Vulnerability of Agriculture, Water and Fisheries to Climate Change

Mohamed Behnassi • Margaret Syomiti Muteng'e
Gopichandran Ramachandran • Kirit N. Shelat
Editors

Vulnerability of Agriculture, Water and Fisheries to Climate Change

Toward Sustainable Adaptation Strategies

Editors
Mohamed Behnassi
Public Law Department,
 Faculty of Law, Economics,
 and Social Sciences
Ibn Zohr University of Agadir
North-South Center for
 Social Sciences (NRCS)
Agadir, Morocco

Gopichandran Ramachandran
Vigyan Prasar, Department
 of Science and Technology
Government of India
NOIDA, UP, India

Margaret Syomiti Muteng'e
Kenya Agricultural Research Institute
Nairobi, Kenya

Kirit N. Shelat
National Council for Climate Change
 Sustainable Development and Public
 Leadership (NCCSD)
Ahmedabad, Gujarat, India

ISBN 978-94-017-8961-5   ISBN 978-94-017-8962-2 (eBook)
DOI 10.1007/978-94-017-8962-2
Springer Dordrecht Heidelberg New York London

Library of Congress Control Number: 2014942852

© Springer Science+Business Media Dordrecht 2014
This work is subject to copyright. All rights are reserved by the Publisher, whether the whole or part of the material is concerned, specifically the rights of translation, reprinting, reuse of illustrations, recitation, broadcasting, reproduction on microfilms or in any other physical way, and transmission or information storage and retrieval, electronic adaptation, computer software, or by similar or dissimilar methodology now known or hereafter developed. Exempted from this legal reservation are brief excerpts in connection with reviews or scholarly analysis or material supplied specifically for the purpose of being entered and executed on a computer system, for exclusive use by the purchaser of the work. Duplication of this publication or parts thereof is permitted only under the provisions of the Copyright Law of the Publisher's location, in its current version, and permission for use must always be obtained from Springer. Permissions for use may be obtained through RightsLink at the Copyright Clearance Center. Violations are liable to prosecution under the respective Copyright Law.
The use of general descriptive names, registered names, trademarks, service marks, etc. in this publication does not imply, even in the absence of a specific statement, that such names are exempt from the relevant protective laws and regulations and therefore free for general use.
While the advice and information in this book are believed to be true and accurate at the date of publication, neither the authors nor the editors nor the publisher can accept any legal responsibility for any errors or omissions that may be made. The publisher makes no warranty, express or implied, with respect to the material contained herein.

Printed on acid-free paper

Springer is part of Springer Science+Business Media (www.springer.com)

About NRCS

The North-South Center for Social Sciences (NRCS) is a research institution founded by a group of researchers and experts from both Global South and North as an independent and apolitical institution. Based in Morocco, NRCS aims to develop research and expertise in many social sciences areas with global and local relevance from a North-South perspective and an interdisciplinary approach. As a Think Tank, NRCS aspires to serve as a reference locally and globally through rigorous research and active engagement with the policy community and decision-making processes. NRCS is currently chaired by Mr. Mohamed Behnassi, Doctor Professor of Global Sustainability and Human Security Politics.

Preface

Human activity is increasingly changing the global environment at an unprecedented rate while humanity is facing a range of complex and interrelated challenges: global warming, ecosystem disruption, biodiversity loss, and for many, increasing difficulty in meeting basic human needs for energy, food, water, and shelter. As a result, environmental issues are inextricably linked to many aspects of local, regional and global development, human security and politics.

A series of recent events have generated interest in food security and food systems, particularly the recent news coverage of high food prices which were variously blamed on biofuels, growing demand for meat and dairy products, commodity speculation, and climate. Other arguments have arisen about the potential impacts of climate change on food availability and water – as the projections of climate change become even more serious – and about the role of integrated policy and governance in shaping food security. The price increases highlighted the connections between food systems in different places – e.g. drought in Australia and demand for meat in Asia, biofuel policy in the USA and Latin America, and between the local food movement in Europe and export farmers in Africa. The challenges facing food systems will accelerate in the coming decades, as the demand for food will double within the next 25–50 years, primarily in developing countries, and with the WTO agriculture talks in disarray, making options for reforming trade policy highly contentious.

Food security and agricultural growth remain high on the science, policy and development agendas. Most research linking global change and food systems focuses solely on the impact of climate change on agricultural production, or the impact of agriculture on land use, pollution and biodiversity. However, interactions with other aspects of the food system – such as food processing, packaging, transporting and consumption, and employment derived from these activities – are often overlooked. There are also important new questions about the interactions between the governance of climate and food such as those associated with carbon trading and labeling, and the role of the private sector in carbon mitigation and in the management of food systems.

Technical prescriptions alone will not manage efficiently the food security challenge. Adapting to the additional threats to food security arising from major environmental changes requires an integrated food system approach, not just a focus on agricultural practices. Many key issues for the research agenda can be highlighted here: adapting food systems to global environmental change requires more than just technological solutions to increase agricultural yields; tradeoffs across multiple scales among food system outcomes are a prevalent feature of globalized food systems; within food systems, there are some key underexplored areas that are both sensitive to environmental change but also crucial to understanding its implications for food security and adaptation strategies; scenarios specifically designed to investigate the wider issues that underpin food security and the environmental consequences of different adaptation options are lacking; price variability and volatility often threaten food security; and more attention needs to be paid to the governance of food systems and to the changing of eating patterns.

Addressing food systems holistically, rather than separate components such as agriculture, markets or nutrition, demands the engagement of multiple disciplines and researchers to understand the causes and drivers of vulnerability. This volume is a contribution to the constructing of this new paradigm.

	The Editors
Agadir, Morocco	Mohamed Behnassi
Nairobi, Kenya	Margaret Syomiti Muteng'e
Noida, India	Gopichandran Ramachandran
Ahmedabad, India	Kirit N. Shelat

Acknowledgements

This contributed volume is based on the best Proceedings of ICCAFFE2011: the International Conference on "Climate change, agri-food, fisheries and ecosystems: Reinventing research, innovation and policy agendas for environmentally- and socially-balanced growth", organized on May 19–21, 2011 in Agadir (Morocco) by the North-South Center for Social Sciences (NRCS) in collaboration with the Deutsche Gesellschaft für Internationale Zusammenarbeit (GIZ) GmbH, Germany, and the Institute for Research and Development (IRD), France.

I have been honored to share the editorship of this book with my colleagues: Margaret Syomiti Muteng'e (Research Scientist, Kenya Agricultural Research Institute-KARI, Kenya), Dr. R. Gopichandran (Principal Research Scientist and Director of Vigyan Prasar, India), and Dr. Kirit N. Shelat (Executive Chairman, National Council for Climate Change, Sustainable Development and Leadership-NCCSD, India) whose commitment and insight made the editing process a wonderful experience and a mutual learning process.

On behalf of my co-editors, I would like to gratefully and sincerely thank the members of the Scientific Committee who have actively contributed to the peer-review of the pre-selected chapters. Deepest thanks go also to all participants in ICCAFFE2011 who made this event possible even if not all could contribute to this volume. We are grateful to the institutions for their support of this book project. In particular, we thank the sponsors of the 2011 Conference, which in addition to NRCS, include the GIZ and the IRD.

While the real value of this volume should be credited to chapters' authors, whose papers have been accepted for publication after a double-blind peer-review, any shortcomings or omissions remain the editors' responsibility. However, the editors and the publisher are not accountable for any statement made or opinion expressed by the chapters' authors.

Mohamed Behnassi

Contents

Part I Agriculture and Climate Change: A Multidimensional Perspective

1 **Mitigation-Adaptation Nexus for Sustainability: Some Important Crosscutting and Emerging Considerations**.......... 3
Gopichandran Ramachandran and Mohamed Behnassi

2 **Climate Change Impacts in the Arab Region: Review of Adaptation and Mitigation Potential and Practices**............ 15
Shabbir A. Shahid and Mohamed Behnassi

3 **Mainstreaming Agriculture for Climate Change Mitigation: A Public Administration Perspective**...................... 39
Kirit N. Shelat and Gopichandran Ramachandran

4 **Learning About Climate Change: A Case Study of Grain Growers in Eastern Australia**..................... 53
Lehmann La Vergne

5 **Economic Impact of Climate Change on Agriculture in SAT India: An Empirical Analysis of Impacts in Andhra Pradesh Using Ricardian Approach**................ 71
Naveen P. Singh, Cynthia Bantilan, and Kattarkandi Byjesh

6 **Vulnerability to Climate Change in Semi-arid Tropics of India: Scouting for Holistic Approach**.................... 89
Naveen P. Singh, Kattarkandi Byjesh, and Cynthia Bantilan

7 **Sustainable Agriculture and Rural Development in the Kingdom of Saudi Arabia: Implications for Agricultural Extension and Education**................... 101
Mirza B. Baig and Gary S. Straquadine

xi

Part II Adapting Agriculture to Climate Risks: Selected Successful Practices

8 Economic Impact of Climate Change on Tunisian Agriculture: The Case of Wheat 119
Ali Chebil, Brian H. Hurd, Nadhem Mtimet, Boubaker Dhehibi, and Weslati Bilel

9 Greenhouse Gas Mitigation Options in Greek Dairy Sheep Farming: A Multi-objective Programming Approach 131
Alexandra Sintori

10 Adaptation Strategies of Citrus and Tomato Farmers Towards the Effect of Climate Change in Nigeria 157
O. Adebisi-Adelani and O.B. Oyesola

11 *Ex-ante* Impact Assessment of 'Stay-Green' Drought Tolerant Sorghum Cultivar Under Future Climate Scenarios: Integrated Modeling Approach 167
Swamikannu Nedumaran, Cynthia Bantilan, P. Abinaya, Daniel Mason-D'Croz, and A. Ashok Kumar

Part III Impacts of Climate Change on Water Resources: Relevant Adaptation Practices

12 Impact of Climate Changes on Water Resources in Algeria 193
Nadir Benhamiche, Khodir Madani, and Benoit Laignel

13 Drought Tolerance of Different Wheat Species (*Triticum* L.) 207
Petr Konvalina, Karel Suchý, Zdeněk Stehno, Ivana Capouchová, and Jan Moudrý

14 Coping with Climate Change Through Water Harvesting Techniques for Sustainable Agriculture in Rwanda 217
Suresh Kumar Pande, Antoni Joseph Rayar, and Patrice Hakizimana

Part IV Impacts of Climate Change on Fisheries and Fishery-Based Livelihoods

15 Climate Change and Biotechnology: Toolkit for Food Fish Security 243
Wasiu Adekunle Olaniyi

16 Climate Change and Fisheries in Chile 259
Eleuterio Yáñez, María Ángela Barbieri, Francisco Plaza, and Claudio Silva

17	**Livelihoods of Coastal Communities in Mnazi Bay-Ruvuma Estuary Marine Park, Tanzania**.......................... 271
	Mwita M. Mangora, Mwanahija S. Shalli, and Daudi J. Msangameno
18	**Livelihood Strategy in Indonesian Coastal Villages: Case Study on Seaweed Farming in Laikang Bay, South Sulawesi Province**............................... 289
	Achmad Zamroni and Masahiro Yamao
19	**Monosex Fish Production in Fisheries Management and Its Potentials for Catfish Aquaculture in Nigeria**........... 301
	Wasiu Adekunle Olaniyi and Ofelia Galman Omitogun
20	**Stock Assessment of Bogue, *Boops Boops* (Linnaeus, 1758) from the Egyptian Mediterranean Waters**................... 313
	Sahar Fahmy Mehanna

Postface.. 323

Notes on Contributors..................................... 325

About the Editors

Dr. Mohamed Behnassi

Mohamed Behnassi, Ph.D. in International Environmental Law and Politics, is Associate Professor and Head of Public Law Department at the Faculty of Law, Economics and Social Sciences of Agadir, Morocco. He is the Founder and Director of the North-South Center for Social Sciences (NRCS) and member of the advisory board of the World Forum on Climate Change, Agriculture and Food Security. His core teaching, research and expertise areas cover the global environmental and human security politics, adaptation to environmental change, and social responsibility. He has published many books with international publishers (such as: *Sustainable Food Security in the Era of Local and Global Environmental Change*, Springer 2013; *Global Food Insecurity*, Springer 2011; *Sustainable Agricultural Development*, Springer 2011; *Health, Environment and Development*, European University Editions 2011; *Climate Change, Energy Crisis and Food Security*, Ottawa University Press 2011) and published numerous papers in accredited journals and communicated several oral presentations in relevant international conferences. In addition, Dr. Behnassi has organized many outstanding international conferences covering the above research areas. On 2011, he had completed the Civic Education and

Leadership Fellowship at Syracuse University and the Maxwell School of Citizenship and Public Affairs (Syracuse, New York). Dr. Behnassi has also managed many research and expertise projects on behalf of national and international institutions such as the Royal Institute of Strategic Studies (IRES) in Morocco (Food and health security face to climate change in Morocco: An adaptation strategy within the perspective of an integrated governance), GIZ (Mainstreaming climate risk in local development planning), Carrfour Associatif and UNICEF (University Social Responsibility).

Margaret Syomiti Muteng'e

Mrs. Syomiti holds a Master Degree in Animal Nutrition and Feed Science from the University of Nairobi, a Bachelor of Science in Animal Production from Egerton University, and she is currently pursuing her Ph.D. in Natural Resources Management from Egerton. Mrs Syomiti began her research career in Kenya Agricultural Research Institute (KARI) on 12 February, 2001, at KARI Muguga South, Animal Production Research Programme. She was promoted from Research I to Research II in 2011. Ms Syomiti is a member of several professional bodies, both local and international. From 2002 to date, she is a member of Animal Production Society of Kenya (APSK). In 2009, she joined the Kenya Professional Association of Women in Agriculture and Environment (KEPAWAE). In 2010–2012, she was competitively selected as an AWARD Fellow, African Women in Agricultural Research and Development, a Gender & Diversity program in the CGIAR. AWARD offers tailored, 2-year fellowships designed to fast-track the careers of African women scientists and professionals delivering pro-poor agricultural research and development that benefits rural communities, especially women. Its goal is to help fellows increase their contributions in the fight against hunger and poverty in sub-Saharan Africa. As an AWARD Fellow,

Ms. Margaret Syomiti has achieved a lot in her research career development: Mentoring Orientation skills in August 2010; AWARDs' Women Leadership and Management Course, in September 2011; and IFS-AWARD science writing, communication and presentation skills course held in February 2011. With AWARD facilitation, she could have the international membership of Women Organizing for Change in Agriculture and Natural Resource Management (WOCAN). During 2011–2014, she is an international Ph.D. Graduate Fellow in Livestock Climate-Collaborative Research Support Program (LCC-CRSP) for Adapting Livestock Systems to Climate Change of Colorado State University. During her research career, Ms. Syomiti has authored and co-authored several scientific papers, which have been published in refereed journals and conference proceedings.

Kenya Agricultural Research Institute
Kari Muguga South
P.O. Box 30148–00100, Nairobi Tel: 020–2519703
E-mail: narcmuguga@yahoo.com
Website: www.kari.org

Dr. R. Gopichandran

Dr. R. Gopichandran is currently the Director of Vigyan Prasar, an autonomous institution of the Department of Science and Technology, Government of India. He has two doctoral degrees in biological sciences specializing in the areas of microbial and chemical ecology. His extensive work on preventive environmental management motivated him to secure a degree in law. Most of his 23 years of scientific, research, communication and capacity building work and output relate to chemical ecological aspects of natural resources management, cleaner production and chemical substitution. The latter two aspects have been through a dynamic association with the Compliance Assistance Programme of the United Nations Environment Programme in particular and with special reference to ozone layer protection, centred on locally relevant action by industry at the national and regional levels. This included innovative work on destruction of ozone depleting substances to twin the objectives of climate and ozone layer protection. His recent contributions to India's 12th Five

Year Plan as member of the Working Group on Natural Resources Management are on the chemical ecological aspects of conservation and sustainable management of agriculture to tackle challenges posed by climate change. He has a large number of publications including books and papers through well-recognized publishers and journals. He has also taught extensively in several centres of higher learning including the Centre for Environmental Planning and Technology and the Indian Institute of Management, Ahmedabad. Prior to joining Vigyan Prasar, he has served in three important institutions over the last 23 years: the Gujarat Energy Research and Management Institute, Gandhinagar, as the Principal Research Scientist; Centre for Environment Education, Ahmedabad, as Scientist SH (Environment and Climate Change Wing); and Environment Management and Entomology Research Institute, Chennai, as Programme Director/Scientist SG. He is also an Alumnus (Year 2000) of the International Visitors Leadership Programme of the Department of State, USA.

Dr. Kirit Nanubhai Shelat

Dr. Shelat holds a Ph.D. in Public Administration. He has served as the Principal Secretary of Agriculture and Cooperation Department, Government of Gujarat, following several other leadership public administration roles in the energy and industry sectors. He recently retired from the Indian Administrative Service. During his career of 40 years, he formulated and implemented many relevant policies covering the stated sectors. This includes his focus on the benefit of poor families, farmers, and micro-entrepreneurs and remote rural areas through micro-level planning for integrated development. He established the new extension approach of meeting with farmers at their door step prior to monsoon by teams led by agriculture scientists. Importantly he introduced soil health and moisture analysis based agriculture systems and provided soil health cards to every farmer in the State of Gujarat. He is also among the Founders of International School for Public Leadership and Executive, Chairman of National Council for Climate Change Sustainable Development and Public Leadership (NCCSD), and Chairman of the Sub Group "Enhancing Preparedness for Climate Change" set up by the Planning Commission, India, for the 12th Five Year Plan. He is author of many books related to rural and agricultural development and public leadership.

Contributors

P. Abinaya Research Program – Markets, Institutions and Policies, International Crops Research Institute for the Semi-Arid Tropics (ICRISAT), Hyderabad, Andhra Pradesh, India

O. Adebisi-Adelani Farming Systems and Extension Research Department, National Horticultural Research Institute (NIHORT), Ibadan, Nigeria

A. Ashok Kumar RP-Dryland Cereals, International Crops Research Institute for the Semi-Arid Tropics (ICRISAT), Hyderabad, India

Mirza B. Baig Department of Agricultural Extension and Rural Society, Faculty of Food and Agricultural Science, College of Food and Agricultural Sciences, King Saud University, Riyadh, Kingdom of Saudi Arabia

Cynthia Bantilan Research Program – Markets, Institutions and Policies, International Crops Research Institute for the Semi-Arid Tropics (ICRISAT), Hyderabad, Andhra Pradesh, India

María Ángela Barbieri Instituto de Fomento Pesquero, Blanco, Valparaíso, Chile

Mohamed Behnassi Public Law Department, Faculty of Law, Economics, and Social Sciences, Ibn Zohr University of Agadir, North-South Center for Social Sciences (NRCS), Agadir, Morocco

Nadir Benhamiche Laboratoire de Biomathématique, Biophysique, Biochimie, et Scientométrie, Faculté des Sciences de la Nature et de la vie, Université de Bejaia, Bejaia, Algérie

Weslati Bilel Department of Agricultural Economics, Mograne High School of Agriculture (ESA Mograne), Mograne-Zaghouan, Tunisia

Kattarkandi Byjesh Research Program – Markets, Institutions and Policies, International Crops Research Institute for the Semi-Arid Tropics (ICRISAT), Hyderabad, Andhra Pradesh, India

Ivana Capouchová Faculty of Agrobiology, Food and Natural Resources, Department of Crop Production, Czech University of Life Sciences, Prague 6, Czech Republic

Ali Chebil Department of Agricultural Economics, National Institute of Research in Rural Engineering, Water and Forestry (INRGREF), Ariana, Tunisia

Boubaker Dhehibi Social, Economic and Policy Research Program (SEPRP), International Center for Agricultural Research in the Dry Areas (ICARDA), Aleppo, Syria

Sahar Fahmy Mehanna Fish Population Dynamics Laboratory, Fisheries Division, National Institute of Oceanography and Fisheries (NIOF), Suez, Egypt

Gopichandran Ramachandran Vigyan Prasar, Department of Science and Technology, Government of India, Noida, UP, India

Patrice Hakizimana USAID, Kigali, Rwanda

Brian H. Hurd Department of Agricultural Economics and Agricultural Business, New Mexico State University, Las Cruces, NM, USA

Petr Konvalina Faculty of Agriculture, University of South Bohemia in České Budějovice, České Budějovice, Czech Republic

Suresh Kumar Pande Department of Soil and Water Management, College of Agriculture, Animal Sciences and Veterinary Medicine, University of Rwanda, Musanze, Rwanda

Lehmann La Vergne University of Ballarat, Ballarat, VIC, Australia

Benoit Laignel UMR M2C CNRS 6143, Département de Géologie, Université de Rouen, Mont-Saint-Aignan, Cedex France

Khodir Madani Laboratoire de Biomathématique, Biophysique, Biochimie, et Scientométrie, Faculté des Sciences de la Nature et de la vie, Université de Bejaia, Bejaia, Algérie

Mwita M. Mangora Institute of Marine Sciences, University of Dar es Salaam, Zanzibar, Tanzania

Daniel Mason-D'Croz Environment and Production Technology Division, International Food Policy Research Institute (IFRI), Washington DC, USA

Jan Moudrý Faculty of Agriculture, University of South Bohemia in České Budějovice, České Budějovice, Czech Republic

Daudi J. Msangameno Institute of Marine Sciences, University of Dar es Salaam, Zanzibar, Tanzania

Nadhem Mtimet Department of Agricultural Economics, Mograne High School of Agriculture (ESA Mograne), Mograne-Zaghouan, Tunisia

Swamikannu Nedumaran Research Program – Markets, Institutions and Policies, International Crops Research Institute for the Semi-Arid Tropics (ICRISAT), Hyderabad, Andhra Pradesh, India

Wasiu Adekunle Olaniyi Biotechnology and Wet Laboratories, Department of Animal Sciences, Obafemi Awolowo University, Ile-Ife, Nigeria

Ofelia Galman Omitogun Biotechnology and Wet Laboratories, Department of Animal Sciences, Obafemi Awolowo University, Ile-Ife, Nigeria

O.B. Oyesola Department of Agricultural Extension and Rural Development, University of Ibadan, Ibadan, Oyo State, Nigeria

Francisco Plaza Instituto de Fomento Pesquero, Blanco, Valparaíso, Chile

Antoni Joseph Rayar Higher Institute of Agriculture and Animal Husbandry, University Crescent Burnaby, Vancouver, B.C., Canada

Shabbir A. Shahid Salinity Management Scientist, International Center for Biosaline Agriculture (ICBA), Dubai, United Arab Emirates

Mwanahija S. Shalli Institute of Marine Sciences, University of Dar es Salaam, Zanzibar, Tanzania

Kirit N. Shelat National Council for Climate Change Sustainable Development and Public Leadership (NCCSD), Navrangpura, Ahmedabad, Gujarat, India

Claudio Silva Pontificia Universidad Católica de Valparaíso, Casilla, Valparaíso, Chile

Centro Andaluz de Ciencia y Tecnología Marina (CACYTMAR), Universidad de Cádiz, Cádiz, Spain

Naveen P. Singh Research Program – Markets, Institutions and Policies, International Crops Research Institute for the Semi-Arid Tropics (ICRISAT), Hyderabad, Andhra Pradesh, India

Alexandra Sintori Department of Agricultural Economics and Rural Development, Agricultural University of Athens, Athens, Greece

Zdeněk Stehno Faculty of Agriculture, Crop Research Institute, Prague 6, Czech Republic

Gary S. Straquadine Chair, Department of Agricultural Communication, Education, and Leadership, Ohio State University, Columbus, OH, USA

Karel Suchý Faculty of Agriculture, University of South Bohemia in České Budějovice, České Budějovice, Czech Republic

Margaret Syomiti Muteng'e Natural Resources Management, Egerton University, Nakuru, Kenya

Masahiro Yamao Graduate School of Biosphere Science, Laboratory of Food Production Management, Department of Bioresource Science, Hiroshima University, Higashi Hiroshima-City, Japan

Eleuterio Yáñez Pontificia Universidad Católica de Valparaíso, Casilla, Valparaíso, Chile

Achmad Zamroni Research Center for Marine and Fisheries Socio Economics, Ministry for Marine Affairs and Fisheries, Jakarta, Indonesia

Abbreviations and Acronyms

ABARE	Australian Bureau of Agricultural and Resource Economics
ADAFSA	Abu Dhabi Agriculture and Food Safety Authority
ADFCA	Abu Dhabi Food Control Authority
ADP	Agricultural Development Programme
AEZ	Agro-Ecological Zones
ANOVA	Analysis Of Variance
ANRH	Agence Nationale des Ressources Hydrauliques
APMC	Agricultural Produce Market Committee
AWC	Arab Water Council
BCM	Billion Cubic Meters
CA	Copenhagen Accord
CC	Climate Change
CCAFS	Climate Change, Agriculture and Food Security
CCI	Climate Change Impacts
CGIAR	Consultative Group on International Agricultural Research
CP	Compromise Programming
CSIRO	Commonwealth Science and Industrial Research organization
EAD	Environment Agency – Abu Dhabi
ECHAM3TR	European Centre Hambourg, Germany
EPA	US Environmental Protection Agency
ESF	European Social Fund
ET	Evapotranspiration
FAO	Food and Agriculture Organization
GCM	General Circulation Models
GEF	Global Environment Facility
GFDL	Geophysical Fluid Dynamics Laboratory
GGA	Grain Growers Association
GHE	Greenhouse Effect
GHGs	Greenhouse Gases
GISS	Goddard Institute for Space Studies
HDI	Human Development Index

HLPE	High Level Panel of Experts on Food Security and Nutrition
HMRDF	Hellenic Ministry of Rural Development and Food
ICBA	International Center for Biosaline Agriculture
ICRISAT	International Crop Research Institute for the Semi- Arid Tropics
IFPRI	International Food Policy Research Institute
ILUC	Indirect land use change
INM	National Meteorological Institute
IPCC	Intergovernmental Panel on Climate Change
IRENA	International Renewable Energy Agency
IRMS	Isotope Ratio Mass Spectrometer
IUCN	International Union for Conservation of Nature
IWASRI	International Water logging and Salinity Research Institute
IWMI	International Water Management Institute
KISR	Kuwait Institute for Scientific Research
KP	Kyoto Protocol
KSU	King Saud University
LCA	Life Cycle Assessment
LUT	Land Utilization Types
MEECC	Ministry of Environment, Energy and Climate Change
MENA	Middle East and North Africa
MOA	Ministry of Agriculture
MOP	Multi-Objective Programming
MREP	Mona Reclamation Experimental Project
NAMA	Nationally Appropriate Mitigation Action
NAPA	National Adaptation Programme of Action
NATCOM	India's Second National Communication
NEPAD	The New Partnership for Africa's Development
NRCS	North-South Center for Social Sciences
NSRF	National Strategic Reference Framework
NSSG	National Statistical Service of Greece
ONM	Office National de la Météorologie
OSU	Oregon State University
PICCMAT	Policy Incentives for Climate Change Mitigation Agricultural Techniques
RCM	Regional Climate Model
RIRDC	Rural Industries Research and Development Corporation
RPWRC	Red Palm Weevil Research Chair
SAC	Space Application Centre
SCAR	Standing Committee on Agricultural Research
SIC	Soil Inorganic Carbon
SLR	Sea Level Rise
SOM	Soil Organic Matter
TCD	Thermo-Conductive Detector
UAE	United Arab Emirates

UKHI	United Kingdom Meteorological Office High Resolution
UNDP	United Nation development Program
UNEP	United Nation Environment Program
UNESCO	United Nations Educational, Scientific and Cultural Organization
UNFCCC	United Nations Framework Convention on Climate Change
WANA	West Asia North Africa
WBGU	German Advisory Council on Global Change
WGP	World Gross Product
WHO	World Health Organization
WR	Water Requirement
WUE	Water-Use Efficiency Evaluation
WUE	Water Use Efficiency

List of Figures

Fig. 3.1	Elements of the integrated outreach for farmers through "Krishi Mahotsav" roughly translated as the grand fair for the farmers and farming	45
Fig. 3.2	Converging themes in "Krishimahotsav"	45
Fig. 4.1	Map of farmer case study locations in the grain belts of eastern Australia	61
Fig. 5.1	Crop wise distribution of agricultural area in Andhra Pradesh (2008–2009)	73
Fig. 5.2	Relationship between net revenue and temperature during the two monsoon seasons	81
Fig. 5.3	Relationship between South West monsoon rainfall and net revenue for rice	81
Fig. 5.4	Relationship between South West monsoon temperature and net revenue for major crops	84
Fig. 6.1	Mapping districts level vulnerability to climate change for Andhra Pradesh and Maharashtra states of India	95
Fig. 8.1	Map of Tunisia showing the regions considered in the study	123
Fig. 9.1	GHG alternative mitigation options of the semi-intensive farm	149
Fig. 9.2	GHG alternative mitigation options of the extensive farm	153
Fig. 10.1	Distribution of categorization of adaptation strategies of respondents	161
Fig. 11.1	World Sorghum area, production and yield (1961–2010)	169
Fig. 11.2	Sorghum production (in metric tons per pixel) and target countries (*shaded* countries) for sorghum technology dissemination	174

Fig. 11.3	The framework of technology development and adoption pathways linked with outcomes and impacts	184
Fig. 11.4	Change in the World price of sorghum under different climate scenarios	188
Fig. 11.5	The net welfare benefits (million US$) for the target countries and non-target countries under no climate change	189
Fig. 12.1	Location of study area	195
Fig. 12.2	Map according to Chaumont for the period 1913–1963	197
Fig. 12.3	Map according to ANRH for the period 1922–1989 (ANRH)	197
Fig. 12.4	Map according to ANRH for the period 1965–2004 (ANRH)	198
Fig. 12.5	Evolution of liquid contributions of catchment areas around Algiers (*Center*) (1965–2003) (ANRH)	198
Fig. 12.6	Evolution of liquid contributions of catchment areas around Constantine (East) (1965–2003) (ANRH)	199
Fig. 12.7	Evolution of liquid contributions of catchment areas around Cheliff (Center-West) (1965–2003) (ANRH)	199
Fig. 12.8	Evolution of liquid contributions of catchment areas around Oran (West) (1965–2003) (ANRH)	200
Fig. 12.9	Decreased volume in the dam of Ksob: area of M'sila (South-East of Algeria) (1932–2006) (ANRH)	200
Fig. 12.10	Decreased volume in the dam of Cheffia: area of Tarf (East of Algeria) (1946–2006) (ANRH)	201
Fig. 12.11	Map of average annual rainfall in Northern Algeria – Decennial frequency **dry** (1965–**2001**) (ANRH)	201
Fig. 12.12	Map of average annual rainfall in Northern Algeria – Decennial frequency **wet** (1965–**2004**) (ANRH)	202
Fig. 12.13	Municipalities victim of floodings during the 3 last years (2003–2005) (ANRH)	202
Fig. 12.14	Localization of the most important floodings during 10 last years (1973–2005) (ANRH) Behlouli (2009)	203
Fig. 12.15	Evolution of the groundwater depth of Mitidja – Area of Algiers (1995–2002) – Piezometer PZ N°1 Hamiz/21 (ANRH 2004)	203
Fig. 14.1	Global water use pattern	219
Fig. 15.1	Microinjection of DNA into salmon eggs	249
Fig. 15.2	Electroporation of fish eggs. Eggs (15–50) are placed in a DNA solution inside the cuvette and then placed between two electrodes to receive rapid pulses of high voltage	249

Fig. 16.1	First axis of the principal component analysis for the time series: capture, fishing effort, Ekman transport and turbulence for anchovy (**a**); recruitment, biomass, sea surface temperature and Ekman transportation for sardine (**b**) (Yáñez et al. 2008a)	261
Fig. 16.2	Jack mackerel CPUE decrease during the period 1983–2008	262
Fig. 16.3	Anomalies of common hake landings and sea surface temperature in central Chile (Yáñez et al. 2003)	263
Fig. 16.4	Integrating conceptual model of local and large-scale phenomena affecting the northern area of Chile and the main fishing resources. The direction and magnitude influences of the phenomena are shown by *arrows* (Yáñez et al. 2008a)	263
Fig. 16.5	Spatio-temporal conceptual model (3 × 3) with planes in the marine, physical, biological and human (fishing) environments, along with processes and forcing from the swordfish ecosystem (Yáñez et al. 2008b)	264
Fig. 16.6	ANN models (10) for anchovy fisheries in northern Chile calibrated (Plaza et al. 2008a, b), and validated with information from 2000 to 2007 (average R2 of 77 %)	266
Fig. 16.7	Monthly anchovy landing projections in northern Chile up to 2100, considering four climate change scenarios	266
Fig. 17.1	Map of the coastal area of Tanzania showing MPAs managed under MPRU. *TMRs* Tanga Marine Reserves System (Kirui, Mwewe, Kwale and Ulenge), *TACMP* Tanga Coelacanth Marine Park, *MIMR* Maziwe Island Marine Reserve, *DMRs* Dar es Salaam Marine Reserves System (Bongoyo, Pangavini, Mbudya, Makatube, Sinda, Kendwa and Funguyasini), *MMRs* Mafia Marine Reserves System (Nyororo, Shungimbili and Mbarakuni), *MIMP* Mafia Island Marine Park, *MBREMP* Mnazi Bay-Ruvuma Estuary Marine Park	273
Fig. 17.2	Map of Mnazi Bay-Ruvuma Estuary Marine Park and surrounding areas	275
Fig. 17.3	Overall percentage distribution of household sizes in the study villages	277
Fig. 17.4	Overall percentage distribution of the reported primary and secondary livelihood occupations in the study areas	279
Fig. 17.5	Household income and expenditure patterns in study villages. a = income variation per different sources, b = income variation per villages, c = expenditure variation per different items, d = expenditure variation per villages	282
Fig. 18.1	Map of Laikang Bay, South Sulawesi Province	292
Fig. 18.2	Income utilization of fishermen at Laikang Bay	294

Fig. 19.1	Effects of genetic manipulation techniques on fertility, hatchability and survival in African catfish (*Clarias gariepinus*)	308
Fig. 19.2	Chromosome spreads (×40) from 24 h larvae of African catfish (*Clarias gariepinus*). (**a**) Haploid control = 28; (**b**) Diploid control = 56; (**c**) Androgenesis = 56; and (**d**) Gynogenesis = 56	309
Fig. 20.1	Egyptian Mediterranean coast	314
Fig. 20.2	Length-weight relationship of *Boops boops* from Egyptian Mediterranean waters	316
Fig. 20.3	Growth curve of *Boops boops* from Egyptian Mediterranean waters	317
Fig. 20.4	Maturation curve of *Boops boops* from Egyptian Mediterranean waters	318
Fig. 20.5	L_c estimation for *Boops boops* from Egyptian Mediterranean waters	319
Fig. 20.6	Yield per recruit analysis of *Boops boops* from Egyptian Mediterranean waters	320

List of Tables

Table 1.1	Differences between mitigation and adaptation	4
Table 3.1	Inputs for locally-relevant and robust micro-level plans	50
Table 5.1	List of crops grown in this study district of Andhra Pradesh	76
Table 5.2	List of variables included in the study	77
Table 5.3	Mean values of area and net revenue of rice and major crops in different districts of Andhra Pradesh	78
Table 5.4	Mean values of the input variables	79
Table 5.5	Ricardian model for net revenue per ha for rice	80
Table 5.6	District wise impact of climate variables on the net revenue per ha for rice	82
Table 5.7	Testing the combined effect of climate variables on net revenue	83
Table 5.8	Ricardian model fit for major crops	84
Table 5.9	Marginal impact of climate variables on the net revenue from major crops in the different districts of Andhra Pradesh	85
Table 5.10	Testing the significance of inclusion of climate variables in the Ricardian model	85
Table 6.1	List and description of indicators used for computing vulnerability (Exposure and sensitivity)	92
Table 6.2	Periodical indexing of vulnerability to climate change for the state of Andhra Pradesh	94
Table 6.3	Periodical indexing of vulnerability to climate change for the state of Maharashtra	97
Table 7.1	Agricultural production for the years 1999, 2003, 2004 and 2008 in the Kingdom of Saudi Arabia (KSA)	105
Table 7.2	An overview of infrastructure and services available to the society in the Kingdom of Saudi Arabia	111

Table 8.1	Cultivated land in the study areas (1,000 ha)	123
Table 8.2	Description of variables used in the analysis (Period: 1990–2010)	125
Table 8.3	Parameters estimates of the Ricardian model	126
Table 8.4	Future projection of climate change under HadCM3 model by 2030	127
Table 8.5	Expected effects of climate change on net revenue in three regions of Tunisia under scenario A2 of HadCM3 model	127
Table 9.1	Population of livestock in Greece and their contribution to CH_4 production (%)	133
Table 9.2	Models used to predict GHG in livestock farms	137
Table 9.3	Pay-off matrix for the semi-intensive farming system	143
Table 9.4	Pay-off matrix for the extensive farming system	143
Table 9.5	Results of the compromise programming method for the semi-intensive farm	145
Table 9.6	Aggregate results of mitigation alternatives for the semi-intensive farming system	146
Table 9.7	Results of the compromise programming method for the extensive farm	150
Table 9.8	Aggregate results of mitigation alternatives for the extensive farming system	152
Table 10.1	Distribution of selected horticultural farmers' personal characteristics	160
Table 10.2	Frequency distribution of zonal categorization of adaptation strategies	161
Table 10.3	Distribution of respondents according to adaptation strategies by selected horticultural farmers and frequency of use	163
Table 10.4	Frequency distributions of respondents according to factors that influence adaptation strategies	164
Table 10.5	Significant difference in the use of adaptation strategies between citrus and tomato farmers across the agricultural zones	165
Table 11.1	Sorghum area, yield and production (Three years average)	170
Table 11.2	Proposed budget allocation for ICRISAT and NARS partners (million US$)	173
Table 11.3	The ceiling adoption level and year of maximum adoption in target countries	173
Table 11.4	Global welfare benefits of drought tolerant sorghum technologies	185
Table 11.5	Net economic benefit drought tolerant sorghum cultivars in the target countries under climate change scenarios	186

Table 11.6	Change in production, consumption and net trade in the target countries in 2050 after the adoption of drought tolerant sorghum in target countries..................	187
Table 12.1	Seasonal climate projections of temperature and rainfall in Algeria by 2020 for two GIEC models (IPCC 1990), (UKHI and ECHAM3TR) (Ministry of Regional Planning and Environment, 2001)...	196
Table 12.2	Projection by 2020 of mobilizable surface water quantities in Algeria (ANRH 1993) according to UKHI (1989) and ECHAM3TR (1995) models................................	196
Table 13.1	List of varieties...	209
Table 13.2	Examination of the contribution of factors (variety, locality, year) and its interactions to $\delta^{13}C$ via the analysis of variance (ANOVA)..........................	211
Table 13.3	Differences in droughtiness in genus *Triticum* L. (mean + SD), LSD test ($P < 0.01$)...............................	213
Table 17.1	Average household sizes in studied villages ($N = 30$ for each village)...	277
Table 17.2	Percentage of households which had at least one member who had attained a given level of education in the villages within study area ($N = 30$ for each village).....................	278
Table 17.3	Village specific percentage distribution of the major household primary and secondary livelihood occupations.......	280
Table 18.1	Socioeconomic data of seaweed farmers in Laikang Bay......	293
Table 18.2	Participation of stakeholder in local-level.......................	297
Table 19.1	Effects of genetic manipulation techniques on fertility, hatchability and survival in African catfish (*Clarias gariepinus*)..	308
Table 20.1	Growth parameters (L_∞, K and t_o) and the Phi index of *Boops boops*...	317
Table 20.2	VIT results...	320

Part I
Agriculture and Climate Change:
A Multidimensional Perspective

Chapter 1
Mitigation-Adaptation Nexus for Sustainability: Some Important Crosscutting and Emerging Considerations

Gopichandran Ramachandran and Mohamed Behnassi

Abstract Managing climate risks can be done via two main strategies: mitigation and adaptation. Yet many studies and practices are increasingly confirming the complementarity of the two strategies. Although available research on mitigation and adaptation synergies remains dispersed and incomplete, examples that demonstrate promising potential for synergies have been identified. Within this perspective, this chapter aims to highlight some important questions that have to be answered in order to foster synergies between mitigation and adaptation approaches with positive implications in terms of sustainability. Researchers in the fields of chemical ecology of crop productivity and environmental management, with special reference to industrial ecology twinning mitigation and adaptation goals, aligned with policy and plan interventions, will gain insights into the unfinished agenda in this regard. Such important facets of integrated management, including peace with implications for quality of life determined by sustained access to information and capacity building on alternatives that are locally relevant, are also discussed.

Keywords Adaptation • Mitigation • Synergy • Sustainable development • Chemical ecology

---

R. Gopichandran (✉)
Vigyan Prasar, Department of Science and Technology, Government of India, A-50, Institutional Area, Sector-62, Noida 201309, UP, India
e-mail: gopi61@yahoo.com; r.gopichandran@vigyanprasar.gov.in

M. Behnassi
Public Law Department, Faculty of Law, Economics, and Social Sciences, Ibn Zohr University of Agadir, North-South Center for Social Sciences (NRCS), Hy Salam, P. O. Box 14997, Agadir 80 000, Morocco
e-mail: behnassi@gmail.com

1 Introduction

Mitigation and adaptation are the two strategies for addressing climate change. Mitigation is an intervention to reduce the emissions sources or enhance the sinks of greenhouse gases. According to IPCC (2001), adaptation is an "adjustment in natural or human systems in response to actual or expected climatic stimuli or their effects, which moderates harm or exploits beneficial opportunities".

According to Locatelli (2011), adaptation and mitigation strategies present some significant differences, especially with regards to their objectives. Mitigation deals with the causes of climate change (accumulation of greenhouse gases in the atmosphere), whereas adaptation addresses the climate change impacts. Both approaches are strongly needed. On the one hand, even with strong mitigation efforts, the climate would continue changing in the next decades and adaptation to these changes is necessary. On the other hand, adaptation will not be able to eliminate all negative impacts and mitigation is crucial to limit changes in the climate system.

For Tol (2005), adaptation and mitigation are also different in terms of spatial scales (although climate change is an international issue, adaptation benefits are local and mitigation benefits are global), temporal scales (adaptation has short-term effects while mitigation has long-term effects), and concerned economic sectors (Table 1.1).

According to a new study (Illman et al. 2013), harnessing adaptation-mitigation synergies can serve as an important component in building the necessary knowledge base, institutional capacity and sectorial collaboration that effective climate policy in the twenty-first century will require. While the state of existing scientific evidence related to mitigation and adaptation synergies remains rather dispersed and incomplete, examples that reveal promising potential for synergies have been identified in several areas. While no quantitative assessments have been made on the magnitude of the potential for synergies, three particular areas are expected to present opportunities: agriculture, forestry, and land use. Developing countries with high (and/or rapidly increasing) emissions and major vulnerabilities are well placed to harness synergies. Promising sectors are those that have a lot of mitigation potential and that have been a focus of national level adaptation plans. In addition, development in urban areas offers ample opportunities to harness synergies in building and infrastructure development, transportation, insurance and waste treatment sectors.

Table 1.1 Differences between mitigation and adaptation

	Mitigation	Adaptation
Spatial scale	Primarily an international issue, as mitigation provides global benefits	Primarily a local issue, as adaptation mostly provides benefits at the local scale
Time scale	Mitigation has a long-term effect because of the inertia of the climatic system	Adaptation can have a short-term effect on the reduction of vulnerability
Sectors	Mitigation is a priority in the energy, transportation, industry and waste management sectors	Adaptation is a priority in the water and health sectors and in coastal or low-lying areas
	Both mitigation and adaptation are relevant to the agriculture and forestry sectors	

Source: Locatelli (2011)

The following analysis attempts to highlight some important questions to be necessarily addressed in order to foster synergies between mitigation and adaptation approaches with positive implications in terms of sustainable development. Researchers in the fields of chemical ecology of crop productivity and environmental management, with special reference to industrial ecology twinning mitigation and adaptation goals, aligned with policy and plan interventions, will gain insights into the unfinished agenda in this regard. Such important facets of integrated management, including peace with implications for quality of life determined by sustained access to information and capacity building on alternatives that are locally relevant, are also discussed.

2 Preparedness of Communities

The importance of sustaining mitigation and adaption outcome, with respect to the quality of life of citizens affected by climate change, cannot over-emphasized (UNEP 2012; UN Global Compact and UNEP 2012; ECPA 2010). While a very large number of mitigation and adaptation initiatives are in progress across the globe, local relevance of these measures and the people's ability to sustain their use have determined their continued success. This aspect of human and institutional development is gaining greater visibility in the development and implementation of such initiatives. The present analysis draws the attention of decision makers to a very important determinant of success in this continuum of knowledge transfer and capacity building, going beyond the normal constructs of technology fix. This is with special reference to the preparedness of stakeholders to receive information and transform learning into real life action.

Nisbet and Scheufele (2009) had highlighted the fact that stakeholders wonder if scientific knowledge will really work especially when initiatives attempt transfers across systems with assumptions that their relevance and feasibility will be mainstreamed anyway. This implies a significant call for local relevance of action. The second dimension, related to the growing body of knowledge regarding this phenomenon, enhances the question whether the projections made by scientists will be all true or will turn up to be false. The third dimension is about duly recognizing all relevant knowledge systems, and hence the need to be open to criticisms by opposite views. Stakeholders tend to also examine the social and other institutional affiliations of scientists and the track record of their trustworthiness. This is also related to scientists' ability to consider prevailing perceptions among the public, and hence duly recognize their opinions regarding the core issues.

Finally, the most important consideration seems to be the preparedness of the scientific community to accept responsibility for unforeseen harm in the future. This last aspect calls for a clear understanding of the sustainability of conceived solutions and a categorical acceptance of the limits and limitations of alternatives proposed. Above all, stakeholders also examine the credibility of the communicators themselves.

These are very important dimensions of public policy in democracies wherein the views of stakeholders – as end users of information and knowledge – are central to the success of interventions for collective good.

The above stated considerations are particularly relevant when the diversity of tools and techniques used for public communication is only apparently growing by the day, reinforcing the value of the framework of knowledge economy. An equally important dimension of public communication is the tendency of the receivers of information to retain the right to not implement the content derived out of the information delivered. This has been yet another critical dimension of translating public policy intent into real life goals. The importance of enabling circumstances cannot therefore be over-emphasized in this context.

Three important determinants of the success of the enabling circumstances include regulations, fiscal and non-fiscal measures to support compliance with regulations, and institutional mechanisms for sustained access to alternatives. Monitoring and reporting on the performance of such institutional mechanisms are crucial to enable mid-course corrections aligned with the availability of alternatives. While governments have normally established such enabling circumstances, the corporate sector is increasingly involved in this interface. A comprehensive presentation by the Global Compact and UNEP captures this emerging trend (UN Global Compact and UNEP 2012).

This emerging role of the corporate sector signifies a business engagement for adaptation; twinning community and business interest for resilience and hence a common perception of risks. Researchers in the fields of community capacity building to mainstream corporate social responsibility initiatives may like to develop locally and culturally specific interventions to harmonize the goals and approaches of these two major stakeholder groups.

3 The Peace Interface

Public policies on climate change mitigation and adaptation imply sustainable development with a bearing on quality of life and hence an over arching peace imperative for communities (Ranjan and Prasad 2011; Afifi et al. 2012). This is an essential component of advocacy for a governance system based on equity and justice (Tripathi and Vasan 2012) and able to link emerging global paradigms for peace with principles of communities' self-governance through the "Swaraj" framework. Such aspects of decentralization, self-sufficiency and cooperation pervade this framework and are directly relevant to efficient management of land, water, and bio-resources, integral to the success of mitigation and adaptation measures.

The Peace Corps and Energy and Climate Partnership of the Americas 2010 (ECPA) Initiative, adopted by the United States of America on 2012 focuses on enhancing capacities of communities to tackle energy related issues in rural Latin America with a special emphasis on energy efficient practices guided by information support, technical training, and mechanisms to access credit. Yet, another initiative taken by the International Organization for Migration in 2011 in Kenya emphasizes environmental stewardship, natural resources management through community action, and resolution of conflicts through solidarity.

The unifying principle of such initiatives is enhanced resilience and access to resources, vulnerability reduction and equity as a part of the larger mitigation/adaptation goal. Tanzler et al. (2010) indicate that climate change accentuates social and economic disparities, and consequently environmental cooperation to tackle common challenges becomes an imperative. However, adaptation has to be perceived as an approach to overcome technical challenges and to enable socio-political transformations. During the adaptation process, we have to ensure that equity is infused in order to avoid conflicts that may arise inadvertently and therefore reinforce adaptation as a means of defusing conflicts.

Koubi et al. (2012) calls for a detailed analysis of climate change determinants, economic developments and conflict pathways in order to develop location-specific preventive adaptation strategies. Mayoral (2012) interestingly defines hot and cold climate wars by linking food and commodity prices with inclement weather systems, impacts, border conflicts and regulatory challenges. Kotite (2012) highlights the importance of education of communities to prevent conflicts and enhance peace. This calls for conflict-sensitive education planning and consulted efforts to build peace, dissolving policies that work at cross-purposes.

4 Crosscutting Adaptation Considerations

The World Bank (2012) highlights the importance of assessing the extent to which financial flows enhance newer local livelihood opportunities in the light of risks they may pose. This is aligned with an emerging holistic understanding of the existing links between development and environmental considerations per se. Melamed et al. (2012) discuss at great length the interrelationship between the two approaches with a historical perspective and within such emerging frameworks as the Human Rights and Millennium Development Goals. The latter have wider social dynamics and it is therefore important to ensure harmonization with political and scientific considerations for sustainability. Brooks et al. (2011) classify adaptation initiatives across a continuum of existing deficits to longer-term manifestations through systemic changes. They further argue that longer-term perspectives are not adequately mainstreamed in adaptation planning, and therefore propose several indicators for holistic interventions; larger the number higher the degree of synthesis.

Qualitative and quantitative dynamics of determinants have to be clearly analyzed in order to develop system specific adaptation strategy that can also tackle emerging challenges. This is also the essence of developing planning as indicated by Lebel et al. (2012). It is important to take note of these emerging perspectives, especially when countries are continually evolving their adaptation strategies. Sectoral and sub/cross-sectoral considerations should gain preponderance over standard approaches, duly considering the constraints they pose on each other. These are expected to reduce policy conflicts and vulnerability while twinning adaptation and mitigation goals.

Sterrett (2011) presents a comprehensive review of adaptation practices in South Asia, further reinforcing the fragmented nature of such approaches with special reference to disaster risk reduction and policies that have bearing on the collective output. Such important aspects – as climate induced migration, consequences of planned/autonomous adaptation approaches and ability of communities to access financial services, including insurance – have to be considered in detailed investigations regarding dynamics. Sterrett (2011) also emphasizes the need to focus on river basin management, including institution mechanisms for just sharing of resource, followed by locally relevant feasible conservation measures. This has been reinforced through a special highlight on the needs of smallholder farmers and their relative food system, health and related economics of markets (Dube et al. 2012).

Dally and Matson (2008) highlighted the importance of integrating ecosystem considerations in the design and implementation of public policies determining the extraction of resources, and hence the health of ecosystems to extend benefits for the society. A special focus on gender strategy has been invited by CCAFS (2012a, b), with respect to adaptation through a pro-poor perspective. This is aligned with the agro-ecology framework for resilient crop production systems with multiple benefits for ecosystems and beneficiaries (Altieri 2012). This was substantiated by Nicholas et al. earlier in 2011.

Some of the other most important leads in the interface of mitigation and adaptation planning and policies have been presented by the FAO Policy Learning Programme (2009); The Hague Conference on Agriculture, Food Security and Climate Change (FAO 2010); Antle (2010); The Climate and Development Knowledge Network (2011); and the Ex-Ante Carbon-Balance Tool of the FAO (2011).

5 The Chemical Ecology Perspective

It is likely that integrated water conservation, land management and related agriculture interventions focus predominantly on the quantum of yield and how to sustain high yields – challenge generally perceived as an impact of climate change. However, climate resilient crop varieties have to be examined from a chemical ecology perspective with special reference to their own biochemical adaptation pathways with significant implications for nutrient dynamics across several trophic levels. For instance, changes in the nitrogen metabolic pathways influence the occurrence and distribution of secondary metabolites that could determine the assemblage of pests and predators including parasites. On the other hand, the nutritive value of a crop could also influence dietary benefits, and hence the processing requirements, with further implications for technologies that would help optimize nutrient benefits.

Yet another, very important dimension of the occurrence and distribution of crops is the nutrient mobilization interface through allelopathic compounds released into the soils determining the dynamics of microbial communities. Nutrients and crops are further influenced by the interactions with weeds in agriculture systems with the cumulative and cross cutting impacts on yields and related biochemical pathways as stated above (Mahajan et al. 2012; Tobin et al. 2011). An understanding of these dynamics should guide the management of landscapes; thereby going beyond quantitative aspects of yields alone. This is true even for fish yields.

Some of the most important insights regarding these chemical ecological dynamics have been highlighted by Parmesan (2006); Wilson et al. (2010); in addition to the policy implications of bio resources management by the 2nd SCAR Foresight Exercise of the EU (2009); Steffen et al. (2009) with reference to Australia; Turral et al. (2011) with respect to water and food security; HLPE Report (2012) integrating food security and nutrition; and CCAFS (2012a, b) including aquatic systems.

Meinwald and Eisner (2008) present and excellent overview of emerging trends in the chemical ecological perspective; which is further substantiated by a recent analysis of the links between atmospheric change, secondary metabolic responses and their ecological interactions by Lindroth (2012). Researchers will find very useful leads to further differentiate individual and synergistic impacts, natural and artificially induced phenomena and the transient from long lasting impacts. Obviously, a comprehensive picture of responses of bio-systems will not emerge without adequate insights regarding the above stated. Policy responses have to therefore duly consider these dimensions and device strategies that can be suitably modified with growing knowledge.

6 The Eco-industrial Development Opportunity

Two important facets of eco-industrial development that are mutually reinforcing are waste minimization and cleaner production aligned with optimal resource recovery and reuse. Several efforts have been made by national and international agencies to reach out to the small and medium scale industrial operations to mainstream these facets and optimize production efficiencies. The special attention to this sector is based on the premise that large industries may be able to fend for themselves. Yet, there is another consideration linked to higher energy efficiencies twinned with emission reduction benefits. Exchange of waste materials across industries in their immediate vicinities has also been attempted; albeit to a limited extent only.

It is essential to upscale such interventions from a preventive management perspective with attractive policy incentives that will encourage waste reduction at source. This will also exert lesser burden on diminishing land resources to secure hazardous wastes. Characterization of wastes to optimize recovery of re-usable portions is a prerequisite and has to be driven by management information systems that can define the origin of such wastes. Training and capacity building of concerned personnel to assess processes are critical to meet these goals. Some of the low hanging fruits in this process include process efficiency gains through careful production scheduling and prevention of contamination. This logic extends to collective management of wastes at the level of the industrial estate with significant improvements in end-of treatment practices. Waste minimization and efficient management of unit processes and operations at such stages too confer significant material and energy benefits.

The most important bottleneck in mainstreaming the preventive management strategies highlighted above is access to alternatives, inadequate technical abilities to diagnose processes, causes of wastes' origin and define management information systems. It is therefore essential to define mutually reinforcing regulations, incentives and disincentives and institutional mechanisms that can trigger and sustain such transitions.

7 Conclusion

Through the above analysis, we have tried to demonstrate the importance of promoting synergies between mitigation and adaptation strategies and the need to address a myriad of questions as pre-requisites within the perspective of sustainable development and the welfare of communities. Illman et al. (2013) state in a recent study that synergies are offering solutions to more efficient, responsive and comprehensive climate policy. They can also result in co-benefits with other goals of sustainable development, especially in developing countries. These countries are increasingly preparing national strategies that aim to integrate both mitigation and adaptation aspects into the development aspiration. Even if this process is still slow but successful mainstreaming of climate change issues into various sectors and the development agenda is continuing. Entering an era of more programmatic and possibly "whole of government" approaches may allow better identifying and harnessing of synergies.

In all cases, and given the importance of this promising area, it is recommended that the concept of synergies is linked with the climate mainstreaming agenda and that more empirical research on synergies is conducted to further define and concretize the benefits and challenges, and hence broaden the spectrum of potential for synergies and trade-offs (Illman et al. 2013).

References

2nd SCAR Foresight Exercise of the EU (2009) New challenges for agricultural research: climate change, food security, rural development, agricultural knowledge systems, KBBE, 130 p. ftp://ftp.cordis.europa.eu/pub/fp7/kbbe/docs/scar.pdf

Afifi T, Govil R, Sakadapolrak P, Warner K (2012) Climate change, vulnerability and human mobility: perspectives of refugees from the East and Horn of Africa, UNU-EHS report no1, UNHCR. reliefweb.int/sites/reliefweb.int/files/resources/East%20and%20Horn%20of%20Africa_final_web.pdf

Altieri MA (2012) The scaling up of agroecology: spreading the hope for food sovereignty and resiliency – A contribution to discussions at Rio+20 on issues at the interface of hunger, agriculture, environment and social justice, SOCLS's Rio+20 position paper 20 p. http://www.weltagrarbericht.de/fileadmin/files/weltagrarbericht/The_scaling_up_of_agroecology_Rio.pdf

Antle JM (2010) Adaptation of agriculture and the food system to climate change: policy issues. Issue brief, 12 p. http://www.rff.org/rff/documents/rff-IB-10-03.pdf

Brooks N, Anderson S, Ayers J, Burton I, Tellam I (2011) Tracking adaptation and measuring development, Climate change working paper 1, November 2011, IIED, 36 p http://pubs.iied.org/pdfs/10031IIED.pdf

CCAFS (2012a) Working paper no 23. In: Thornton P, Cramer L (eds) Impacts of climate change on the agricultural and aquatic systems and natural resources within the CGIAR's mandate, 201 p http://cgspace.cgiar.org/handle/10568/21226. Accessed 6 Aug 2012

CCAFS [Kristjanson AJ, Thornton P, Campbell P, Vermeulen S, Wollenberg E] (2012b) CGIAR Research program on climate change, Agriculture and food security CCAFS, Copenhagen, Denmark, 29 p. http://www.ccafs.cgiar.org

Dally GC, Matson PA (2008) Ecosystem services: from theory to implementation. PNAS 105(28): 9455–9456. http://www.pnas.org/content/105/28/9455.full.pdf+html. Accessed 06 Aug 2012

Dube L, Pingali P, Webb P (2012) Paths of convergence for agriculture health and wealth. PNAS 109(31): 12294–12301. http://www.pnas.org/content/109/31/12294.full.pdf+html

FAO (2009) Climate change and agricultural policies, how to mainstream climate change adaptation and mitigation into agriculture policies. EASYPol, 48 p http://www.fao.org/docs/up/easypol/779/policy-clim_change_240en.pdf

FAO (2010) The Hague conference on agriculture, food security and climate change with a special emphasis on climate-smart agriculture. Climate-smart agriculture policies, practices and financing for food security adaptation and mitigation. http://www.fao.org/fileadmin/user_upload/newsroom/docs/the-hague-conference-fao-paper.pdf

FAO (2011) The Ex-Ante carbon balance tool: climate change and agriculture policies, how far should we look for synergy building between agriculture development and climate mitigation? http://www.fao.org/docs/up/easypol/868/synergy_building_agric_dev_climate_mitigation_098en.pdf

HLPE (2012) Report 3, food security and climate change. A report by the High Level Panel of Experts on Food Security and Nutrition, June 2012. http://www.fao.org/fileadmin/user_upload/hlpe/hlpe_documents/HLPE_Reports/HLPE-Report-3-Food_security_and_climate_change-June_2012.pdf

Illman J, Halonen M, Rinne P, Huq S, Tveitdal S (2013) Scoping study on financing adaptation-mitigation synergy activities, Nordic Council of Minister, Nordic working papers, 902. www.diva-portal.org/smash/get/diva2:702352/FULLTEXT01.pdf

IOM (2011) Mitigating the impact of climate change among pastoralist communities. https://www.iom.int/jahia/webdav/shared/shared/mainsite/activities/countries/docs/kenya/Mitigating-Resource-Based-Conflict-among-Pastoralist-Host-Communities.pdf

IPCC (2001) Climate change 2001. Synthesis report, Cambridge University Press

Kotite P (2012) Education for conflict prevention and peacebuilding meeting the global challenges of the 21st Century, IIEP occasional paper, IIEP France. http://www.iiep.unesco.org/fileadmin/user_upload/info_services_publications/pdf/2012/pkotite_conflict_prevention.pdf

Koubi V, Bernauer T, Kalbhenn A, Spilker G (2012) Climate variability, economic growth and civil conflict. J Peace Res 49:113

Lebel L, Li L, Krittasudthacheewa C, Juntopas M, Vijitpan T, Uchiyama T, Krawanchid D (2012) Mainstreaming climate change adaptation into development planning, Bangkok, Adaptation Knowledge Platform and Stockholm Environment Institute. http://www.indiaenvironmentportal.org.in/files/file/mainstreaming%20climate%20change_2.pdf

Lindroth RL (2012) Atmospheric change, plant secondary metabolites and ecological interactions. In: Iason GR, Dicke M, Hartley SE (eds) The ecology of plant secondary metabolites: from genes to global processes, Cambridge University Press, pp 120–153. http://labs.russell.wisc.edu/lindroth/files/2012/05/Lindroth-Iason-et-al-chapt-2012.pdf. Accessed 06 Aug 2012

Locatelli B (2011) Synergies between adaptation and mitigation in a nutshell. http://www.cifor.org/fileadmin/fileupload/cobam/ENGLISH-Definitions%26ConceptualFramework.pdf

Mahajan G, Singh S, Chauhan BS (2012) Impact of climate change on weeds in the rice-wheat cropping system. Curr Sci 102(9):1254–1255. http://www.currentscience.ac.in/Volumes/102/09/1254.pdf

Mayoral A (2012) Climate Change as a conflict multiplier, United States Institute of Peace Brief 120, 01 Feb, http://www.usip.org/sites/default/files/PB%20120_0.pdf. Accessed 07 Aug 2012

Meinwald J, Eisner T (2008) Chemical ecology in retrospect and prospect. PNAS 105:12, 4539–4540, http://www.pnas.org/content/105/12/4539.full.pdf+html

Melamed C, Scott A, Mitchell T (2012) Separated at birth, reunited in Rio? A roadmap to bring environment and development back together, Background Note, May 2012, Overseas Development Institute. http://www.odi.org.uk/resources/docs/7656.pdf

Nicholas A, Birch E, Begg GS, Squire GR (2011) How agro-ecological research helps to address food security issues under new IPM and pesticide reduction policies for global crop production systems. J Exp Bot 62(10):3251–3261. doi:10.1093/jxb/err064

Nisbet MC, Scheufele DA (2009) What's next for science communication? Promising directions and lingering distractions. Am J Bot 96(10):1767–1778. http://climateshiftproject.org/wp-content/uploads/2012/01/Nisbetscheufele2009_whatsnextforsciencecommunication_promisingdirectionslingeringdistractions_americanjournalbotany.pdf

Parmesan C (2006) Ecological and evolutionary responses to recent climate change. Annu Rev Ecol Evol Syst 37:637–669

Ranjan R, Prasad V (2011) Exploring social resilience in state fragility: a climate change perspective. In: Hamza M, Korendea C (eds) Climate change and fragile states: rethinking adaptation, UNU-EHS and Munich Re Foundation, pp 22–42. http://www.ehs.unu.edu/file/get/9717.pdf. Accessed 7 Aug 2012

Steffen W, Burbidge A, Hughes L, Kitching R, Lindenmayer D, Musgrave W, Smith MS, Werner P (2009) Australia's biodiversity and climatic change – A strategic assessment of the vulnerability of Australia's biodiversity to climate change, Technical synthesis, http://www.climatechange.gov.au/publications/biodiversity/~/media/publications/biodiversity/biodiversity-technical-synthesis.pdf

Sterrett C (2011) Review of climate change adaptation practices in South Asia, Oxfam research reports, Climate concern, Melbourne, Australia, November 2011, Oxfam GB for Oxfam International UK. http://www.oxfam.org/sites/www.oxfax.org/files/rr-climate-change-adaptation-south-asia-161111-en.pdf

Tanzler D, Mass A, Carius A (2010) Climate change adaptation and peace. WIREs climate change, Wiley. http://www.adelphi.de/files/uploads/andere/pdf/application/pdf/taenzler_maas_carius_climate_change_adaptation_peace.pdf

The Climate and Development Knowledge Network (2011) Agriculture and climate change, Policy brief, Main issues for UNFCCC and beyond, Meridian Institute. http://ccafs.cgiar.org/sites/default/files/assets/docs/accpolicy_english.pdf

The Peace Corps Energy and Climate Partnership of the Americas (ECPA) (2010) Initiative. http://www.ecpamericas.org/initiatives/?id=35

The World Bank (2012) Carbon livelihoods social opportunities and risks of carbon, Finance 69163, http://www.ddrn.dk/filer/forum/691630WP00PUBL0on0Livelihoods0WEB20.pdf

Tobin PC, Berec L, Liebhold AM (2011) Exploiting allee effects for managing biological invasions. Ecol Lett 14:615–624. doi:10.1111/j.1461-0248.2011.01614x http://www.isarmc.org/wp-content/uploads/2012/07/nrs_2011_tobin_001.pdf

Tol RSJ (2005) Adaptation and mitigation: trade-offs in substance and methods. Environ Sci Policy 8(6):572–578

Tripathi S, Vasan A (2012) Global peace a climate change conflict: a Gandhian approach. In: Proceedings of Gandhirama 2012: a feast of ideas and a festival of art, Indian Council of Philosophical Research, New Delhi, India, 17–22 Aug 2012, p 169

Turral H, Burke J, Faures JM (2011) Climate change, water and food security, FAO water reports. http://www.fao.org/docrep/014/i2096e/i2096e.pdf

UN Global Compact and UNEP (2012) Business and climate change adaptation: toward resilient companies and communities, UN Global Compact Office

UNEP (2012) UNEP year book 2012: emerging issues in our global environment. http://www.unep.org/yearbook/2012/pdfs/UYB_2012_fullreport.pdf

Wilson SK, Adjeroud M, Bellwood DR, Berumen ML, Booth D, Marie Bozec Y, Chabanet P, Cheal A, Cinner J, Depczynski M, Feary DA, Gagliano M, Graham NAJ, Halford AR, Halpern BS, Harborne AR, Hoey AS, Holbrook SJ, Jones GP, Kulbiki M, Letourneur Y, De Loma TL, McClanahan T, McCormick MI, Meekan MG, Mumby PJ, Munday PL, Ohman MC, Pratchett MS, Riegl B, Sano M, Schmitt RJ, Syms C (2010) Crucial knowledge gaps in current understanding of climate change impacts on coral reef fishes. J Exp Biol 213:894–900 http://jeb.biologists.org/content/213/6/894.full.pdf

Chapter 2
Climate Change Impacts in the Arab Region: Review of Adaptation and Mitigation Potential and Practices

Shabbir A. Shahid and Mohamed Behnassi

Abstract This paper aims at presenting a comprehensive review of mitigation and adaptation efforts being made to cope with climate change impact in the Arab region. The review was completed through consulting already published literature (such as official reports, books, scientific papers, conference proceedings, flyers, pamphlets, newspapers, newsletters, and websites). In addition to these, efforts made by Dubai-based International Center for Biosaline Agriculture – to which the key author belongs – will also be shared. The focus is being made on the Arab region with some examples from around the world. It is revealed that climate change (CC) is old phenomenon and the most discussed topic of the present time, and the management of this challenge extends the individual capacity of concerned countries. The CC is impacting all continents, but significantly the water-scarce developing countries, including the Arab region where the major concern is linked to the increase in temperature and rainfall decline leading to increase in evapotranspiration and changes in water cycle depleting the groundwater resources respectively, crucial for both farming and survival of nature, living beings and biodiversity, and in combating desertification. It is envisaged that effective mitigation and adaptation actions, as well as communication of related achievements, can pave the way to slow CC impacts. However, there are pragmatic views expressed by scientists and businesses regarding mitigation and adaptation efforts like the shift to alternate energy sources, biofuels, organic farming, change in land use, deforestation, using set aside or marginal lands, no till or low till farming, chemical

---

S.A. Shahid (✉)
Salinity Management Scientist, International Center for Biosaline Agriculture (ICBA),
P.O. Box 14660, Dubai, United Arab Emirates
e-mail: s.shahid@biosaline.org.ae

M. Behnassi
Public Law Department, Faculty of Law, Economics, and Social Sciences,
Ibn Zohr University of Agadir, North-South Center for Social Sciences (NRCS),
Hy Salam, Agadir 80 000, Morocco
e-mail: behnassi@gmail.com

fertilizers and leguminous crops, livestock management, rangelands, food security, etc. Each component has its own *pros* and *cons* under a set of environmental and geographical conditions. It is believed that adaptation practices can't be generalized to all vulnerable countries; hence the relevance of such adaptation practices to vulnerable country resources and needs must be carefully out looked and understood prior to enacting any adaptation action. It is also assumed that "business as usual" will increase GHG, whereas, adherence to global climate action (such as Kyoto Protocol and all related subsequent decisions) will reduce emission of GHG. It is clear that water scarcity – and not only land – will be a limiting factor to increase agriculture production, an issue that will be exacerbated in the Arab region by the predicted trends of climate change. It is visualized that limited efforts are made and implemented in the Arab region to meet climate change challenges, especially with regard to agriculture and biodiversity. Scientifically-determined climate patterns record barely exists while economic considerations are merely ignored.

Keywords Arab region • Climate change • Mitigation • Adaptation • Vulnerable countries

1 Introduction

The Arab region occupies 10 % of the planet and contains 1 % of the world's freshwater resources, 80 % of which is used in agriculture. The Arab world comprises 22 countries ranging from oil rich nations to poor least developed countries, and divided into five hydrological regions: The Mashreq (Iraq, Syria, Lebanon, Jordan and Israel); Al Maghreb (Libya, Tunisia, Algeria, Mauritania and Morocco); Nile Basin (Egypt and Sudan); Arabian Peninsula (Saudi Arabia, Kuwait, United Arab Emirates, Qatar, Oman, Bahrain and Yemen); and Sahel (Somalia, Djibouti and Comoros Islands). Developing sustainable land use and water practices and addressing climate change impact (CCI) in the Arab region requires a detailed knowledge of the system dynamics. A way to establish the changing trend of these complex systems is through using simulation models. In the Arab region, some efforts have been made to project and predict changing trends of rise in temperature, reduction in precipitation etc. and some mitigation and adaptation efforts are being made – based on local conditions and available resources – and communicated.

Arab region is considered the world most water scarce, with high water demand but less recharge and supply; where the climate models are predicting a hotter, drier and less predictable climate. It is speculated that CC will affect rainfall amounts, frequency and patterns, and duration – rainfall becomes less reliable – leading to increase in floods, hurricanes, storms, and drought while disturbing the water balance in the region as well as affecting the quantity and quality of the water. Higher temperature and less rainfall will reduce the flow of rivers and streams, slow the rate at which aquifer recharge, progressively raise sea levels and make the entire region more arid leading to reduction in biodiversity, and significant impact on agriculture, infrastructure, and water management. Enhanced release of CO_2 to the atmosphere from soil organic carbon as a result of increased temperature is likely to occur (Jones et al. 2005). It is noted that limited efforts are being made and implemented in Arab region to meet CC challenges, and these efforts remain insufficient to cope with relevant impacts, especially for agriculture and biodiversity. Scientifically determined, climate patterns record barely exists and economic considerations are merely ignored. The missing link of GCC countries, including UAE (no GCC country is member of Global Environment Facility-GEF), therefore restrains funding to UAE for local environment projects (Gulf News September 14, 2010). The GEF provides fund for projects covering biodiversity, climate change, international waters, land degradation, the ozone layer and persistent organic pollutants. This paper reports a review of efforts made in the Arab region with respect to CCI on environment, agriculture, biodiversity, adaptation and mitigation. The basic concepts of CC as well as introduction of Kyoto Protocol and Copenhagen Accord are briefly given.

1.1 Climate and Climate Change

Climate is the average weather conditions, such as temperature, rainfall and wind usually taken over a 30 year-period (Shahid 2010), whereas the CC is the long-term alteration in global weather patterns, especially increases in temperature and storm activity and the potential consequences of the greenhouse effect. Both atmospheric temperature and rainfall are linked to climate change impact (CCI) in terms of melting glaciers, sea level rise (SLR), hurricanes, floods and drought. It is wiser to state that the CC is an environmental issue of peripheral concern but has recently reached the realm of high politics.

1.2 Greenhouse Effect (GHE)

The GHE is heat buildup near the Earth's surface atmosphere. Part of the sun's heat reflected from earth is trapped by greenhouse gases (GHGs) in the atmosphere (which normally exit into space) and then re-radiated back toward the Earth's surface, causing rise in temperature of lower atmosphere, inflicting potentially catastrophic climate damage. The GHG absorbs infrared radiation in the atmosphere. The GHGs comprise water vapor, carbon dioxide (CO_2), methane (CH_4), nitrous oxide (N_2O), hydrofluorocarbons (HFCs), perfluorocarbons (PFCs) and sulfur hexafluoride (SF6). The atmosphere used to hold 280 parts per million (ppm) CO_2 in 1750 before the industrial revolution; rising to 315 ppm in 1955 and 430 ppm today, expected to reach 550 ppm by 2035, and global average temperature may rise by more than 2 °C. With business as usual (BAU), Tolba and Saab (2009) warned that the stock of GHG may be more than triple by the end of twenty-first century, with 50 % risk of 5 °C temperature rise in the coming decades. The contribution of the Arab region to global GHG emissions from major sources is about 4.2 % (Tolba and Saab 2009). Worldwide agricultural land occupies 37 % of earths' land surface and account for 52 and 84 % of global CH_4 and N_2O emissions respectively (Smith 2007). If we look at agriculture from mitigation aspect, it reduces its own emission, offsetting emissions from other sectors by removing CO_2 from atmosphere through photosynthesis and storing carbon in soil. The section below presents an overview of how CC issue initiated through Kyoto Protocol and led to Copenhagen Accord in 2010.

1.3 Kyoto Protocol (KP) and Copenhagen Accord (CA)

The CC is an old phenomenon but the most discussed topic of the twenty-first century. To address this important issue, the United Nations Framework Convention on Climate Change (UNFCCC) was adopted at the Earth Summit in Rio de Janeiro in 1992. Later, the KP was adopted in Kyoto (Japan) on 11 December 1997, ratified by 192 parties, and entered into force on 16 February 2005. The KP's first commitment period runs out in 2012. The ultimate objective of UNFCCC is to achieve "stabilization of GHGs concentrations in the atmosphere at a level that would prevent dangerous anthropogenic interference with the climate system". The KP sets emission limits for 37 industrialized countries. The UNFCCC encouraged industrialized countries to stabilize GHG emissions whereas the KP commits them to do so. The G-8 Summit in 2008 & 2009 agreed to reduce global GHG emission up to 50 % by 2050 and to limit the rise in temperature to no more than 2 °C. However, developing countries have serious concerns over this reduction. Similarly, the European Union (EU) can commit 20 % reduction by 2020 from 1990 levels and can go to 30 % if others make the same commitment. This situation proves that no one is strictly volunteering in making commitment with regard to mitigation. Within the same framework, Japan would reduce its emissions up to 25 % by 2020 from 1990 levels. On the other hand, US legislation, if it becomes law, would result in a 17 % reduction by 2020 from 2005 levels, *which was a dream to come true*.

As a follow-up to above, delegates from 194 countries met in Copenhagen (7–18 December 2009) in an effort to negotiate a binding accord on reduction of GHGs to tackle global warming beyond 2012. The major achievement was a compromise voluntary pact titled *Copenhagen Accord (CA)*, which states that US, China, India, Brazil and South Africa agree that CC is one of the greatest challenges of our time. The chief provision of CA is: the increase in global temperature should be below 2 °C; and developed countries should provide financial and technological assistance to developing countries, particularly to the most vulnerable ones. This includes 30 billion US$ for 2010–2012 and a goal of providing US$ 20 billion a year by 2020. Substantial funds to reduce deforestation and forest degradation are to be provided. A Copenhagen Green Climate Fund will be established to provide funds. Developing countries, especially those that now have low emissions, should be provided with incentives to develop with low emissions. All developing countries should take mitigation actions, report these and allow international verification of their emission data.

2 Climate Change Impacts Projection in the Arab Region

This section describes CCI projections in the Arab region, in the context of sea level rise (SLR) and effects on infrastructure and development, water resources, coastal areas, agriculture, biodiversity, and food security, etc.

2.1 Climate Change Impact on Rainfall and Water Resources

Annually about 2000 billion m^3 rain falls over Arab countries with significant losses; however rainfall is likely to be decreased due to CCI. The Regional Circulation Models (RCM) projection for precipitation change in the Gulf region predicts the region to get drier, with significant rainfall declines in the wet season outweighing slight increases during the drier summer months (Cruz et al. 2007; Hemming et al. 2007). Meanwhile, the rainfall distribution will change (moving to north), the weather is also likely to become more unpredictable with the region experiencing an increase in extreme rainfall events (Hemming et al. 2007). Such events have been witnessed in Kuwait in 1998 (70 mm rain in an hour) and the government had even declared emergency; in Saudi Arabia in 2009, rainfall inundated large urban areas of Jeddah with some death tolls.

Let us assume that rainfall does not decrease, which is less likely, the rise in temperature will still have significant impact on evapotranspiration and water balance, reducing available resources up to 15 % in a context marked by the increase in agricultural water demand by the year 2020 arising to 6 % (Bou-Zeid and El-Fadel 2002). A high water exploitation index (annual total abstraction of fresh water divided by long term average of fresh water resources) reaching 83 % (Tunisia), 92 % (Egypt), 169 % (Gaza) and 644 % (Libya) has been reported by Ragab and Prudhomme (2000). They predicted, through climate model in Mediterranean and Middle East countries, the percent change in rainfall with respect to mean monthly values, that is 20–25 % less rainfall in the dry season (April to September) by the year 2050 (North Africa, some part of Egypt, Saudi Arabia, Iran, Syria and Jordan). This decline is accompanied by rises in temperature between 2 and 2.75 °C. In winter, rainfall will decrease by about 10–15 %. Winter temperatures in the coastal areas will also increase by 1.5 °C on average, while inside the region it will increase by 1.75 °C to 2.5 °C. In this regards, the non-conventional water resources, such as reclaimed or recycled water and desalinated brackish water or seawater, are expected to play an important role. An innovative super-high-resolution (20-km) global climate model was employed (Kioth et al. 2008) which accurately reproduces the precipitation and the stream-flow of the present-day Fertile Crescent revealing loss of current shape of Fertile Crescent, which could disappear by the end of the century. Jorgensen and Al-Tikiriti (2002) working on archaeological sites (4,500 years old) in the UAE developed aridity trends (desertification) and linked the drop of ground water levels to trends on increasing aridity (climate change). The increased aridity trends are contemporaneous with reported increased atmospheric CO_2 trends. The flow in major rivers like Blue Nile, the white Nile, the Euphrates, the Tigris (Turkey, Syria and Iraq), the Jordan, and the Yarmuk River in Southern Lebanon will be affected, become variable and non dependable. Under moderate temperature increases, some analysts anticipate that the Euphrates River would carry 30 % less water than it currently does and the Jordan River could shrink by up to 80 % by the end of century (Brown and Crawford 2009). The UN warns of CC effects on water shortages and the message took worldwide attention (www.un.org/News). The report warns that an extra 155–600 million peoples could experience additional water stress if temperatures increase by few degrees, as well as water scarcity, food insecurity and erratic economic growth may be exacerbated by CC. The Middle East as a whole withdraws the world's highest proportion of its total renewable water resources (75 %); South Asia is second with 25 % (Raphaeli 2007). The lack of water resources means reliance mainly on other resources (ground water, wastewater and desalinated waters). The possible implications of CC in MENA region (Saleh 2010) are the vulnerability of water resources projects, the sea level rise in coastal regions coupled with more frequent extreme events like tropical cyclones add to design consideration especially for infrastructure. The risk of lower water availability, especially in the Middle East, could affect the food production, and hence socio-economic implications of CC could be severe in some countries of the region, such as Syria where around 20 % of all active workers are employed in agriculture sector.

2.2 Temperature and Sea Level Rise (SLR) – Consequences for the Arab Region

The chief provision of *Copenhagen Accord* (CA) is: the increase in global temperature should be below 2 °C and this requires effective mitigation and adaptation efforts to meet the commitment, otherwise temperature may extend 2 °C. To achieve this target and maintain current mean temperature, the energy balance needs to be maintained between lands, sea and atmosphere. For the time-being, this balance does not exist, temperature is increasing and hence global warming is melting glaciers worldwide resulting into SLR. The past century has witnessed a 17 cm in SLR (IPCC 2001) at a mean rate of 1.75 mm per year (Miller and Douglas 2004), the IPCC (2007) predicts SRL up to 59 cm by 2100. Taking the full likely rise of temperature, SLR could even increase to 1.4 m by 2100 (Rahmstorf 2007), and even between 5 and 6 m in the event of West Antarctic Ice sheet collapse (Tol et al. 2006). The SLR witnessed in the last century was for the first time observed to be driven primarily by human-induced warming (Cruz et al. 2007). However, an increased infusion of melt water into the world's ocean and a thermal expansion caused by warming sea temperatures both contributed to the rise, which since 1993 have averages 3 mm per year with variations (WBGU 2007). The earth's mean global temperature rose by $0.6 \pm 0.2\,°C$ during the second half of the twentieth century, at an average rate of $0.17\,°C$/decade (IPCC 2001; Lal 2007; Khan and Lal 2007). This climate change is attributed mainly to the increase in concentration of GHGs (CO_2, CH_4, N_2O) by fossil fuel combustion, land use change and deforestation, soil degradation because of inappropriate land use and soil management practices (Lal 2004).

The Levant region-a mix of arid and semi-arid zones due to low rainfall (Syria, Lebanon and the occupied Palestinian territory) is the world most water scarce, where water demand exceeds supply in many places, and climate models are broadly predicting a hotter, drier and less predictable climate (IPCC 2007). The RCM predictions of average temperature (°C) and precipitation (%) changes across the Gulf region for the 2020s, 2040s and 2070s, by reference to 1990 levels have been presented (Hemming et al. 2007). These predictions show that the region is expected to get hotter across all seasons by 2050: models predict an increase of temperature between 2.5 to 3.7 °C in summer, and 2.0 to 3.1 °C in winter (Hemming et al. 2007; Cruz et al. 2007). In the Middle East, the average temperature rose by 1.5–4 °C during the course of the twentieth century (Alpert et al. 2008) with aggravation in summer heat levels (Ziv et al. 2005). The climate predictions in the Middle East for the next 100 years foresee droughts that are ten times more frequent (Weib et al. 2007), temperature will rise by 4–6 °C (Alpert et al. 2008), rainfall will decrease by up to 15 % (Ragab and Prudhomme 2000) and the Jordan River flow predictions show a decrease between 23 and 73 % (Kitoh et al. 2008).

These predictions are higher than CA commitment, and therefore, extreme temperature rise forces the policy makers to mitigate GHG and implement policies for innovative adaptations. It has been predicted (Freimuth et al. 2007) that a rise in temperature of just 1.5 °C is expected to shift the Mediterranean climate zones (or biomes) 300–500 km northwards, making the region as a whole more arid. If temperature warms beyond 2–3 °C, there is a growing risk that the climate system will pass critical thresholds. Positive feedback loops could then trigger runaway changes to our climate – so called "*non-linear*" events or "*climate surprises*", such as the melting of Greenland ice sheet at large-scale, die back of the Amazon rainforest (WBGU 2007), though these events may seem outside of the immediate concern of the region, all carry significant global consequences. Should the Greenland ice sheet melt, for example, global sea levels could rise by 7 m with serious consequences for the entire region, every country of which has a densely populated coastline (WBGU 2007). If the energy balance is maintained (less likely) and the present rising temperature trend continues, a drastic increase in global temperature is projected by the end of the twenty-first century and the consequences will be SLR and accelerated meltdown of polar ice sheets. Scientists recently observed that a shelf of floating ice, which was larger than 100-km^2, and which jutted into the

Arctic Ocean for 3,000 years from Canada's northernmost shore, broke away in the summer 2005 because of sharply warming temperatures. It is predicted that global SLR will be between 0.1 and 0.3 m by 2050 (Cruz et al. 2007). The IPCC's Fourth Assessment Report (IPCC 2007) posits an upper boundary for global sea level rise by 2,100 of 0.59 m; however, it does not include ice sheet dynamics, as per IPCC the SLR will continue rising for centuries, even if climate forcing is stabilized (IPCC 2001), up to 5 m SLR if the melting of the Antarctic ice sheet is taken into consideration, and even up to 10 m SLR may be possible based on emission scenario, albeit over a long time frame (IPCC 2007).

The AFED report (Tolba and Saab 2009) summarizes CCI in Arab World, stating that the projected CC impacts on the fragile environment (rising temperature, less rainfall, SLR) as well as aridity, drought and water scarcity in the region is immense and call for effective mitigation and adaptation measures. A simulation carried out by Boston University revealed that 1 m SLR would directly impact 41,500 km^2 of the Arab Coastal land (Egypt, Tunisia, Morocco, Algeria, Kuwait, Qatar, Bahrain, and the UAE). Almost 11 % of the land area of Bahrain will be lost due to 50 cm SLR if no action is taken for protection (Al Janeid et al. 2008). In Egypt, 1 m SLR will put 12 % agricultural land at risk, and directly affect 3.2 % population compared to 1.28 % worldwide. A 2 °C rise in temperature will make extinct up to 40 % of all species, and 1 °C rise in average global temperature economic growth will drop by 2–3 %.

According to World Economic and Social Survey released by UN in 2009, if mitigation and adaptation actions are not taking place while keeping a business as usual state of mind, it is predicted that World Gross Product (WGP) loss could be as high as 20 %. The unique Cedar forests in Lebanon and Syria, mangroves in Qatar & UAE, the red marshes of Iraq, the mountain ranges of Yemen and Oman, and coastal mountain ranges of the Red Sea remain highly vulnerable to CCI. 75 % of the buildings and infrastructure (water and energy generation networks) in the region are also at high risk of CCI (SLR, higher intensity and frequency of hot days and storm surges). The newly-built small islands will be first to disappear through SLR due to their small size and perhaps low elevation. An increase of mean annual temperature up to 4.5 °C, and 25 % decreases in mean annual precipitation in the mountainous part of the upper Jordan catchment has been reported through regional climate-hydrology simulation (Kunstmann et al. 2007). In a worldwide scale, the coastal areas are at high risk of SLR. Of 34,000 km coastline in the Arab World, 18,000 km is inhabited. The landmass of Qatar, UAE, Kuwait and Tunisia will be most vulnerable and in these countries under different projections and scenarios, 3 % of land will be affected (1 m SLR), 8 % (3 m SLR) and 13 % (5 m SLR). Qatar is by far the most exposed country. Over 2 % of the GDP of Qatar, Tunisia and UAE is at risk in case of 1 m SLR, increasing to between 3 and 5 % for 3 m SLR. The low-lying coastline exists in most of the GCC countries (UAE, Kuwait, Qatar, Oman, Bahrain) which may be within the reach of few meters sea level rise, e.g., in the UAE which has nearly 1,300 km of coastline, about 85 % population and 90 % of infrastructure are located within several meters of sea level (EAD 2009b).

Two scenarios (EAD 2009b) of SLR were used in the Abu Dhabi Coastal Ecosystem base map: (1) no accelerated ice cap melting and by 2050 1 m above mean sea level (MSL); (2) accelerated ice cap melting, 1 m above MSL by 2050 and 9 m above MSL by 2100. Much of Abu Dhabi is above 1 m, therefore most of the area is not vulnerable to 1 m rise, but for the low lying areas and under scenario 2, with 3 m SLR, several offshore islands, mangrove village and industrial city south of main land will be under water; with 9 m of sea level rise, Abu Dhabi will look fundamentally different as a city and society as of today. The extent of inundation in Abu Dhabi Emirate range from 344 km^2 (1 m SLR), 805 km^2 (3 m SLR) and 1,672 km^2 (9 m SLR), where as in Dubai it is 14 km^2, 147 km^2 and 221 km^2 inundation area with 1, 3 and 9 m SLR scenarios respectively (Tolba and Saab 2009). The derived Digital Elevation Model (DEM) from topographic maps shows 332 km^2 area below 10 m in three emirates (Sharjah, Ajman, Umm Al-Quwain) and highly vulnerable to SLR. It is projected that 1 m SLR would inundate 8.1 % (Ajman), 1.2 % (Sharjah) and 5.9 % (Umm Al-Quwain) area, with 8 m SLR the inundated area would be 24 %, 3.2 % and 10 % respectively.

2.3 Climate Change Impact on Coast, Soil Salinity, Agriculture and Biodiversity

Increased temperature and decreased rainfall will lead to predominantly upward water movement in soil, as currently seen in most of the arid-region soils, this will increase soil salinity. It has long been recognized that CC will increase SLR (explained above) resulting into inundation and salinity along coastal regions worldwide (Pezeshki et al. 1990). Overall worldwide twin problem of soil salinity and water logging lowering the productivity of 25 % of the world's irrigated cropland (Brown and Yong 1990) and potentially will have significant impacts on future agricultural production under a changed climate. Recently Shahid et al. (2010) has presented case studies on soil salinity mapping in the Middle East, the estimates reveal 11.2 % soils are salinized in the entire Middle East (Shahid et al. 2010; Hussein 2001), 12.1 % area affected to varying degrees of salinity in Kuwait (Shahid et al. 2002), and 36 % of soil map units (partially or completely) affected by salinity Abu Dhabi Emirate (EAD 2009a; Abdelfattah et al. 2009), it is most likely that the changing climate and subsequent SLR will increase soil salinization in the coastal zones to a significant extent.

About 600 million people living worldwide in low lying coastal areas need to be worried about estimated projections of SLR, which may lead to population displacement and infrastructure destruction. The menace of global warming is already taking a heavy toll, and low-lying countries such as Maldives and Bangladesh are on the brink owing to SLR. One third of Bangladesh will be flooded with 1 m SLR affecting nearly 20 million people. In 2005, Hurricane Katrina temporarily displaced 1.5 million people. Similar to non-Arab countries, the 2007 hurricane Gonu in Oman has left its footprint in the coast as well as Muscat city of Oman. In 2010, at least 160 million people living in coastal areas – a very conservative estimate – might be at risk of flooding from storm surges, recently 20 million peoples are affected and displaced internally in Pakistan due to devastating floods of July 2010 which has also significantly affected agriculture and food supply. In Africa, coastal areas are already facing problems – such as coastal erosion, flooding and soil subsidence – while the coastal areas – Gaza and Shat-al-Arab such as Iraq, Iran and Kuwait – are the most vulnerable to the projected SLR and its impact on environment, agriculture and biodiversity (flora and fauna) as well as the coastal community migration to other areas. SLR inundates coastal areas, causes soil erosion and soil subsidence, increases sea water intrusion into coastal aquifers, of particular concern is the Gazan coastal aquifer, upon which 1.5 million Palestinian depend for their drinking water. This aquifer is particularly vulnerable to seawater intrusion as a result of over-pumping (both inside and outside Gaza), which has lowered the water table (Brown and Crawfield 2009), affect flora and fauna, and destroy infrastructures and community migration. Any added salinity from SLR will further compromise the water quality in the aquifer, which is already polluted. When it comes to the agricultural sector, Egypt will be most impacted by SLR. More than 12 % of Egypt's best agricultural lands in the Nile Delta are at risk (Tolba and Saab 2009).

As climate changes, the value of biodiversity for food and agriculture will increase. It is anticipated that the modest climate change till 2020 could be beneficial to agriculture (early season supply to markets), while climate change as anticipated till 2100 will be detrimental, therefore, a reduction of 20 % in state-wide annual agricultural net-revenues is projected compared to 2002 (Trondalen 2009). Rise in temperature and decrease in rainfall due to GHE would have great impact on water losses through evapotranspiration, thus crop water demand will definitely be increased, leading to ultimate change in cropping patterns and yield declines. Immediate impacts will be on dryland farming in Africa, specifically in Ethiopia where less than one per cent of total cultivated lands are irrigated and the rest is rain-fed, therefore, dry areas are likely to even get drier and too hot for certain crops. By 2020, yields from rain-fed agriculture in some African countries are projected to reduce up to 50 %; thus increasing food insecurity and hunger; and 75 to 250 million peoples are predicted to be exposed to water stress due to CC.

The Arab region, due to high aridity and hot climatic conditions, has already low biodiversity by reference to global standards; and this will further decline due to intensifying climate change, a 2 °C rise in temperature will make an extinct up to 40 % of all species (Tolba and Saab 2009). Using the IUCN threat categories in Yemen, there are 159 threatened plant species, while Somalia and Sudan have 17 each. Other countries – Egypt, Jordan, Morocco, Saudi Arabia, Somalia, Sudan, Yemen – have more than 80 threatened animal species, with Egypt more than 108, affecting the entire ecosystem. In sub-Sahara Africa, the combination of historical crop production and weather data into a panel analysis predicts a yield decline of maize, sorghum, millet, groundnut and cassava by 22, 17, 17, 18 and 8 % respectively by 2050 (Burke et al. 2009). A research by CGIAR based on distribution models of wild relatives of three staple crops of the poor – peanuts, cowpea and potato – suggests that by 2025, 16–25 % of wild species will be threatened by extinction. For a better vision about biodiversity and CCI, it is basic requirement to match biodiversity distribution mapping with different CC scenarios to develop national and regional conservation strategies. While describing the biodiversity and climate change in Kuwait, Omar and Roy (2010) stressed on developing strategic plan for climate change mitigation and adaptation in Kuwait including identifying vulnerable species through field survey, monitoring and *ex-situ* and *in-situ* conservation, increasing protected areas, and through awareness and capacity building, as biodiversity in Kuwait is under severe threat to various natural and anthropogenic factors, where land impacted by anthropogenic activity is reaching 10.47 %. They further pointed out that the potential rise in seawater temperature would affect spawning period of fish and shrimp and relocation to other suitable areas, disturbing the fish industry in Kuwait and the region. Increase seawater temperature and acidity will cause coral reefs bleaching, and thus affect tourism in Jordan and Egypt.

3 Mitigation and Adaptation

Mitigation is a human intervention to reduce the anthropogenic impacts on climate system; it includes all strategies aiming at reducing GHG sources and emissions and enhancing GHG sinks. Within the perspective of IPCC (2007), *adaptation* is the adjustment in natural or *human systems* in response to actual or expected climatic stimuli or their effects, which moderates harm or exploits beneficial opportunities. Various types of adaptation can be distinguished, including anticipatory, autonomous and planned adaptation: (1) adaptation that takes place before the observation of CCI (anticipatory or proactive adaptation); (2) adaptation that does not constitute a conscious response to climatic stimuli but is triggered by ecological changes in natural systems and by market or welfare changes in human systems (autonomous or spontaneous adaptation); (3) adaptation that is resulting from a deliberate policy decision, based on awareness that conditions have changed or are changing and that action is required to return to, maintain, or achieve a desired state (planned adaptation). Examples of adaptation activities have been defined by Klein et al. (2006).

3.1 Adaptation in the Vulnerable Area of Arab World

In order to enact effective adaptation to CC and to mitigate GHG impact, it is essential to produce relevant information and projections. In this regard, it is necessary to set up mechanisms aiming at collecting climate-related information and to build early warning systems. It has been concluded that major sources of GHG emissions in the Arab region don't extend a merely 4.2 % to the global emission potential. However, the impact of climate change on the fragile environment of the region and its people is expected to be immense, and this demands urgent planning for adaptation measures (Tolba and Saab 2009). The SLR rise is the main threat to coastal area; therefore, it is essential to be careful in building new infrastructures (buildings, industries, agricultural activities, etc.), slowly displace/migrate the community and agricultural activities to inland. Adaptation projects should address core issues through better water management, agricultural development and disaster prevention. Once it is guaranteed that water shortage is not an issue, adaptation efforts can be achieved easily compared to water-stressed regions. For quick results, innovative approaches for adaptation tested in similar conditions should be reproduced while considering the local specifities. An important, although obvious, conclusion is that adaptation does not reduce the inherent vulnerability of concerned territories; rather it serves to enable communities to withstand, bounce back from or absorb the effects of vulnerability to climate change (Briguglio 2010).

It is speculated that water and not land will be a limiting factor to increase agriculture production, an issue that will be exacerbated by the predicted trends of climate change. Indeed, agriculture sector is the greatest consumer of water and it is not a matter of absolute use but relative use. It is therefore important to find ways as how to increase the efficiency of water for agriculture (Shetty 2006) and how to reduce the overall use. In this perspective, some countries – such as Tunisia, Morocco and Jordan – have already started to address the issue of water reallocation (Adams et al. 1999).

It has been estimated (Alcamo et al. 2005) that global demand for ecosystem services will substantially increase by 2050: cereal consumption will increase by a factor of 1.5–1.7; fish consumption by a factor of 1.3–1.4 up to 2020; water withdrawals by factor of 1.3–2.0; and biofuel production by a factor of 5.1–11.3. With the scenario of business as usual (BAU), and without significant mitigation efforts, local agriculture production will be at high risk due to increased temperature and water scarcity, especially in countries like UAE, Kuwait, Qatar, etc. One way many government are saving precious water resources and assuring food security is through *"Land Resource Grab"* in other countries. During 2008, the emergence of "land-grabbing" – the purchase or long-term lease of vast tracts of land from mostly poor, developing countries by wealthier, food-insecure nations as well as private entities to produce food for export – has raised deep concern over food security and rural agricultural development. The foreign interests were seeking or securing between 37 million and 49 million acres of farmland between 2006 and the middle of 2009 (Buying Farmland Abroad 2009). The issue of Land Grab has been described in great length by many authors (Daniel 2011; Behnassi and Yaya 2011) with three main trends driving the land grab movement: the rush to secure food supply by increasingly food-insecure nations, the surging demand for agrofuels and other energy and manufacturing demands, and the sharp rise in investment in both the land market and the soft commodities market. The Gulf States, with scarce water and soil resources on which to grow food, but vast oil and cash reserves, have watched their dependence on food imports become increasingly uncertain and ever more expensive, their total food import bill ballooning from

US$ 8 billion to US$20 billion from 2002 to 2007 (GRAIN 2008). These states have moved quickly to extend control over food-producing lands abroad. Qatar, with only 1 % of its land suitable for farming, has purchased 40,000 ha in Kenya for crop production and recently acquired holdings in Vietnam and Cambodia for rice production and in Sudan for oils, wheat, and corn production (Capital Business 2009). There has been strong criticism in Kenya over land garb (Kilner 2009). The United Arab Emirates (UAE), which imports 85 % of its food, purchased 324,000 ha of farmland in the Punjab and Sindh provinces of Pakistan in June 2008 (Kerr and Farhan 2008). Other emerging nations such as China, Japan, and South Korea are also seeking to acquire land as part of a long-term strategy for food security.

The FAO Director General Jacques Diouf, while having clearly expressed his concern about the potential consequences of swift land grabbing on political stability, has supported the proposed Gulf food deals *as a means of economic development for poor countries. If the deals are constructed properly,* he said, *they have the potential to transform developing economies by providing jobs both in agriculture and other supporting industries like transportation and warehousing* (Coker 2008). However, a dangerous element of the land grab trend is the shift from domestic to foreign control over food resources and food-producing lands. Where food production is still feasible in vulnerable areas, adaptation strategy should include innovative technologies and those best suited to the vulnerable community in the region, improved irrigation and fertilizer techniques, water saving techniques, integrated water resources management, such as rainwater harvesting, recycling, increasing promoting water conservation and efficiency, using alternate water sources (waste water) and using drought resistant and salt tolerant varieties. With the BAU, there is a risk of 50 % decline in food production. Adaptation strategies should be mainstreamed in national policies and supported by governments in terms of implementation.

The environmental costs associated with increased food production may rise with globalization, for example due to increased GHG emissions, especially N_2O and CH_4 which are relatively more damaging than CO_2 and for which agriculture is the major source (Stern 2007). The emissions of GHG from agriculture are mostly related to three main activities: (1) CO_2 emissions from fertilizers (urea and ammonium bicarbonates), land conversion to cropping, use of agricultural machinery and livestock production; (2) nitrous oxide (N_2O) emissions from nitrogen fertilizers, manures and nitrogen-fixing legumes, as well as microbial conversion of other nitrogen sources in agricultural soils; and (3) CH_4 emissions from livestock and irrigated rice production. The livestock industry produces about 7.1 billion tons of CO_2 equivalents, which represents 18 % of total anthropogenic GHG emissions (mainly CH_4). Relative contributions along the food chain are 36 % land use and land use change, 7 % feed production, 25 % animals, 31 % manure management and 15 % processing and transport (Gerber and Steinfeld 2008). The GHG emissions from livestock rearing are mainly CH_4 emissions from enteric fermentation, animal manures and respiration. It is essential to change cattle fodder to reduce CH_4 emissions from enteric fermentation, manure management and management of livestock population.

According to recent FAO estimates, the livestock industry is responsible for 18 % of total anthropogenic GHGs. The GHG contribution from the livestock industry would be reduced by vegetarian diets, supported by the fact that the conversion efficiency of plants into animal matter is approximately 10 %, and thus there is a prima facie case that more people could be fed from the same amount of land if they were vegetarian, as a third of global cereal production is currently being fed to animals (FAO 2006). In reality the trend is opposite, as in the past five decades, due to rapidly increasing meat and dairy consumption, the number of animals – cattle, sheep and goats – increased approximately 1.5 fold, with equivalent increases of 2.5 and 4.5 fold for pigs and chickens respectively (FAO 2009). Other options to reduce GHGs are improvement in rice varieties and cultivation practices (introduction of drainage in paddy field), rationale use of fertilizers, introduction of biofertilizers, green manuring, solid waste composting rather than incineration.

In Abu Dhabi Emirate, total water consumption (EAD 2009b) by various sectors is agriculture (57 %), forestry (18 %), domestic (15.5 %), amenity (7 %), and industrial (1.5 %). As a part of adaptation to CCI, the Government of Abu Dhabi has taken serious steps to save water resources through banning Rhodes grass cultivation to save huge amounts of water (Gulf News 24 August 2021a). Gulf News further states that this move is significant as the government is concerned about unplanned and uncontrolled ground water withdrawals of more than two billion cubic meters annually, especially in the agriculture and forestry sectors. The Abu Dhabi Food Control Authority (ADFCA) revealed a plan to phase-out the cultivation of Rhodes grass (the phase-out will eventually lead to a ban), a principal feed in the region and a water-intensive crop (rhodes grass uses 24,000 m^3 water per hectare annually). The intention is to reduce water usage in agriculture by 40 % over existing levels through changing cropping pattern (less water consumption), adoption of improved water application techniques and updating of old and inefficient on-farm water systems currently practiced, and improved marketing strategies. Abu Dhabi Agriculture and Food Safety Authority (ADAFSA) has signed a 5-year contract with ICBA (ADAFSA-ICBA 2010; ICBA 2010a) to accomplish above activities. The major objective is to replace Rhodes grass (using high quantity of water) with alternate grasses such as buffle grass (*Cenchrus ciliaris*), blue panic (*Panicum antidotale*) and tuman (*Pennisitum divisum*) and *Distichlis spicata, which* have been tested successfully at ICBA – these grasses use less water and are well adapted to the harsh desert environment – as well as importing feed from other countries.

Other efforts, such as the use of *organic farming*, zero or low tillage, biodiesel crops should be selected prudently after careful considerations of *"pros and cons"*, based on the country economic and food situation. *Organic farming* in the context of changing climate was deliberated and reported in the Synthesis Report (Shahid et al. 2009): *"Improvements in organic farming techniques can make a contribution to food security, particularly in situations where capital is scarce or expensive and labor is plentiful and inexpensive. In these cases, the yield penalty is likely to be small and can be more easily made up through improved farmer skills"*. In this context, the use of organic farming as a replacement of conventional farming based on chemical fertilizers is a risky shift, because in poorly developed countries the yield gap between inorganic and organic farming cannot be matched. Using chemical fertilizers and pesticides have impact on GHG emissions through their preparation and uses; biofertilizers also emit GHGs through microbial decomposition in mineralization process. The rich countries importing most of their foods should check whether such a practice could be used from food quality point of view. Recently a similar effort is being made in Abu Dhabi Emirate where more than 700 organic farms are presently functional, especially that the country imports around 83 % of its food from other countries. The begging question here: Is it REALLY necessary to grow and buy organic? And the number one factor fueling that question is the fact that organic products are typically more expensive than non-organic, and poor consumers cannot afford such a high quality food. However, the organic produce tastes better and is healthier than commercially cultivated produce. It contains higher levels of vitamins C and A, antioxidants and essential minerals like calcium, magnesium, iron, selenium and chromium. It also has no residue from the more than 500 chemical pesticides routinely used in conventional farming.

Another interesting area to sustain food security is the use of *Precision Agriculture*, which refers to the use of a series of technologies that allow the application of water, nutrients and pesticides only to the places and at the time they are required, thereby optimizing the use of inputs (Day et al. 2008). However, this requires investigation and investment for its implementation.

Conservation agriculture (CA) aims to conserve, improve and make more efficient use of natural resources through integrated management of available soil, water and biological resources combined with external inputs. Intensification of agriculture production has been an important factor influencing GHG emissions and affecting water balance. Agricultural activities contribute to CO_2 emissions released to the atmosphere through the combustion of fossil fuel, soil organic matter (SOM) decomposition and biomass burning. Improved CA practices, especially in water-limited areas, have great potential to increase soil organic carbon sequestration, available water storage, and decrease in net emissions of CO_2 and other GHG. Lal's (2004) estimates of the potential of soil C sequestration in the Central Asia countries indicate a range of 10–23 Tg C year^{-1} over 50 years, and to achieve this improved management practices require less fallowing. To reduce GHG emissions and for better adaptation, no-tillage and conservation agriculture (CA) are experiencing a persistent and steady growth in the world (covering 95 million ha worldwide). Traditional intensive tillage cultivation systems lead to soil degradation and loss of productivity. There are advantages and disadvantages of tillage and no- or low-tillage cultivation. No tillage is a practice of planting crop on untilled soil by opening a narrow slot, trench or band sufficient size for proper seed coverage. It has benefits: soil remains covered by crop residues from previous crops or green manure to protect soil from erosion (reduce soil degradation), reduce moisture loss (mulch), conserve organic matter, and improve nutrients (green manuring). However, it has disadvantage, as weeds will grow in the field, which requires herbicides/weedicides for eradication, which involves GHG emission for their preparation. The long-term gains from widespread conversion to no-tillage could be greater than from any other innovation in third world agricultural production (Warren 1983), and no technique yet devised has been anywhere near as effective at halting soil erosion and making food production truly sustainable as no-tillage (Baker et al. 1996). Before selecting the option of no- or low tillage, the soil conditions, farmers' economic situation and market demand have to be investigated carefully.

As climate changes, the value of biodiversity for food and agriculture will increase. For a better adaptation and CCI management, it is crucial to conserve genetic resources to adapt food and agricultural production to changing needs. Maintaining and using this reservoir of genetic diversity will be the foundation for potential climate change management plans. The Dubai-based ICBA has unique salt-tolerant crops Gene bank at its Headquarter, which houses over 11,000 accessions from 251 species in 134 countries. Significant efforts are being made to multiply new germplasm accessions and seed conservation, trials, nurseries, and distribution. Recognizing the importance of gene bank and climate change, Environment Agency – Abu Dhabi with ICBA has prepared a proposal (EAD-ICBA 2010) to establish a Gene bank and Botanical Garden. The Abu Dhabi Executive Council has approved the funds for project implementation by UAE University. The scientists at ICBA endeavor to establish a science-based integrated strategy, which minimizes losses due to CC and takes advantage of adaptation opportunities through the implementation of a US$ 2.4 million project "Adaptation to climate change in marginal environments through diversification of forage/livestock system (ICBA 2010b) in WANA region".

3.2 Mitigation Efforts and Soil Carbon Pool

According to Lal (2001), the major global concerns of the twenty-first century are food security, soil degradation by land misuse and soil management, and anthropogenic increase in atmospheric greenhouse gasses. The extent to which any ecosystem can withstand land-use pressure or its resilience or potential to recover from a degraded state is a function of the ecosystem in question, with semi-arid and arid regions being particularly vulnerable (Stewart and Robinson 1997). The extent to which any land can be sustainably farmed depends on soil quality from the physical, chemical and biological perspectives (Karlen et al. 2001). Depletion of the soil organic matter pool exacerbates emission of CO_2 into the atmosphere. Soil degradation decreases CH_4 uptake by agricultural soils. Indiscriminate use of nitrogenous fertilizers and water logging also accentuate N_2O emissions from crop land (Lal 2007).

Mitigation of GHGs means implementing policies and measures to reduce anthropogenic GHG emissions from sources – such as power plants, industrial facilities and the transport sector – as well as enhance natural GHG sinks – such as forests, land use change and carbon capture and storage. Increasing energy efficiency and moving to renewable resources of energy generation would bring important economic benefits for energy-poor regions. The decrease in GHG emissions is possible, particularly with regard to CO_2 from terrestrial ecosystem

(Khan and Lal 2007) by: (1) increasing C sinks in soil organic matter (SOM) and the above ground mass; (2) avoiding carbon emissions from farm operations by reducing direct and indirect energy use; and (3) increasing renewable-energy production from biomass that either substitutes for consumption of fossil fuels or replaces inefficient burning of fuel wood or crop residue (Pretty et al. 2002). The total soil C pool comprises of SOM and Soil Inorganic Carbon (SIC). The SOC pool, comprising of highly active humus, and relatively less active charcoal carbon, is a major source or sink for atmospheric CO_2 and a key detriment of soil quality (Lal 2004). The SIC pool includes elemental C and carbonate minerals – such as calcite and dolomite – and is an important constituent of the soils of the arid and semi-arid regions such as those in Kuwait (KISR 1999) and Abu Dhabi Emirate (EAD 2009a). Formation of pedogenic carbonates ($CaCO_3$, $MgCO_3$ or Ca, Mg $(CO_3)_2$) plays a significant role in C-sequestration. Dissolution and leaching of carbonates under irrigated or high rainfall conditions thus enhance atmospheric CO_2 sequestration, temporarily.

The total amount of global SOC has been estimated at 1,500Pg, which is three times more than the biotic pool (living plants), and two times more than the atmospheric pool (Landi and Mermut 2007). The estimation on the total amount of SIC varies between 700 and 1,700 Pg. This is at least partly due to the difficulty of separation between pedogenic and lithogentic carbonates (Lal et al. 1998). Various factors affecting SOC in Arab region soils are: (1) high soil temperature which mineralizes OM; (2) infrequent and low rainfall; (3) water scarcity; (4) low biodiversity and vegetative cover; (5) poor inherent soil fertility to support vegetation; and (6) soil degradation. Mermut and Eswaran (2001) believe that carbon sequestration can be highly cost effective and environmentally friendly mitigation technique. A brief overview of SOC in the Mediterranean's WANA regions revealed a number of important generalization, such as, most soils have relatively low SOM levels, mainly 1–2 %, with some notable exception at local level, such a low level is assumed to be due to low quantity of crop residues returned to the soil and favored SOM mineralization due to high temperature (Parton et al. 1987), extractive farming practices, conventional tillage, removal of crop residue for fodder and fuel, use of dung as household fuel, and climate aridity. There is strong correlation between low SOM and soil physical degradation (Lal 2004) such as desert sandy soils have low resistance to wind erosion. The current SOM levels are probable no more than half of what they were originally before pressure of land use in the Middle East today (Lal 2001), the use of fertilization increased SOM and relative properties (Ryan 2002). Conservation and shallow tillage enhanced SOM accumulation by reducing condition for mineralization. However, conservation tillage (minimum, reduced, no-till) is still in the experimental stage in the Middle East Region (Bessam and Mrabet 2003). In the Arab region in general, there is little or no information available on carbon sequestration, and therefore, a potential area for future research.

It should be appreciated by the policy makers and politicians that mitigation and adaptation costs are to be born immediately, with the benefit to come after significant time. Investing in renewable sources of energy generation and increasing energy efficiency can bring economic benefits to this energy deficit region.

The Arab region is invited to develop a joint consensus and commitment to reduce GHG emissions without any conflict. As a part of mitigation efforts to join the world community, the Arab region – due to its geographical location – have abundant renewable energy resources (solar, wind) that can stress the development of clean energy technologies. The rich countries are also planning for the nuclear energy. Such mitigation efforts include: improved efficiency of windmills and commercialization of wind energy (Egypt); solar heating (Palestine, Tunisia and Morocco); CNG as transport fuel (Egypt); and concentrated solar power project (Egypt, Tunisia, Morocco and Algeria). Other initiatives deserve to be reported: first two Arab Green building councils (UAE & Egypt); forestation "greening the desert in UAE"; the first zero-carbon emission city in Abu Dhabi (Masdar); carbon capture and storage project in Algeria; import of hybrids cars and tax exemption (Jordan); and the set up of IRENA (International Renewable Energy Agency) headquarter in Masdar City in Abu Dhabi, UAE. Another step taken by Environment Agency, Abu Dhabi jointly with the Arab Water Council, is the launching in 2008 of Arab Water Academy (AWA) as a regional *"Center of Excellence"* for executive education in water. The AWA is initiated by Arab Water Council (AWC) and is based in the EAD headquarter in Abu Dhabi. The AWA's focus is on strengthening the knowledge and skills of the MENA's decision-makers to address and manage effectively the region's water challenges. The Abu Dhabi water master plan (EAD 2009c), soil survey of Abu Dhabi emirate (EAD 2009a), strategy for the reuse of wastewater Abu Dhabi Emirate (EAD 2010), and water conservation plan of UAE (MoEW 2010) are other initiatives for soil and water management and governance in the UAE. These initiatives set examples for other Arab countries to follow regarding water management in terms of mitigation and adaptation.

Biofuel as a substitute to fossil fuel has been prudently adopted as approach by many countries. Fuel alcohol, biodiesel, hydrogen and biofuel-cell are some of the current interests and technologies being improved to exploit these products economically. In recent years, there is a growing interest in biofuel as a substitute to fossil oils in order to counter GHG emissions and global climate change. The *Jatropha curcas*, a tropical shrub is attracting the interests of policy makers and the energy industry as one of the promising sources of high quality biofuel, extracted from the seeds. *Jatropha* is a perennial plant that can grow on a wide range of soils, including marginal lands, which are of poor quality in terms of agricultural use. Currently biodiesel feedstock is dominated by rapeseed and other food crops, posing challenges to future food security. Thus, using marginal land for *J. curcas* cultivation is therefore attractive since it would not displace food-producing crops. Although *Jatropha* has been publicized as a wonder biofuel with unlimited potential, the key issue is the productivity in the dry, degraded lands. Evidence suggests that the tree will grow far more productively on higher quality land with more rainfall or irrigation and there are warnings against overestimating Jatropha's potential to produce economically viable yields on severely degraded lands. The adoption of crops – like *Jatropha* – for biodiesel production has to be carefully made in an area of interest, and even though it may grow well or may have an option for economical production, it should not be grown on good productive lands in order to maintain to assure food security. Otherwise, the competition would impact food security significantly. Indirect land use change (ILUC) has become a major biofuels topic, ever since Searchinger (2008) pointed out that calculations of potential GHG savings through adoption of biofuels had not taken into account the effects caused by the crops displaced by biofuels. Such approach may be suitable on marginal lands and those set aside due to low quality (degradation), and these soils can be brought into *Jatropha* cultivation for biodiesel production. To get income during the establishment period (3–5 years), intercropping (pulses and vegetables) can be practiced in *Jatropha* plantation. From the climate change perspectives, it is to be noted that *Jatropha* can fix up to 10 t ha^{-1} year^{-1} CO_2 (NDAU 2010), which can be traded international, however, this system is still in its infancy and there is no way to know how much producers can realistically expect to make from selling carbon credits. Carbon sequestration can be traded under the KP or the World Bank at EU price of $20 Mg^{-1} of CO_2, however and since KP expired in 2012, the sustainability of C trading is a big question.

Another issue to worry about is the unintended impact of energy on increased nitrate contamination from biofuels production (Twomey et al. 2010). The amount of carbon that can be sequestered and the amount of emissions that can be reduced through the adoption of different agricultural production systems still require further investigation and research; however, it is well recognized that agriculture plays a great role in carbon sequestration. The *Jatropha* has been grown extensively in Brazil, Nigeria, Ghana and the local community has some concerns about economics, toxicity, nitrate pollution and concerns over food security due to shift from conventional farming to biofuel crops. For all national biofuel' plans to be implemented, 30 million ha more land and more than 180 km^3 more water globally is needed; many countries like China and India will not meet food and biofuel water demand. The question is, is it ethical to use crops to produce energy when 860 million peoples are undernourished? (Smakthin and co-workers presentation from IWMI). Climate change related interventions (mitigation) like biofuels have implications for water and agriculture management; they need to be evaluated carefully to assure food security. Most of these actions are fragmented (individual country efforts) and are not part of a comprehensive national strategy. Arab region, especially GCC countries, have fossil fuel energy resources and this will remain major energy resource for many years ahead; therefore, above mitigation efforts are important initiatives and could become part of global mitigation efforts.

4 Capacity Building, Knowledge Sharing and Media Coverage

The vulnerable communities (especially farmers) have the largest stake in developing strategies to cope with climate change. The awareness to climate change phenomenon and impact does lead to significant behavioral change among stakeholders (Halady and Rao 2010). Therefore, ensuring permanent access of farmers and other stakeholders to relevant information is crucial. Awareness may help to encourage a culture of conservation and efficiency in the Arab region. There is a great difference in the region regarding the media coverage of issues related to climate change; this is traced back to the financial resources differences in concerned countries and institutional research interests. Some institutes are taking great interest to aware the large public and youth on climate change, such as the Environment Agency – Abu Dhabi in the United Arab Emirates. It is essential to aware youth at elementary level and other stakeholders (the public) about water scarcity, saving, recycling, conservation techniques, and increasing energy efficiency. Effective communication between parties in the region about climate change adaptation results into transfer of technology to vulnerable areas. At smallholder farming level, it is important to aware farmers about existing innovations such as rainwater harvesting or the use of small-scale solar panels as source of energy. There is a great need to share and communicate skills and technologies in the region to select best adaptation practices as a part of technology transfer. Effective communication in the region will help develop efficient and relevant plans for adaptation.

5 Conclusions

The Arab region is highly vulnerable to climate change impact. It is most likely that water resources over time will be diminishing, climate will become warmer, the resources will be degraded, ultimately affecting the resource capacity to produce food for the existing and ever increasing population. It is therefore essential to take necessary actions to mitigate the impact of climate change and work seriously on the adaptation strategies. The land-surface properties such as vegetation, soil type, and the amount of water stored on the land as soil moisture, and groundwater, all strongly influence climate, particularly through their effects on surface albedo and evapotranspiration. Land-surface schemes calculate the fluxes of heat, water, and momentum between the land and the atmosphere. It is therefore important to use improved models to downscale the predictions to map communities highly vulnerable to climate change, identify major climate change issues and develop/adopt highly productive agriculture systems. Among other options, innovations in agriculture and technology transfer to climate change vulnerable communities can be a way forward for successful adaptation.

References

Abdelfattah MA, Shahid SA, Othman YR (2009) Soil salinity mapping model developed using RS and GIS – A case study from Abu Dhabi, United Arab Emirates. Eur J Sci Res 26(3):342–351

Adams B, Calow R, Chilton PJ (1999) Groundwater management in the Middle East and North Africa (MENA) Region: a review. Technical report WD/99/5R, British Geological Survey, Natural Environment Research Council, Keyworth, Nottinghamshire, UK

ADASFA-ICBA (2010) Memorandum of agreement. Signed between ADAFSA & ICBA, 2010

Al Janeid S, Bahanacy M, Basr S, Raey ME (2008) Vulnerability assessment of the impact of sea level rise on the Kingdom of Bahrain. Mitig Adapt Strateg Glob Chang 13(1):87–104

Alcamo J, van Vuuren D, Ringler C, Cramer W, Masui T, Alder J, Schullze K (2005) Changes in nature's balance sheet: model-based estimates of future worldwide ecosystem services. Ecol Soc. Resilience Alliance Publications, Waterloo, Canada

Alpert P, Lrichak KO, Shafir H, Haim D, Osetinsky I (2008) Climatic trends to extremes regional modeling and statistical interpretation over the E. Mediterranean. Glob Planet Change 63:163–170. Elsevier, Amsterdam

Baker CJ, Saxton KE, Ritchie WR (1996) No-tillage seeding, science and practice. CAB International, Wallingford/Oxon, 158 pp

Behnassi M, Yaya S (2011) Land resource governance from a sustainability and rural development perspective. In: Behnassi M, Shahid SA, D'Silva J (eds) Sustainable agricultural development: recent approaches in resources management and environmentally-balanced production enhancement. Springer, Dordrecht/New York, pp 3–23

Bessam F, Mrabet R (2003) Long-term changes in soil organic matter under conventional tillage are no-tillage systems in semi-arid Morocco. Soil Use Manag 19:139–143

Bou-Zeid E, El-Fadel M (2002) Climate change and water resources in Lebanon and the Middle East. J Water Resour Plan Manag 128(5):343–355

Briguglio LP (2010) Defining and assessing the risk of being harmed by climate change. Int J Clim Change Strateg Manag 2(1):23–34

Brown O, Crawfield A (2009) Rising temperatures, rising tensions – Climate change and the risk of violent conflict in the Middle East. International Institute of Sustainable Development (IISD), Winnipeg, p 40

Brown LR, Young JE (1990) Feeding the world in the nineties. In: Brown LR et al (eds) State of the world 1990, Worldwide Institute report on progress toward a sustainable society. WW Norton and Co., Inc., New York, pp 59–78

Burke MB, Lobell DB, Guarino L (2009) Shift in African crop climates by 2050, and the implications for crop improvement and genetic resources conservation. Glob Environ Change (JGEC) 699:9

Buying farmland abroad: outsourcing's third wave (2009) The Economist, May 23, 2009

Capital Business (Nairobi) (2009) Kenya, Qatar land deal questioned, 19 May 2009

Coker M (2008) UN Chief Warns on Buying Farms. The Wall Street Journal. September 10

Cruz RV, Harasawa H, Lal M, Wu S (2007) Chapter 10: Asia' In: Parry ML, Canziani OF, Palutikof JP, van der Linden PJ, Hanson CE (eds) Climate change 2007a: impacts, adaptation and vulnerability. Contribution of working group II to the fourth assessment report of the intergovernmental panel on climate change, Cambridge University Press, Cambridge

Daniel S (2011) Land-grabbing and potential implications for world food security. In: Behnassi M, Shahid SA, D'Silva J (eds) Sustainable agricultural development: recent approaches in resources management and environmentally-balanced production enhancement. Springer, Dordrecht/New York, pp 25–42

Day W, Audsley E, Frost AR (2008) Philos. Trans R Soc Biol Sci 363:527

EAD (2009a) Soil survey of Abu Dhabi Emirate. Environment Agency, Abu Dhabi, 5 Volumes

EAD (2009b) Climate change – Impacts, vulnerability and adaptation. Environment Agency, Abu Dhabi

EAD (2009c) Abu Dhabi water master plan. Environment Agency, Abu Dhabi

EAD (2010) A strategy for the reuse of wastewater for Abu Dhabi Emirate. Environment Agency, Abu Dhabi

EAD-ICBA (2010) Establishment of Abu Dhabi Gene Bank and Botanical Garden. A joint proposal of EAD and ICBA

FAO (2006) World agriculture towards 2030/2050. FOA, Rome

FAO (2009) The State of Food and Agriculture-livestock in the balance. http://www.fao.org/docrep/012/i0680e/i0680e.pdf

Freimuth L, Bromberg G, Mehyar M, Khateeb NAl (2007) Climate change: a new threat to Middle East Security. Friends of the Earth Middle East: Amman, Bethlehem and Tel Aviv

Gerber P, Steinfeld H (2008) Livestock's role in global climate changes. Agriculture Department-Animal and Health Division-FAO. Presented by Irene Hoffman, Hammamet, Tunisia, 17 May

GRAIN (2008) Seized! The 2008 land grabbers for food and financial security, October 2008

Gulf News (2010a) Grass ban to save huge amount of water. Abu Dhabi to phase-out cultivation of crop used as principal animal feed in the region, Gulf News, UAE, 24 August, Tuesday

Gulf News (2010b) Local environment projects miss funding. UAE not a member of key organization. Gulf News, UAE, 14 September, Tuesday

Halady IR, Rao PH (2010) Does awareness to climate change lead to behavioral change? Int J Clim Change Strateg Manag 2(1):6–22

Hemming D, Betts R, Ryall D (2007) Environmental stress from detailed climatic model simulations for the Middle East and Gulf Region (Ref. MOEN/04/04/02/02b) completed by the Met Office Hadley Center for UK Ministry of Defense Project 'Defense and Security Implications of Climate Change'

Hussein H (2001) Development of environmental GIS database and its application to desertification study in middle east. A remote sensing and GIS application. PhD thesis Graduate School of Science and Technology, Chiba University, Japan

ICBA (2010a) Proposal for Farmers Service Center (FSC). Potential cooperative projects for farming systems, technology transfer and capacity building. Proposal submitted to Farmers Service Center (FSC), Abu Dhabi, p 19

ICBA (2010b) Adaptation to climate change in marginal environments through diversification of forage/livestock system in WANA region. Proposal submitted to IFAD and Arab Fund for Economic and Social Development (AFESD)

IPCC (2001) Climate change 2001: the scientific basis. Inter-governmental panel on climate change. Cambridge University Press, Cambridge

IPCC (2007) Climate change 2007: impacts, adaptation and vulnerability. Retrieved 10 Oct 2010 from www.ipcc.wg2.org/

Jones MC, Coleman K, Cox P, Falloon P, Kenkinson D, Powlson D (2005) Global climate change and soil carbon stocks; predictions from two contrasting models for the turnover of organic carbon in soil. Glob Change Biol 11(1):154–166. Blackwell Publishing, Oxford

Jorgensen DG, Al-Tikiriti WY (2002) A hydrologic and archaeological study of climate change in Al Ain, United Arab Emirates. Glob Planet Change 35(1):37–49. Elsevier, Amsterdam

Karlen DL, Andrews SS, Doran JW (2001) Soil quality: current concepts and applications. Adv Agron 74:1–40

Kerr S, Farhan B (2008) UAE investors buy Pakistan farmland. Financial Times, May 11

Khan AUH, Lal R (2007) Potential of carbon sequestration in the soils of Afghanistan and Pakistan. Chapter 18. In: Lal R, Sulaimenov M, Stewart BA, Hansen DO, Doraiswamy P (eds) Climate change and terrestrial sequestration in Central Asia. Taylor & Francis, Balkima, pp 235–249

Kilner D (2009) Kenyans have concerns over land grab deal in times of hunger. VOANews.com, January 13

Kioth A, Yatagi A, Alpert P (2008) First super-high resolution model projection that the ancient 'Fertile Crescent' will disappear in this century. Hydrol Res Lett 2:1–4. Japan Society of Hydrology and Water Resources

KISR (1999) Soil survey for the State of Kuwait. In: Reconnaissance soil survey. Kuwait Institute for Scientific Research, Kuwait

Klein RJT, Alam M, Burton M, Doughtery W, Ebi K, Fernandes M, Huber L, Rahman A, Swartz C (2006) Application of environmentally sound technologies for adaptation to climate change. Prepared for the UNFCCC Secretariate, FCCC/TP/2006/2

Kunstmann H, Suppan P, Heckl A, Rimmer A (2007) Regional climate change in the Middle East and impact on hydrology in the Upper Jordan catchment. In: Quantification and reduction of predictive uncertainty for sustainable water resources management. IAHS Publication, Wallingford, pp 141–149

Lal R (2001) Managing world soil for food security and environmental quality. Adv Agron 74:155–192

Lal R (2004) Carbon sequestration in soils of Central Asia. Land Degrad Dev 15:563–572

Lal R (2007) Preface. In: Lal R, Sulaimenov M, Stewart BA, Hansen DO, Doraiswamy P (eds) Climate change and terrestrial carbon sequestration in Central Asia. Taylor & Francis, Balkima, p XI

Lal R, Kimble JM, Follet R (1998) Pedospheric processes and the carbon cycle. In: Lal R, Kimble JM, Eswaran H, Follett T, Stewart BA (eds) Soil processes and the carbon cycle. CRC Press, Boca Raton, pp 1–8

Landi A, Mermut AR (2007) Carbon dynamics in Saskatchewan soils: implications for the global carbon cycle. Chapter 14. In: Lal R, Sulaimenov M, Stewart BA, Hansen DO, Doraiswamy P (eds) Climate change and terrestrial carbon sequestration in Central Asia. Taylor & Francis, Balkima, pp 189–197

Mermut AR, Eswaran H (2001) Some major developments in soil science since the mid-1960s. Geoderma 100:403–426

Miller L, Douglas B (2004) Mass and volume contributions to twentieth-century global sea level rise. Nature 428:406–409

MoEW (2010) UAE water conservation strategy. Ministry of Environment and Water, United Arab Emirates

NDAU (2010) Jatropha production technology. Flyer, Tamil Nadu Agricultural University, Coimbatore, 641003, India, p 4

Omar SAS, Roy WY (2010) Biodiversity and climate change in Kuwait. Int J Clim Change Strateg Manag 2(1):68–83

Parton WJ, Scimel DS, Cole CV, Ojima DS (1987) Analysis of factors controlling soil organic matter levels in Great Plains Grassland. Soil Sci Soc Am J 51:1173–1179

Pezeshki SR, De Laune RD, Patrick WH (1990) Flooding and saltwater intrusion: potential effects on survival and productivity of wetland forests along the US Gulf coast. For Ecol Manag 33–34:287–301

Pretty JN, Ball AS, Li XY, Rivindranath NH (2002) The role of sustainable agriculture and renewable resource management in reducing greenhouse-gas emissions and increasing sinks in China and India. Philos T Roy Soc A Math Phys Eng Sci 360(1797):1741–1761

Ragab R, Prudhomme C (2000) Climate change and water resources management in the southern Mediterranean and Middle East Countries. The Second World Water Forum, The Hague, 17–22 March

Rahmstorf S (2007) A semi-empirical approach to projecting sea level rise. Science 315:368–370

Raphaeli N (2007) 'Potential water conflicts in the Middle East' Capitol Hill briefing on June 6, 2007, to staff of the Senate Foreign Relations Committee on water poverty and potential water conflicts in the Middle East. The Middle East Media Research Institute. http://memri.org/bin/latestnews.cgi?ID=IA36707. Accessed 20 Jan 2009

Ryan J (2002) Available soil nutrients and fertilizer use in relation to crop production in the Mediterranean area. In: Krishna KR (ed) Soil fertility and crop production. Science Publisher, Inc., Enfield, pp 213–246

Saleh AW (2010) Climate change in Middle East and North Africa (MENA) region and implications for water resources project planning and management. Int J Clim Change Strateg Manag 2(3):297–320

Searchinger T (2008) http://www.scienceexpress.org/7 February 2008/page 3/10.1126/science. 1151861

Shahid SA (2010) Climate change – Its impact and biosaline agriculture. Farm Outlook 9(2):24–27

Shahid SA, Abo-Rezq H, Omar SAS (2002) Mapping soil salinity through a reconnaissance soil survey of Kuwait and Geographic Information System. Annual research report, Kuwait Institute for Scientific Research, Kuwait, KISR 6682, pp 56–59

Shahid SA, Behnassi M, Silva JD (2009) Synthesis report of the international conference. In: The integration of sustainable agriculture, rural development, and ecosystems in the context of climate change, the energy crisis and food insecurity, Agadir, Morocco, 7–10 Nov 2010, 16 p

Shahid SA, Abdefattah MA, Omar SAS, Harahsheh H, Othman Y, Mahmoudi H (2010) Mapping and monitoring of soil salinization-remote sensing, GIS, modeling, electromagnetic induction and conventional methods – case studies. In: Ahmad M, Al-Rawahy SA (eds) Key note paper published in the Proceedings of the International conference on soils and groundwater salinization in Arid Countries, Sultan Qaboos University, Muscat Sultanate of Oman, Keynote papers, vol 1, pp 59–97

Shetty S (2006) Water, food security and agricultural policy in the Middle East and North Africa Region. Working paper series no 47, The World Bank Group, Washington, DC

Smith P (2007) Greenhouse gas mitigation in agriculture. Encyclopedia of Earth. Website at: http://www.eoearth.org/article/greenhouse_gas_mitigation_in_agriculture. Accessed 14 June 2009

Stern N (2007) The economics of climate change. Cambridge University Press, Cambridge

Stewart BA, Robonson CA (1997) Are Agro-ecosystems sustainable in semi-arid regions? Adv Agron 60:191–228

Tol RSJ, Bohn M, Downing TE, Guillerminet ML, Hizsnyik E, Kasperson R, Lonsdale K, Mays C (2006) Adaptation to five meters of sea level rise. J Risk Res 9:467–482

Tolba MK, Saab NW (2009) Arab environment climate change – Impact on climate change on Arab countries. 2009 report of the Arab Forum for environment and development, Technical Publications and Environment & Development Magazine, Beirut, Labanon, p 159

Trondalen JM (2009) Climate change, water security and possible remedies for the Middle East. The United Nations World Water Assessment Programme side publication series, Scientific paper

Twomey KM, Stillwel AS, Webber ME (2010) The unintended energy impact on increased nitrate contamination from biofuels production. J Environ Monit 12(1):218–224

Warren (1983) Technology transfer in no-tillage crop production in the third world agriculture. In: Proceedings of symposium "No-tillage crop production in the tropics", Monrovia, Liberia, International Plant Protection Center, Oregon State University, Corvallis, OR, p 25031

WBGU (2007) Climate change as a security risk. German Advisory Council on Global Change (WBGU)

Weib M, Florke M, Menzel M, Alcamo J (2007) Model based scenarios of Mediterranean droughts. Adv Geosci 12:145–151. Katlenburg-Lindau, Germany, Copernicus Publications

Ziv B, Saaroni H, Baharad A, Yekutieli D, Alpert P (2005) Indications for aggravation in summer heat conditions over the Mediterranean basin. Geophys Res Lett 32(12). American Geophysical Union, Washington, DC

Chapter 3
Mainstreaming Agriculture for Climate Change Mitigation: A Public Administration Perspective

Kirit N. Shelat and Gopichandran Ramachandran

Abstract The present chapter shares insights on micro level planning; essential to optimize top-down and bottom-up approaches. The chapter accordingly reveals some of the predominant strands that have to be supported through locally adapted multi-pronged strategies including institutional mechanisms that foster public leadership to sustain action. Interesting lessons from some multipronged strategies, including farmers outreach on soil health, water quality and productivity management, nutrient management, local knowledge and preparedness to support conservation through an applied biodiversity perspective and integrated river basin management interventions, are presented from an Indian perspective. These insights will be useful for countries with comparable circumstances of growth and externalities that determine development priorities.

Keywords Agriculture • Outreach • Climate change • Mitigation • Adaptation

1 Introduction

Four important dimensions of agriculture-centered management of climate change impacts have significant implications for sustainable development. These are policies to enhance climate-resilient agriculture through locally adapted best practices, access to appropriate information on alternatives in a timely manner, an enabling institutional

---

K.N. Shelat (✉)
National Council for Climate Change Sustainable Development and Public Leadership (NCCSD), Patel Block, Rajdeep Electronic's Compound, Near Stadium Six Road, Navrangpura, Ahmedabad 380 0014, Gujarat, India
e-mail: drkiritshelat@gmail.com

R. Gopichandran
Vigyan Prasar, Department of Science and Technology, Government of India,
A-50, Institutional Area, Sector-62, Noida 201309, UP, India
e-mail: gopi61@yahoo.com; r.gopichandran@vigyanprasar.gov.in

mechanism for farmers to exert their influence on decisions regarding choice of practices through well-informed community leadership, and robust use of tools of knowledge economy for knowledge transfer and hybridization. It will be useful to understand the interplay of these four dimensions from a public policy perspective to improve preparedness of farmers to overcome related challenges at the local level.

2 Emerging Facets of Agriculture-Centered Mitigation

The past decade has seen significant developments in the field of agriculture-centered climate change mitigation. It is important to take stock of some of the major developments and ask if the extent of progress through real life transformations has been indeed significant. Some of the useful indicators of such transformations will be a large number of medium to long term initiatives, almost ubiquitous; delivered tangible mitigation and adaptation benefits and modified policy settings through learnings to enhance intended benefits for production systems and people. The start-up segment of such a continuum of success could be in the form of initiatives and country frameworks still emerging. This seems to be fairly robust with the development of missions and related National Adaptation Programme of Action (NAPA) and Nationally Appropriate Mitigation Action (NAMA) in several countries. The link could be strands of predominant thinking that define thrust areas for action, growing spread and depth of parameters and convergence through initiatives at the grass-root level. These appear to be embedded in the former and are emerging in a continual manner.

Policy settings and programs that foster community leadership through locally adapted practices, sustain the structural and functional dynamics of systems, and generate verifiable indicators of progress will represent the continually evolving and open-ended segment of the continuum. This particular aspect is probably lagging behind and has to be galvanized through significant political will and not be embroiled in eternal debates.

Two other important considerations are central to the focus on public leadership and need based outreach in the context of climate change mitigation through agriculture. The first aspect is about the unique role the farmer plays in crop production. The farmer assumes the role of an individual producer responsible for the productivity of her/his farm guided by one's own decisions about the choice of farming practices. It is essential to target the farmer almost on an individual basis, going beyond mechanisms that interact with them on a collective mode. The second important consideration is the bottom-up approach to complement macro policy-based interventions. Farmer-centered bottom-up approaches are essential bearing in mind the diversity of information needs and extreme specificity in crop choice and menu of production practices. This is especially so in the case of small landholdings, characteristic of several developing countries. These are distinctly different from the large corporate approaches seen in the West. Public policy and administration systems should ensure congruence between bottom-up and top-down interventions.

3 Important Focal Areas of Immediate Relevance at the Grass-Roots Level

We indicate four important areas that have not received adequate attention in this context. Some prominent publications in this regard have been cited in the framework that follows the listing below. The four areas are:

- The precise quantification of long-term consequences of loss of agricultural land to urban development and other land use preferences on sustainable development of communities and natural resources that will be adversely affected: while it can be argued that the intensity of agriculture activities/outputs compensate for land area loss, several externalities – including the loss of community livelihoods and the tenure of ownership – remain unanswered. On the other hand, it is more likely that higher intensity output is also not achieved to the extent desired due to various constraints. The opportunity to absorb CO_2 through photosynthesis is also lost to the extent vegetation is not present/fostered in barren lands. It is therefore essential to appropriately enhance locally-adapted agriculture by recruiting larger tracts of land that are not covered by vegetation/agriculture.
- Micro-level planning tools for sustained community involvement considering changing environmental and resource quality: aspects such as institutional mechanisms to optimize delivery of extension messages with support services and integrating local knowledge, tenure of support and technical assistance to sustain transitions and feasibility of local interventions, become important.
- Community/public leadership to be fostered in a systematic manner by delivering appropriate information in a timely manner: These interventions can be a logical extension of the traditional extension and training programs with a precise understanding of institutional mechanisms that can help communities exert their rights and influence decision making at the local level. Focused technical/compliance assistance should follow and guide awareness to action.
- From a public policy perspective focused on the role of the farmer to sustain agriculture production, it is important to consider the transience of crop systems and accordingly quantify the immediate carbon sequestration benefits achieved through crops. These cannot however be equated with industry-centered carbon avoidance/capture interventions because the scales of time/technology and impacts are markedly different. It is equally important to consider the technical preparedness levels of farmers to contribute through related protocols. The lack of appropriate facilitators at this point in time is a case strong enough to initiate appropriate capacity-building efforts and mainstream agriculture-centered mitigation through photosynthesis more emphatically. These have a bearing on the costs, benefits, barriers and tradeoffs that determine the quality of life of farmers in particular.

It will be further clear from the following that large-scale and long-term interventions are needed in the areas of public leadership, land use planning, conservation of natural capital and watershed functions to optimize mitigation potential while twining conservation and economic goals. These are essential to actually prevent backsliding of the most affected. These include administrative capacities to handle climate-proof investments and related ecosystem functions in a scientific manner. Is it therefore appropriate to ask if mitigation action has not yet reached the much needed threshold to deliver tangible benefits? If so, how do we ensure implementation of policies and mutually reinforcing initiatives?

We take note of some of the most important references on the photosynthetic and related mitigation potential of agriculture (Paustian et al. 2006; Smith et al. 2007; Altieri and Koohafkhan 2008; Rodale Institute 2008; PICCMAT 2008; Murphy et al. 2009; Scherr and Stapith WWI 2009; Johnson et al. 2010; FAO 2010a, b; Branca et al. 2011; Ramirez et al 2011; Hoffman 2011; Kumar et al. 2011) and deliberately refrain from re-stating the facts and figures indicated therein. On the other hand, we focus particularly on statements regarding information/knowledge gaps and investigations needed to generate empirical evidences; learnings from assessments of community preparedness to adopt and adapt appropriate practices and leadership from a public policy perspective. This is to consolidate the clarion call for immediate and focused attention of decision makers in the public policy interface through a micro-planning perspective. The present stock taking is important when the call for action has progressed from a focus on the photosynthetic potential of crops as a mitigation tool to such integrated aspects as food security, trade, development and voluntary standards to benchmark and promote mitigation efforts. While these focal-thrust areas could undoubtedly synergize and foster mitigation and conservation for climate-smart/efficient development, the scope for immediate focus on capacity building of farmers to initiate and sustain appropriate action appears large due to the fact that such initiatives are either few and far between or are only emerging at this point in time.

Even when countries moved towards Qatar and took note of the emerging initiative of the Committee on World Food Security (CFS 2012) voluntary guidelines and responsible governance (Potts 2012), several system-related challenges remain. These challenges pertain to carbon sequestration/removal within overall GHG accounting, related methodological and technical challenges, intensity targets related to growing demand, regulations, monitoring and related institutional mechanisms and the confluence of market and economic conditions. These challenges remain generally unanswered despite the call for action almost consistently through the previous decade and probably reflect an inadequate response from decision makers. It is essential to take these on board on a priority basis.

The 2007 Assessment Report of the IPCC (Smith et al. 2007) indicated the need to enhance sink functions through soil carbon sequestration implying links with crop-based photosynthesis efficiencies. Estimates about per area sequestration as a function of specific mitigation measures and their influence on economic conditions and social behavior adaptations had to be assessed in greater detail. Rosegrant et al. (2008) reinforced the need for pro-poor investments to complement incentives for sustainability practices that justify CO_2 mitigation potential. These are also related to land use patterns in forestry and may assume significance in post 2012 negotiations to mainstream agriculture-centered mitigation (Murphy et al. 2009). Scherr and Stapith (2009) reinforced the photosynthetic potential through a "Carbon-rich farming" framework, that will also cater to animal feed needs wherein agroforestry is integrated.

The fundamental premise of sustainable development is equitable access to resources for future generations. This is indeed a dynamic perspective duly considering emerging needs of developing countries. Intra-generational equity too is a well-recognized imperative especially when the dependence of communities on agriculture and related incomes is inextricably linked to survival. The present generation, particularly in areas highly vulnerable to climate change impacts, has to be enabled to secure sustained access to resources and enhanced resilience. This highlights an augmentation approach and calls for continual improvements in mitigation and adaptation strategies aligned with their development aspirations. A logical extension of this framework is to ensure a gradual increase in income and safeguards to prevent reduction. In this context, it is important to recognize that agriculture has not received its due attention in international deliberations. This is particularly so with reference to the photosynthetic function of crops that makes agriculture a dynamic natural tool to combat increasing levels of carbon dioxide. Substantial efforts are needed to mainstream this aspect in order to enhance the efficiency of agriculture production systems.

4 Some Integrated Local Solutions: An Indian Perspective

Since the 1950s, India has progressively tackled high levels of poverty, recurring droughts, water scarcity and dependence on food imports. She has duly recognized the cross-cutting nature of mitigation/adaptation across water, land, bio resources, nutrients and market thrust areas, in response to such realities as inclement weather and related vagaries that in turn determine production and extraction practices. The stressed quantitative and qualitative profiles of resources create the context for dynamic and evolving strategies in response to emerging challenges. Five decades on, India is striving to reduce the spread and depth of externalities through several national and state-level initiatives. Integrated management of resources and systems through people's initiatives has also largely been responsible for this emerging positive trend.

India has to be commended for her recent efforts to align with global efforts to tackle climate change impacts. This is notwithstanding the debates and considerations on such aspects as equity, caps on emissions, historic emission, and monitoring and verification commitments. The National Mission on Sustainable Agriculture and the National Initiatives on Climate-Resilient Agriculture represent a robust intent further substantiated with action at the ground level (Sunita Narain et al. 2009). India's NATCOM and the recent Results Framework (DARE 2012–2013) have also highlighted her efforts and initiatives to tackle knowledge gaps and define the way forward. The present chapter restricts its focus on some of the grassroot level aspects of implementation and indicates the scope to enable optimal delivery through policies, plans, programs, and projects. Aspects of equity and

India's position in international negotiations are beyond the scope of the present chapter. We accordingly trace the consolidation of our understanding of knowledge gaps as stated above to draw the attention of decision makers on some thrust areas and maverick initiatives to complement India's growing preparedness.

One of the best public policy and public administration-centered initiative is titled KrishiMahotsav from the State of Gujarat in western India. It is a classic representative of political will to respond and tackle sustainability-related challenges in agriculture with implications for livelihood. An analysis of the process and outcome of KrishiMahotsav creates the context for an integrated approach that can simultaneously enhance resilience across water, soil, yield and market linkages that in turn influence production practices.

5 The Gujarat Experience: Sustainable Agricultural Development

Gujarat is a state situated on the west coast of India. Diverse in its topography, it boasts of a 1600 km coastline and is home to the largest desert in the country known as Rann of Kutch. The State has all possible handicaps faced by agriculture such as 70 % being rain-fed, recurrent droughts, untimely/irregular rainfall and some areas receiving rain only 3–4 days in a year. In normal years, the agricultural growth rate used to be 2–3 %. Agriculture was not sustainable in many parts of the state due to recurrent crop failures. However, this is a story of the last millennium. A recent review by Dholakia and Datta (2010) indicated that the State's initiatives were bolstered by political will and leadership supported by a dynamic public administration initiative. This was in response to a felt need to tackle many challenges such as depleting water tables, deteriorating soil and water conditions, especially due to salinity ingress compounding the impacts of irregular rainfall and drought-like situation.

Shankar Acharya (2011), based on the work of Dholakia and Datta cited above, highlighted the fact that a dynamic synergy was established between water management and conservation programs, realigned rural electricity distribution, support for livestock development complementing an emphasis on non-food crops. This turnaround, which became possible due to certain successful experiences, did not remain specimen or model projects, but became a base to launch an overall initiative in all 18,600 villages in the state, known as "KrishiMahotsav" (Fig. 3.1). This was also related to a change from a agro-climatic zone approach to farmer- and village-level interventions with implications for farmers' sustainable livelihood in addition to resource management imperatives.

The first author of this chapter was responsible for developing this new extension approach that helped transform related intentions into reality reinforcing bottom-up and top-down people-centered approaches (Fig. 3.2). Appropriate information on preventive and remediation measures was delivered in a timely manner (Shelat 2007).

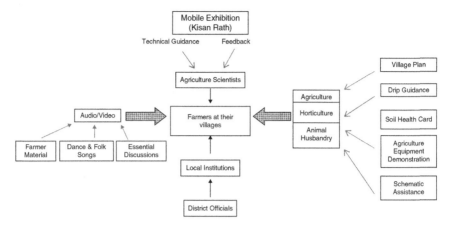

Fig. 3.1 Elements of the integrated outreach for farmers through "Krishi Mahotsav" roughly translated as the grand fair for the farmers and farming

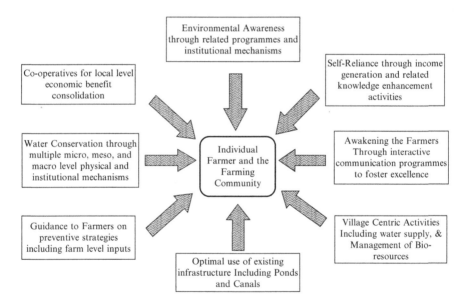

Fig. 3.2 Converging themes in "Krishimahotsav"

This sets the context for effective micro-level planning to determine alternative cropping patterns. From a public administration point of view, this called for a massive coordination across nearly 18 governmental departments, involving nearly 100,000 personnel every year and consistently so.

Shah et al. (2009) discussed the dynamics of this intervention in detail. They refer to the 9.6 % rate of growth in agricultural state domestic product. This is significantly higher than the nearly 3 % in the national GDP from agriculture and allied sectors. This has to be considered commendable due to the fragile nature of the State's agriculture production systems. This macro-level quantification was however attributed to growth in cotton, wheat, livestock, fruits, and vegetables in particular. A concomitant rise in farmers' income is also evident aligned with increase in area under cultivation. The former was also possible due to remunerative minimum support prices as incentives for higher production.

Some of the related policy initiatives that accentuated Gujarat's agriculture progress were improved market access and focused extension services, research support and input supply. The Government revived several extension activities and delivered through the "KrishiMahotsav" campaign. This took the form of a direct door-to-door extension program to guide farmers at village level at the pre-Kharif (pre-monsoon) stage. This signified an intensive festival-like initiative that strengthened the sense of belonging and ownership between the government, farmers and agriculture-related stakeholders.

The key to this success was direct involvement of public leadership – both elected and non-elected members of the public governance system – in close proximity with scientists, extension staff, officers from the veterinary department, co-operatives, irrigation department, rural development department, and local financial institutions. District-Level Management Committees, including the Minister and the Secretary of the Department, were to provide outlines of the programs and ensure the involvement of all concerned stakeholders. The District Implementation Committees comprised officers for coordination, including securing administrative and financial approvals. The Block-Level Committees were responsible for implementation and consultations with local institutions. The Village-Level Implementation Committees included volunteers, principal of the local school, health workers, milk cooperatives etc. with the strongest local connect. Farmers were given soil health cards and guidance on management of soil moisture with respect to the best suited crops; complemented with information on market price based on a 5-year trend. This accordingly included information about soil and water parameters based on which it was possible to understand the suitability of the soil for the proposed crops. Rainfall data was used to derive useful correlates on the possibility of access to water; availability of the resultant moisture in the soils and its links with expected productivity.

The farmers were guided about using certified seeds, quantity of fertilizers and pesticides and comprehend prices through the Agricultural Produce Market Committees (APMCs) before selling their farm produce. Farmers subsequently sowed crops that gave them higher return and were sustainable in the soil of their farms. Cattle were vaccinated and kits on animal husbandry techniques were also provided with a special focus on benefits for poor farmers on a priority basis. This exemplified a value-added micro-level extension system that enabled scientific agriculture. An important indicator of success was the reduction in the volumes of nitrogen and phosphorus used in the soils. These aspects helped farmers reduce costs and optimize production, only to improve their net income.

Information and Communication Technology oriented material was made available to farmers in local language for crop management, including the use of fertilizers and pesticides. Free telephonic help lines were introduced to answer the queries of farmers. The E-Gram project helped thousands of villages with computer facilities. All these initiatives were backed by total involvement of public leadership, both elected and non-elected. Chief Minister, Village Sarpanch, the Chief Secretary village-level worker, voluntary agencies, input dealers and co-operatives and bankers – representing the rank and file of institutions – were involved in the projects. This accordingly enabled multiplier effects and linkages for success at the micro-level.

The "KisanRath" – the grand vehicle meant for reaching out to farmers – assumed the form of a mobile exhibition including demonstration facilities. A team of scientists and officers of the agriculture cooperation department moved as part of the entourage with the mobile facility; hence delivering information at the door step of the farmer. Two hundred and twenty nine such raths served the outreach needs of the State simultaneously, reaching out all the nooks and corners. Figure 3.1 presents some of the thrust areas that were integrated in this initiative.

On the water front, more than 100,000 check dams were constructed and enhanced access through a network of canals. In its rain-starved areas, such as North Gujarat and Kutch, a special scheme for irrigation known as "Sujalam-Sufalam" was introduced. Micro-irrigation techniques were promoted. Consistent quality and timing of electricity was ensured to assist efficient use of energy in the farming sector. Importantly, the initiative had to take note of the diverse agro-ecological characteristics across the regions of the State that differed also in terms of enterprise, conjunctive use of ground and canal water, semi-arid and humid climate, and of course soil types.

As a result of this multipronged approach, a positive water balance is also seen in some areas. This has implications for sustainable use of resources in times of adversity through appropriate conservation and extraction practices. The Bhaskaracharya Institute For Space Applications and Geo-Informatics, an institute set up in collaboration with Space Application Centre (SAC), Ahmedabad, by the Government of Gujarat, prepared a micro-level plan for land use by identifying sites for check dams and village ponds for every village (Fig. 3.2).

The Junagadh Agriculture University assessed the impacts of this extensive intervention and observed that:

- farmers changed over to improved varieties in the case of groundnut, castor, bajra, and sesame in particular;
- they comprehended the importance of inter-cropping/mixed cropping and improved their returns;
- production costs were reduced with respect to nutrients and pesticides in particular and avoided unscientific use;
- this was true even of improved irrigation and water use; and
- improved grading and packaging practices for better marketing.

It will be obvious from the above that the multipronged strategy called for a systematic linking of initiatives and has improved resilience of farmers and strengthened the natural resource base to a certain extent. From a public policy perspective, it is equally important to recognize the individual and synergistic impacts of policies across sectors to reduce distortions and mutually exclusive impacts. Such linkages need not always to be linear/unidirectional. These could pertain to subsidies, product pricing and trading as modulators of output that in turn could determine extraction practices. While the impacts have been largely consistent over the last 5–6 years, the growing impact of climate change has to be factored to develop more robust preventive strategies. Crop technologies, congruence of water/moisture and nutrient use to ensure adequate and appropriate access to the crop in various stages of growth, have to be aligned. Infrastructures for post-harvest storage and transport also influence gains further reinforcing the link between institutions, delivery, and access.

The initiative highlighted the fact that communities have to be oriented to documentation and reporting on observations they make regarding local-level changes in resource quality; with special reference to yields of crops. It is however important to understand the links between agronomy and determinants of output, including variations in other related biota (weeds, pests, predators and other natural enemies) and allelochemical and allelopathic interactions. The institutional mechanisms – including periodic reporting, leadership to facilitate knowledge and technology transfer, documentation of local innovations – the architecture of fiscal and non-fiscal mechanisms, the market-based measures (productivity related incentives), and capacity building opportunities could be assessed and documented to articulate, policies, plans, programs, and projects. It is therefore essential to describe and assess the main natural and anthropogenic interactions at a watershed scale, their aggregation at the river basin scale and the role of water in sustaining different land uses, including ecosystems.

Livestock management and related issues have not secured adequate attention. It is important to also examine the feasibility of traditional systems of optimal grazing intensity, renovation of grasslands, precision farming systems and tillage practices considering the link between cultivar performance and soil types, regulate and enable implementation of soil quality/nutrient management practices, reduce erosion and intrusion of salt and chemicals, and provide incentives to farmers for climate-efficient action.

From a public policy point of view, it is equally important to involve community stakeholders by building on their insights to manage externalities, enhance their capacity to comprehend issues and options for collective and well-informed mitigation and adaptation actions. Communities should accordingly know about the effect climate change would have on them and whether authorities engage in appropriate action. The continuum in this context is signified by simple involvement in initiatives to start with the leadership to tackle challenges. Communities and their leaders should know about policy commitments for long-term management of causes and effects of vulnerability and contribute to decision making through their insights. This will help develop appropriate development plans and actions.

Policy and plan initiatives have to also be familiar with temporal and spatial scales of issues in hydrology and management approaches, including economic instruments considering watershed and river basin scales. Recently, Harsha (2012) emphasized the need for coordinated management of land and water resources with stakeholder engagement, catchment treatment and management, including watershed scale assessments and their aggregation at the river basin scale. Earlier, the Preliminary Consolidated Report on effects of climate change on water resources (Ministry of Water resources India 2012) pointed out that Regional Circulation models do not give basin-level scenarios and that India is planning to develop appropriate models for prediction of flow under varying climate conditions. Interestingly, the updated recommendations on environmental standards set by the United Kingdom Technical Advisory Group on the Water Framework Directive for the period 2015–2021 cover standards for pollutants, implications of proposed standards and the need for remediation and prevention strategies, thresholds for assessing risks due to influents, and the influence of alien species and impacts on landscapes. The assessments of the Bioforsk and IWMI reveal interesting dynamics of conservation.[1]

The Draft National Water Policy (2012) of India lists a large number of important concerns that have to be addressed on a priority basis. This includes the need to enhance stakeholder participation in land-water-soil management strategies in a holistic manner. Management information systems could incorporate valuable information on water available for use, opportunities to improve water use efficiency, preservation of river corridors, related infrastructure and flood forecasts. Public leadership is central to the success of these initiatives supported by information support and sustained communication among stakeholders.

6 The Way Forward Through Enhanced Leadership at the Local Level to Support and Upscale Positive Action: Framework for Specific Interventions

We present a framework of eight important dimensions and 12 cross-cutting aspects in the table below. The framework recognizes the heterogeneous nature of these thrust areas and the need for an indepth attention to enable action. Empirical inputs on these parameters are essential to develop locally relevant micro – level plans (Table 3.1).

Insights on the interrelated dynamics of these aspects will be of direct benefit to the national mission on agriculture and cross-cutting aspects of other missions through an understanding of knowledge gaps, limits, and limitations. These could substantiate convincing evidences and help link institutions for value-added assistance for the way forward. The efforts of the Anand Agricultural University, Gujarat, India (Patel et al. 2012) has to be commended in this context.

---

[1] www.iwmi.cgiar.org/News_Room/Archives/Water_and_Climate_Change_Book/PDF/Delhi_CC_book_launch_final_300112.pdf

Table 3.1 Inputs for locally-relevant and robust micro-level plans

Eight important dimensions	Twelve cross cutting aspects
Soil systems and their responses (changes in the physical, chemical and biological profiles)	Individual and synergistic impacts of the stated individual aspects including allelopathic and crop-weed-tri trophic interactions
Water availability, access and quality	Natural variations vis-à-vis changes in response to micro-climate profiles
Crop varieties	
Specific species that appear to be resilient	Predominant signals of stress due to impacts of climate change in all systems stated
Most susceptible to changes and reasons	
Productivity patterns	Well-known adaptation strategies and the scope to strengthen them for tangible benefits
Main crops and associated crops within a cropping schedule	Spread and depth of data gaps
Sustainability of productivity enhancement measures	Data sources/expertise that have to be tapped
	Incongruent policies/plans
Weeds	Plan for integrated preventive and remedial measures to sustain and enhance productivity
Occurrence and distribution patterns	
Well known and expected impacts on crops	Important lessons for management from programs and projects
Influence on productivity	
Sustainability of control measures	Scope for value addition to traditional knowledge for preventive action
Diseases in crops and weeds	
Microbials	Information about sites that can be visited for detailed discussions and documentation as part of the present project.
Vector relations	
Crop-weed-soil-water linkages	
Nutritional status of the host plant (weed and crop)	Information and training needs of communities to sustain well-informed action, including communication with institutions for better-enhanced flow of support and rewards.
Sustainability of control measures	
Pest infestations in crops and weeds	
Crop-weed host dynamics	
Predators & Parasites	
Influence of host plant nutritional status	
Animal resources	
Stress responses	
Access to nutrients	
Diseases	
Links with productivity	
Local level institutions and community initiatives	

7 The Clarion Call Revisited

From a public policy point of view, it is essential to sustain the interest of citizens to participate in collective actions. While debates can rage on regarding the causes and perpetrators of climate change, the need to tackle present and emerging impacts in an expedited manner cannot be over-emphasized. This is especially so in the context of agriculture that can also help solve related issues through a multi-sectoral framework. The eclectic analysis of a still emerging agenda presented in this chapter highlights the fact that the scope for immediate attention and support for long-term action that can demonstrate the feasibility of locally-suited adaptation and mitigation measures through well-informed public leadership is quite large.

In countries with circumstances of development and local level institutions similar to India's, it is possible to involve non-elected leaders and other members of public governance – such as farmers, animal holders, fishermen, entrepreneurs, village-level workers, voluntary organizations, youth and officers of government institutions – in a collective action closely linked with networks of agriculture research institutes etc. It is also essential to unravel the communities' potential to tackle local-level problems, especially with the involvement of youth in a significant manner. The need for focused information on tools and techniques for productivity enhancement and institutional mechanisms that will help people exert their rights cannot be over-emphasized in this context. The central theme has to be poverty alleviation through climate-resilient agriculture with the aim to enhance livelihood options. This will ensure a twin advantage of sustainable livelihood/community development and climate-efficient management of natural resources. The scope is indeed quite significant and has to be realized through concerted efforts.

References

Acharya S (2011) Agriculture: be like Gujarat. Other states have much to learn from Indian agriculture's star performer. Business Standard, July 11 2012

Altieri MA, Koohafkan P (2008) Enduring farms: climate change, small holders and Traditional Farming Communities. Third World Network, Malaysia, 63p

Branca G, McCarthy N, Lipper L, Joleiole MC (2011) Climate smart agriculture: a synthesis of empirical evidences of food security and mitigation benefits from improved cropland management. MICCA, FAO, 35p

CFS (2012) Consensus reached on guidelines for land tenure and access to fisheries and forests Negotiating group sends text to Committee on World Food Security for final approval 13/3/2012. http://www.fao.org/news/story/en/item/128907/icode/

Department of Agricultural Research and Education (2012–2013) Results framework Document (RFD) for the Department of Agricultural Research & Education. http://www.icar.org.in/files/RFD-2012-13-DARE.pdf

Dholakia RH (2010) Has agriculture in Gujarat shifted to high growth path? In: Dholakia RH, Datta SK (eds) High growth trajectory and structural changes in Gujarat Agriculture, Centre for Management in Agriculture, Publication no 234, Indian Institute of Management, Ahmedabad, 215 p

FAO (2010a) Conference on ecological agriculture: mitigating climate change, providing food security and self reliance for rural livelihoods in Africa, Conclusions and recommendations, African Union headquarters, Addis Ababa, Ethiopia, 2008, 10 p

FAO (2010b) Soil for food security and climate change adaptation and mitigation, Committee on Agriculture, 22nd session, Rome

Harsha J (2012) IWRM and IRBM concepts envisioned in Indian water policies. Curr Sci 102 (7):986–990. http://www.currentscience.ac.in/Volumes/102/07/0986.pdf

Hoffmann U (2011) Assuring food security in developing countries under the challenges of climate change: key trade and development issues of a fundamental transformation of agriculture. No 21. UNCTAD, 43 p

Johnson R, Ramseur JL, Gorte RW (2010) Estimates of carbon mitigation potential from agricultural and forestry activities. Congressional Research Service, 7–5700, http://fas.org/sgp/crs/misc/R40236.pdf, R40236

Kumar NS, Aggarwal PK, Rani S, Jain S, Saxena R, Chauhan N (2011) Impact of climate change on crop productivity in Western Ghats, coastal and northeastern regions of India. Curr Sci 101(3):332–341

Ministry of Water resources, Government of India (2012) Draft national water policy 2012. http://www.indiawaterportal.org/sites/indiawaterportal.org/files/revised_draft_national_water_policy_june_2012_english.pdf as recommended by the National Water Board in its 14th meeting held on 7th June 2012, 11 p

Murphy D, Vit De C, Nolet J (2009) Climate change mitigation through land use measures in the agriculture and forestry sectors. Background paper IISD, 31 p

Narain S, Ghosh P, Saxena NC, Parikh J, Soni P (2009) Climate change perspectives from India, 67 p. UNDP, India. http://www.undp.org/content/dam/india/docs/undp_climate_change.pdf

Patel RH, Usadadiya VP, Shah SN (2012) Contingency plan for weather aberrations (Ahmedabad, Anand, Dahod, Kheda, Panchmahal & Vadodara), Directorate of Research, Anand Agricultural University, Anand, Gujarat, India, 340p

Paustian K, Antle JM, Sheehan J, Paul EA (2006) Agriculture's role in greenhouse gas mitigation. Pew Centre on Global Climate Change, Arlington VA, USA, 76 p

PICCMAT (2008) Policy Incentives for Climate change mitigation agricultural techniques http://www.climatechangeintelligence.baastel.be/piccmat/; http://www.climatechangeintelligence.baastel.be/piccmat/draft.php

Potts J (2012) Mitigating climate change: leveraging the potential of voluntary standards in the agriculture and forestry sectors. Issue Brief ENTWINED 2012/03/01 http://www.iisd.org/pdf/2012/mitigating_climate_change_ag_forestry.pdf

Ramirez VJ, Lau C, Kohler AK, Signer J, Jarvis A, Arnell N, Osborne T, Hoober J (2011) Climate analogues: finding tomorrow's agriculture today. Working paper no 12. CCAFS, 40p

Rosegrant MW, Ewing M, Yohe G, Burton I, Huq S, Santos RV (2008) Climate change and agriculture, threats and opportunities. GTZ, Gmbh, Germany

Scherr SJ, Sthapit S (2009) Mitigating climate change through food and land use. Ed Lisa Mastny, Worldwatch report 179, Worldwatch Institute

Shah T, Gulati A, Hemant P, Shreedhar G, Jain RG (2009) Secret of Gujarat's agrarian miracle after 2000. Econ Polit Wkly XLIV(52):45–55

Shelat KN (2007) What ails our agriculture? The Gujarat Experience, 171 p. Shree Bhagwati Trust, Ahmedabad, India

Smith PD, Martino Z, Cai D, Gwary H, Janzen P, Kumar B, Carl M, Ogle F, O'Mara C, Rice B, Scholes O, Sirotenko O (2007) Agriculture. In: Metz B, Davidson OR, Bosch PR, Dave R, Meyer LA (eds) Climate change 2007: mitigation, contribution of working group III to the fourth assessment report of the IPCC. Cambridge University Press, Cambridge/New York

The Rodale Institute (2008) Regenerative 21st century farming: a solution to global warming, 12 p. http://grist.files.wordpress.com/2009/06/rodale_research_paper-07_30_08.pdf

United Kingdom Technical Advisory Group on the Water Framework Directive, Updated Recommendations on Environmental Standards, River basin management 2015–2021, 77 p. www.wfduk.org/sites/default/files/Media/UKTAG%20Summary%20Report_final_260412.pdf

Chapter 4
Learning About Climate Change: A Case Study of Grain Growers in Eastern Australia

Lehmann La Vergne

Abstract "Farmers are the only indispensable people on the face of the earth." — Li Zhaoxing, Ambassador, China. "The diligent farmer plants trees, of which he himself will never see the fruit." — Cicero. These two quotes encapsulate the challenge of educating farmers about issues relating to climate change and adapting to farming in drying conditions. As a society we need farmers to provide the sustenance for the increasing global human population while at the same time taking on the role of an environmental steward to the land that they farm. These are tasks that need to be done within an economic framework that ensures that food is and remains affordable and in many regions, a climate that has become increasingly dry. After more than a century of progressively industrializing agriculture, particularly in developed nations, thereby making the practice of farming less labor intensive through the use of increased technology, many farmers are now faced with the challenge of adapting farming practices to drier climatic conditions. This is particularly the case with broad acre cropping and mixed farmers across the grain growing regions of eastern Australia. This case study research identifies how farmers choose to obtain information about climate change and which sources they find most trustworthy. It also identifies that a belief in human induced climate change may not be critical in getting farmers to adapt their farming practices.

Keywords Climate change • Adaptation • Agriculture • Farming • Sustainability

---

L. La Vergne (✉)
University of Ballarat, Lydiard St S, Ballarat, VIC 3350, Australia
e-mail: l.lehmann@ballarat.edu.au

1 Introduction

While there has been a rich history of formal agricultural education in Australia from secondary colleges through to tertiary and post-graduate education, there has also been a significant traditional of informal education within the Australian farming community. Australian farmers have long been active participants in field days, workshops and crop walks and field trial days. However as we enter the second decade of the twenty-first century there are two major issues that will impact on both the way Australian grain farmers operate and how they access information to improve their operations. Climate change is already impacting on how farmers operate, even if they do not believe in or recognize human induced climate change theory. The growth of digital communications through the Internet and mobile phones is also impacting on how they access information. While this research was initially primarily concerned with how farmers adapted to climate change, the emergence of technology as a significant medium for deriving that information has added a new dimension to this research.

In looking at the rich history of agricultural colleges in southeastern Australia it was clear that traditionally studies in agriculture have been delivered in a face-to-face mode with an emphasis on practical applications rather than classroom theory. In Australia the formal agricultural education market has largely been young men, although in recent years there has been increasing interest shown in the industry by young women as well. TAFE level Agricultural Colleges located in many regional areas in Australia have thrived on live-in cohorts of young student farmers. At the higher education degree level, Dunn and Wolfe (2001, p. 278) acknowledged that face-to-face delivery of agricultural programs needs to broaden to include more distance education and web delivery to ensure equity and accessibility of education in the agriculture sector in the longer term.

How do established farmers learn about adapting farming practices in a drier environment is the primary question in this research. Subsidiary research questions that were also explored included how they derive information about climate change, the different responses to climate change and the types of skills farmers will need to adapt for future farming in a changing climate.

In endeavoring to answer these questions, a path for the development of farmer orientated climate change education programs, either formal or informal, has emerged. In developing future education programs it was critical that the learning preferences of all farmers, regardless of age, gender, geographic location and level of education and socialization be considered.

2 Literature Review – Farmer Learning Preferences for Formal or Informal Learning Activities

This literature review focuses on how farmers actually learn as they operate their farming enterprise. This means that one of the purposes of my research is to look at agricultural education and learning beyond specialist agricultural secondary schools, the TAFE and higher education sector. Farmers and agricultural professionals, like most other occupations now have to participate in life-long learning activities. Indeed for farmers, the need to continuously update their knowledge and skills has been challenged, not only by increasing levels of technology, regulation and structural change, but also by the need to adapt to changing environmental conditions.

Like many adult learners, farmers do not restrict their learning practices to formal education activities. In my experience, as a teacher at an agricultural TAFE college, early career farmers (aged under 25) will often seek some formal qualification either through the TAFE system or in higher education, however, once established they are more likely to undertake non-formal or informal learning. In this context non-formal learning is undertaken through a program that is not usually evaluated and does not lead to certification (Halliday-Wynes and Beddie 2009). Informal learning results from everyday work-related, family or leisure activities (Halliday-Wynes and Beddie 2009). This is borne out by activities such as Landcare, field days, extension activities and farmer or industry groups. The challenge with understanding the value of these forms of adult learning is the difficulties in assessing how valuable it is for the learner and what effect it has in promoting increased understanding and change. Golding et al. (2009) also make some interesting points about the lack of value placed on informal learning in workplace environments because it is not part of a structured educational approach. Understanding the types and range of these informal learning activities were critical in answering my research questions.

It was some initial research by Kilpatrick (1997) which suggested that those farmers who attended formal or non-formal education activities were able to operate their farm businesses more profitably and correspondingly increased their willingness and ability to make successful changes to their farming practices. This suggests that farmers could be more successful if they participate in lifelong learning. It is likely that many farmers may not recognize informal learning activities, such as field days, as a form of education because they may also include significant social and community elements. This had implications in the design of the interview question guide for my research in order to ensure that all forms of informal learning were captured.

The most significant report on how farmers learn was commissioned by the Rural Industries Research and Development Corporation (RIRDC) in 1999. Undertaken by Kilpatrick et al. (1999), this report synthesized two significant Australian studies researching the learning styles of farmers. While the context was principally about farmer learning in the management and marketing areas, it may be applied to other farm management issues. The report by Kilpatrick et al. (1999) is very comprehensive and highlights numerous recommendations. Of most interest in my research project are the following:

- Not all farmers learn in the same way;
- Progressive farmers are the most likely to undertake specific training programs;
- Multiple learning sources are common;
- Attendance at training activities increase with the level of formal education;
- A need for localized training opportunities is identified;
- Farmer-directed groups and agricultural organizations are able to deliver effective training programs for adult learners;

- Many farmers are not aware of the benefits of learning and rely on experts to assist when they are contemplating changes;
- Less progressive farmers struggle to keep up with large amounts of information provided to them;
- Farmers prefer more interactive training programs;
- Easy access to support networks following training results in an increased likelihood of change after training;
- Older farmers are less likely to access formal training opportunities because they do not expect to be in farming long enough to reap sufficient returns for the time invested in the training;
- Farmers find new practices easier to apply if they can see a practical example of success;
- Other farmers are an important learning source for all farmers; and
- A wide range of farmers attend field days.

Kilpatrick and Rosenblatt (1998) observed that while many Australian farmers are happy to seek and acquire information, they are less ready to participate in formal training programs. While most educators would argue that the acquisition of information is at least part of learning and hence part of training, Kilpatrick and Rosenblatt indicated that for some farmers the activity of training has negative connotations. This would suggest that for some farmers, at least, optimal learning does not necessarily involve a 'training program'. Extrapolating from Kilpatrick and Rosenblatt (1998), five reasons why farmers might prefer to seek information rather than undertake training include:

- A preference for independence;
- Familiarity with a highly contextual learning mode;
- Lack of confidence in working in training settings;
- A preference for information from known sources; and
- A fear of being exposed to new knowledge and skills.

In a later study, Kilpatrick et al. (1999) identified that informal learning is an important aspect of education for many farmers. The main sources of informal knowledge include experts such as agronomists and meteorologists, other farmers, family and other workers, industry organizations, field days along with print and electronic media. A farmer survey conducted by the Australian Bureau of Agricultural and Resource Economics (ABARE 1998) that the number of farmer directed groups such as Landcare and commodity based groups such as the Grain Growers Association had increased significantly in recent years. At this stage there is no evidence to suggest that this trend has not continued over the subsequent decade. Indeed, farming groups focusing on more sustainable farming practices, such as the Victorian No Till Farming Association, which was established in 2003, have become more common as well. Benefits of farming groups include interaction between participants, a variety of perspectives, the practical application of new practices, collective problem solving and recognition of the importance of local knowledge (Millar and Curtis 1997).

It has also been important to recognize that farmer learning styles are not homogeneous across different sectors. Reeve and Black (1998) identified that for the grazing and horticulture sectors that training rather than extension activities relying on social networks were more likely to bring about the adoption of sustainable management practices. Farmer directed groups were also identified as an important source of motivation for change.

Confirming other reports about farmer learning, the 2002 National Land and Water Resources Audit highlighted in the key points relating to Australian farmers and natural resource management:

> Farmers do not all learn about sustainable practices in the same manner. Styles of farmer learning vary from reliance on a few key informants to the use of a wide range of personal and indirect sources. No delivery system will be appropriate for all farmers. Dissemination of local knowledge will remain a key feature of any successful training program (Commonwealth of Australia 2002).

A more recent piece of research that explores the questions of how farmers learn, how they construct knowledge, and how learning informs their practice was produced by Allan (2005). Conducted in New Zealand, the research report by Allan (2005) comes from the stance that farmers are now living in a time where things are changing much faster than they had in the past and now needed to deal with globalization, the 'New Economy and the Knowledge Age'. As with earlier Australian research, Allan's case studies also highlighted the lack of value placed on formal training in education and professional development.

Allan's paper also addressed the issue of the conflict between farmers themselves and 'outsider experts' involved in agricultural training, who appeared to have little or no understanding of the needs of farmers, the culture of farming and the value of informal learning in a farming context. This appears to have impacted on the perceived relevancy of the training programs for farmers.

The nature of farming as an activity and its geographical location are significant issues in relation to education for farmers. Farmers are often perceived as 'practical' people who learn through trial and error. This has implications for how such informal learning is 'valued' in the education sector. There also appears to be some conflicting evidence about the correlation between formal agricultural education and the profitability of a farm. This might be considered to be a both a blank spot and a blind spot in Allan's research as there are so many variables that impact on farm profitability, that formal education may be quite a minor issue in certain circumstances (Wagner 1993).

The idea of a community of practice (Lave and Wenger 1991) among farmer learners is also highlighted in Allan's research. The issue of social isolation among farmers is considered and may well be one reason why many farmers tend to congregate for other activities including Landcare, farming organizations, field days, trials and social occasions. This leads to the need for my research to consider learning as a social construction in this context rather than individual learning. In effect it could mean that for farmers learning is only effective or possible in a social context rather than as an individual learning context.

The emergence of computerized communications technology and the corresponding social media that may provide social connectedness for learning purposes through websites, blogs, Facebook and twitter also has the potential to increase opportunities in this area (Wagner 1993). However, given the timeframe and resources available for my research, consideration of these alternative forms of informal learning or information gathering has not been possible. This is an area that should be considered in future research on farmer learning activities.

It is also apparent in Allan's case studies that those farmers who are resource-rich and have extensive work practices that involve a broad range of peer and advisory influences are more innovative and successful than those who rely on historical practices alone. The overall conclusion of Allan's study was that the learning of farmers was reflected by the richness or poorness of their extended communities of practice and their dispositional knowledge was strengthened or weakened by their interactions with the community. Previous formal qualifications appeared to have little impact on the performance of the farmer, although the nature of the study, with just six case studies, would not make that outcome conclusive.

More recently the focus for the farming community has been on capacity building. Aslin et al. (2006) prepared a report for the Cooperative Venture for Capacity Building that focused on capacity building in a rural Australian context using the following definition for capacity building originally from:

> ...externally or internally initiated processes designed to help individuals and groups associated with rural Australia to appreciate and manage their changing circumstances, with the objective of improving the stock of human, social, financial, physical and natural capital in an ethically defensible way (Macadam et al. 2004, cited in Aslin et al. 2006)

Aslin et al. (2006) highlighted that learning, either as an individual or a social process, is central to building capacity in the rural sector. Aslin et al. (2006) also argued that there was a need to move away from conventional ideas of learning because of the adult learning context that exists when referring to the farming community. The learning process was considered more in the context of:

- A lifelong, experiential process, which is not often a formalized process;
- People or groups being able to engage in a broad range of formal and informal processes and information sources;
- A way in which people are presented with new ways to undertake their work but are also able to critically evaluate the assumptions used in developing new work methods;
- Learning directed at a very practical level to deal with everyday issues.

Aslin et al. (2006) evaluated a number of different learning scenarios for farmers, both formal and informal, and came to many of the same conclusions that Kilpatrick et al. (1999) came to in her report on farmer learning. In terms of learning activities, it was clear that contextual, short, practical group activities were favored by farmers.

These preferences are likely to be due to a number of factors including availability of activities in the local area, costs incurred, and time. Preferred activities are ones that are likely to take the farmer away from the farm for only relatively short periods and that are either free of charge or relatively inexpensive to attend.

The implication here is that training providers need to use these types of activities as much as possible and perhaps investigate other kinds of informal activities as well. Applying adult learning principles and providing activities that allow farmers to interact directly and learn from one another in relatively informal settings are likely to foster participation (Aslin et al. 2006).

Among the recommendations in the Aslin et al. (2006) report was the need to support further work on developing better methods of identifying the contribution made by informal capacity building activities, and encourage use of these methods in future surveys of farmer learning activities. In endeavoring to address my research question, the role of informal learning among farmers was a critical issue to be addressed in terms of establishing the types of informal learning activities and the effectiveness of them.

Having considered the current range of research literature in the areas of farmer learning styles it is already clear that there is no one answer that will work for or suit all farmers. Indeed it is likely that not only will most farmers establish their own way of learning about changing environmental conditions but they will also individualize their own adaptation practices. The problem with individualization is that it makes it very difficult for government agencies, industry groups and education institutions to develop appropriate strategies for communicating the information that will inform adaptation practices. This research aims to establish what learning methods or combination of methods will best reflect the way that Australian farmers operate.

The types of methods and tools that could be used to inform or educate farmers about climate change adaptation are likely to vary from web-based learning for those who are more computer savvy to the formation of farmer learning groups based around community halls (often located in the middle of nowhere) to participate in a social learning environment.

Allan's study (2005) promoted the idea of a community of practice (Lave and Wenger 1991) among farmer learners. Alan considered the issue of social isolation among farmers suggested that is why many farmers tend to congregate for other activities including Landcare, farming organizations, field days, trials and social occasions. This leads to consideration of learning as a social construction in this context rather than individual learning. Does this mean that for farmers learning is only effective or possible in a social context rather than as an individual learning activity? The emergence of more advanced computerized communications technology and social media has also provided new opportunities for farmer learners to gain information in ways that have not previously been researched.

3 Research Approach and Method

Two of the leading paradigms in social science research are the positivist and phenomenological paradigms. While positivist research is most closely aligned with quantitative methods of data collection and analysis, phenomenological research mostly relies on qualitative data collection and analysis (Collis and Hussey 2008). The phenomenological paradigm has a primary focus on the understanding of the meanings of human experience in a particular context. In-depth and semi-structured interviews are often associated with the phenomenological paradigm.

The use of case studies in education research is reasonably widespread. Case studies can be particularly useful in a broad ranging socially complex field such as education, which is not always well served by being limited to quantitative research (Yin 2003). This research has been undertaken as a case study of grain or mixed farmers in the grain growing regions of Australia. A sample of eight farmers from grain growing regions in Queensland, New South Wales and Victoria, who are all facing similar issues in terms of climate change adjustment across their farming activities were selected to participate in an interview.

For this research a case study approach allowed for the collection of comprehensive and in-depth information (Collis and Hussey 2008). It produced rich data for theorizing and conducting detailed analysis of the way in which farmers obtain information and education about issues such as climate change and adapting farming practices to drier climatic conditions. It was especially useful as exploratory research used in an area that has been identified as lacking in a significant knowledge base (Collis and Hussey 2008).

In discussing the presentation of case study reports Yin (2003) emphasizes that there is no fixed or hard and fast method of presentation. Yin discusses a different range of approaches to presenting a series of case studies, such as those in this research, and in weighing up those approaches, a short narrative has been prepared for each case study, highlighting the main points of discussion in each as they relate to the research questions. This approach then allows for cross-case analysis after all the case study narratives are presented.

4 The Case Studies – Location and Data Collection

Many of the grain growing regions in eastern Australia have been experiencing ongoing drought conditions for more than a decade. While there is still debate about the impact of climate change on the agricultural sector, there are a significant number of farmers who have demonstrated an interest in gaining a greater understanding of issues relating to climate change. This was demonstrated by the Grain Growers Association obtaining funding in 2009 to develop a climate change education program for their members over a 3-year period. This provided an opportunity to work with the GGA through their membership to find out how they currently obtain their information on climate change and adaptation.

The Grain Growers Association is made of 17,000 members located predominantly in New South Wales, Queensland and Victoria. Using the electronic communications network within the grain Growers Association membership, volunteers were sought in December 2009 for interviews to be conducted in early 2010. A number of GGA members responded and 7 members were selected and

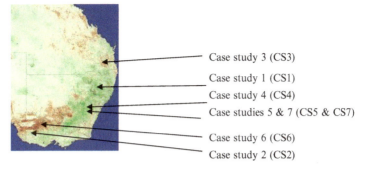

Fig. 4.1 Map of farmer case study locations in the grain belts of eastern Australia

contacted for telephone interviews. These interviews took place between February and April 2010. Figure 4.1 is a map identifying the spread of the case study farmer volunteers for this research project.

This research consisted of one data collection method; in-depth telephone interviews with Grain Grower Association member volunteers. A semi-structured interview approach through the use of an interview guide or questionnaire was used in order to ensure that there was a framework for the information collected (Remenyi et al. 1998). Other than for demographic questions, the question guide was designed to elicit open-ended responses and to encourage the interviewees to express honest responses regarding their views on climate change, how they gain information, who they trust and the types of learning activities they choose to participate in.

It is important to note that all participants were male and although there has been an increasing participation of females in the agriculture in Australia this trend has been less obvious in broad acre grain cropping and given the small sample size not surprising. The age range of participants is from 27 through to 56 with an average of 44. Three participants were aged under 40 and three over 50.

In terms of education, all those under the age of 40 had acquired some form of post secondary education to at least diploma level. The only one who had postgraduate qualifications was in the 50 plus age group with the remaining three above the average age had no post secondary qualifications.

In order to maintain anonymity among each case study they are referred to with a shortened version of Case Study? As CS? To avoid confusion.

4.1 Case Study 1 – Climate Change Is an Inconvenient Truth

Some things are inevitable and changing and we have to adapt...

Case Study 1 describes his understanding of climate change he used an example of a 2,000 year old village in England, which was once on the coastline and is now nowhere near it. CS1 also cited imagery used by Dorothea Mackellar in describing the Australian landscape and climate to emphasize that it has a changing climate that can be variable and erratic.

In outlining CS1's reaction to the science of climate change he expressed the view that people only had to look at the Bureau of Meteorology graphs of the Australian climate and the change was very clear. CS1 cited Al Gore's *An Inconvenient Truth* saying that he took the information from that at face value and if we ignored CO_2 levels then we '*just have our heads in the sand*'.

In terms of how CS1 acquired his knowledge on issues such as climate change, he stated that main sources were industry magazines and newsletters and newspapers, both paper and web-based publications. CS1 explained that the Australian media had been very negative about the drought and that drought conditions were really just part of the climate in Australia. Other knowledge on climate change came from local observations of his farm, field day activities and other industry meetings.

CS1 stated that there was not one single credible source for information on climate change and that he sought information across a variety of sources. CS1 highlighted that it was really local observed information that had the greatest impact in that there has been little rain for the last 6 years. This has led to a significant change in his approach to farming and although CS1 is optimistic that water allocations will return, this needs to be balanced with realistic expectations. Recent farming practices adopted by CS1 included no-till farming and stubble retention as well as improved crop varieties.

4.2 Case Study 2 – *Farming with the Environment in Mind*

We have to be practical and learn to live with our environment…if it is changing then so do we…

Case study 2 acknowledged that he is a climate change believer and thinks that we need to do more to adapt and mitigate than what is currently being done. CS2 has drawn most of his information on climate change from reading books, magazine articles, industry newsletters and websites and listening to those he believes are credible sources.

CS2 said that he had a preference for practical learning activities such as field days, workshops, crop walks, trials and demonstrations because they are usually timed to fit in with the farming cycle and do not generally involve any more than a day or two as a time commitment.

CS2 expressed concern about the future of family farming because of the difficulty in improving profitability without increasing the farm size overall. CS2 indicated that while he has four children he did not expect any of them to take over the farm in the future and hence was uncertain about the future of his farm.

CS2's farming philosophy involved developing niche markets such as organically accredited grain and livestock products. CS2 had also previously dabbled in specialized marketing activities of his own lentils but found it too difficult to maintain. Recent farming practices adopted by CS2 included no-till farming, stubble retention, increased native vegetation for shelter belts and trialling of different crop varieties.

4.3 Case Study 3 – Skepticism Combine with Best Practice Farming

I can't see any real change and everyone says that each season is different...

Case study 3 expressed the view that those who believed in climate change accepted that human interaction on earth was changing the weather on the planet. CS3 acknowledged that he was skeptical about human impacts on climate change. CS3 stated that he was certainly not convinced that there are any significant changes relating to the climate that will impact on his farm or farming activities. CS3 said that he does believe that his farm will be impacted by government policy and legislation that relates to perceived climate change.

Most of the information that CS3 has gained about climate change issues has been derived from the media, mainly newspapers, radio and sometimes from the Internet. CS3 acknowledges that the most reliable source for information on climate issues is from industry related organizations.

CS3's main farm-related learning activities involve field days, workshops and trials that he would attend approximately once a month. CS3 said he believes that the most effective learning activities are field days and workshops but highlights the importance of communicating about farm-related activities with his neighbors. Recent farming practices adopted by CS3 included a combination no-till farming and minimum till farming as well as precision agriculture using improved technology and machinery.

4.4 Case Study 4 – Sustainability Is the Key

History has proven that we have had similar dry events in the past... is it permanent or is it cyclical...as savage as it has been, we know the CSIRO (Commonwealth Science and Industrial Research organization) has identified global warming as a problem....

Case study 4 explained that even if it was not human induced climate change, we have to do something because we cannot rely on the current sources of energy such as coal and petroleum because they are not sustainable in the long term. CS4 expressed an interest in issues relating to climate change and stated that we have to be smarter in how we manage our energy sources in the future.

Despite following the climate change debate closely CS4 said he was undecided as to the impact on his farming operation. CS4 explained that he was aware that climate change was a naturally occurring phenomenon and that there was also some longer term cyclical patterns in climate.

CS4 said that the impact of climate change on livestock farming was less noticeable apart from the massive reduction in livestock numbers on farms in recent years because of the dry conditions and a lack of feed. CS4 observed that Australia had lost 100 million sheep in recent years and that it would take at least a decade to recover the industry as long as the seasons were good.

CS4's main sources of information on climate change have been the internet, in particular Farm Online and television programs and debates. CS4 indicated that he also reads widely and carefully on the subject. He has also been involved in a climate change taskforce with the Grain Growers Association.

CS4 stated that his view on farmer learning activities was that one day workshop style activities were the most useful and effective learning activities for farmers. *'Farmers are more likely to pick up a brown snake than a pen...'* was how he described the learning style of farmers. Recent farming practices adopted by CS4 included no-till farming, improved crop varieties, increased attention to technological advances in precision agriculture and weather forecasting.

4.5 Case Study 5 – Climate Is More Than Just a Seasonal Trend

I am not sure if we caused it but it seems like this is more than just a temporary change...this is long term...

Case study 5 stated that is clear in his understanding of climate change that it involves more than just seasonal variations and that it is a much longer-term trend in changing of climatic conditions. CS5 admitted that while he is not entirely convinced about anthropogenic climate change, he acknowledges that it is probable that there will be some changes in seasonal conditions in the overall climate.

CS5 explained that he has gained knowledge on climate change issues through his involvement in scientific and industry management along with wide reading of articles from the popular press, the internet, agricultural journals, the Australian Farm Institute publications, Australian Grain magazine and the rural press. CS5 has also had significant contact with leading scientists in the field through his involvement in agricultural research. CS5 indicated that he has found that the CSIRO is the most credible source of information relating to climate change issues.

In terms of learning activities relating to his farming operation, CS5 has maintained membership of two farming systems groups; Central West Farming Systems and the Conservation and Natural Farming Association based in central west NSW. CS5 stated that he has also been prepared to consult with relevant experts such as agronomists to assist in better managing his farm. Recent farming practices adopted by CS5 included no-till farming, stubble retention and improved soil testing practices and trialling of different crop varieties.

4.6 Case Study 6 – Serious Climate Change Research Is Important

Climate change is too important to sensationalise in the media...even the CSIRO is guilty of climate change sensationalising...

Case study 6 understands that climate change involves increased temperatures which impact around the world with some places receiving more rain while others will receive less. CS6 does acknowledge that the climate has changed and highlights the increasing number of dry springs since 2000 and believes that there is more to it than just climate variability, but is unsure of what those factors are. CS6 expresses concern about the impact of farming in his region (Mallee) by 2030 and that a significant reduction in rainfall will significantly impact productivity by reducing yields by as much as 30 %.

CS6 expressed some frustration with sensational media reports by people who do not understand what they are talking about. But he also indicated that CSIRO is guilty of sensationalizing climate change issues as well. CS6 said that he does believe in anthropogenic climate change but he does like to read what skeptics have to say on the subject. CS6 highlighted issues such as duplication in research funding involving climate change these days and consequently he said he had become very frustrated with the CSIRO.

CS6 said that the most credible sources of information are industry bodies such as the Grain Growers Association and the World Grain magazine produced in Kansas because it always has quite a bit of information about climate change and variability. CS6 also derives information from a variety of sources such as the internet, the Financial Review, Citigroup sharebrokers newsletter, the GRDC Groundcover publication, The Weekly Times, Australian Grain Magazine, the Victorian Department of Primary Industry, skeptics such as Dr Jennifer Marohasy. CS6 gives qualified support to CSIRO publications but expresses concerns about the politics and self-interest in the organization. CS6 said *'in terms of climate change issues, the general media such as newspapers were complete rubbish ...'*

Recent farming practices adopted by CS6 included a combination of minimum and no-till farming, composting trials and different crop varieties.

4.7 Case Study 7- Keeping Farming Options Open

I have become a lot more reactive to climatic conditions and now keep my options on the farm open...

Case study 7 explains that climate change involves the long term change in climatic conditions. CS7 acknowledges that his farming operation has been impacted by climate change and will continue to do be impacted on into the future. CS7 stated that he has become much more reactive to climatic conditions in his farming practices each season.

CS7 says he has gained most of his climate change information from industry organizations such as the Grain Growers Association, the Kondinin Group, the Australian Farm Institute and NSW Farmers. CS7 has also gained information from the general media such as the radio, television, newspapers and the Internet. CS7 says that he believes the most credible source of information was the Australian Farm Institute because of the quality of work they have done in the area of climate change. Recent farming practices adopted by CS7 included no-till farming precision agriculture using improved technology and use of the Internet for weather forecasts.

5 Analysis and Discussion

The principle purpose of this section is to apply what has been discovered from the seven case studies and apply it to the research questions in light of what is already understood from the literature review. In answering the main question posed in this research project; *How do established farmers learn about adapting farming practices in a drier environment?*, it was clear that the case study farmers understood that climate change was an issue for the future of agriculture and in particular grain and mixed farming enterprises.

Some of the factors that emerged in this research that have implications for how farmers learn include the importance of informal learning and what we can now include in that, given ongoing advances in information technology, the importance of risk management in a farming enterprise and the role of peer group communication.

It became apparent through the case study interviews that there are really two separate learning issues in this research. While the principal research question focused on learning about farming practices to adapt to a drier environment, there was a significant underlying question relating to how farmers learnt about climate change itself. This second question aligns with one of the subsidiary research questions; *how do farmers learn that the environment is changing?* Learning about climate change itself can be a very different activity compared to learning about farming practices. The science behind climate change is essentially an academic learning activity. It is not a 'hands-on' type of learning activity. It is the type of learning activity that will involve listening to someone speak, almost all of this will be in an informal learning setting. In most cases learning about the science behind climate change involved listening to popular media, accessing the internet or exchanging information through farming peer groups such as the GGA. Broad acre grain farmers also often spend a considerable amount of time operating farm machinery such as trucks, tractors and headers, and participants in my study reported listening to farm specific programs such as the ABC radio lunchtime program, *The Country Hour* or the early morning *Rural Report*. This type of learning activity also tends to line up with Kilpatrick and Rosenblatt's (1998) five reasons why farmers seek information rather than undertake formal training and Allan's (2005) lack of value placed on formal training by established farmers.

Kilpatrick (1997) identified the importance of all forms of education, both formal and informal learning, as a contributor to the success of a farming enterprise. In the 14 years since that study, there has been a significant change in the role of information technology and communications across our whole society, and farming is not immune from these developments. In the context of an issue like climate change this has some important implications because all farmers are now exposed to a much greater range of communications than they would have been in the previous century. It also highlights the need to consider what is informal learning now. Where once informal learning would only have included activities such as field days or workshops, the advent of the Internet has opened up the definition of informal learning for farmers in this context. What is apparent in my research is that all of the case study farmers had actively used modern communication technology such as the Internet, along with traditional media to educate themselves about climate change issues, regardless of the source of the information.

In addition the issue of risk management was highlighted by virtually all participants as a critical issue for future farming profitability and sustainability. Indeed the risk management issues in the area of climate change was becoming increasingly critical even for traditional family farming enterprises I think the risk management factor also plays a very large role in some of the decision made by farmers concerning learning about and adopting new farm practices. As Allan (2005) highlighted in her research, it was those farmers who were more resource rich and involved a broad range of peer influences that tended to be more successful.

There was considerable knowledge expressed by all participants on improved practices that will assist their farming operation to adapt to changing climatic conditions even if they did not agree that climate change was happening. These practices included direct drilling of seed, controlled traffic farming, soil moisture retention through better stubble management and acceptance of the need to look at new crops and crop varieties. All of these reflected some of the adaptive practices listed by Howden and Stokes (2010). Much of the knowledge concerning new and innovative farming practices came from attendance at traditional informal learning activities such as field days, crop walks and workshops. Most of these were run by peer group farming organizations such as the GGA, No-Till or Conservation farming groups. It was clear that the adoption of improved practices did not rely on the belief that climate change was happening but rather that these practices would better suit their particular farming operation.

Farm size was also another focus on future profitability and sustainability. All participants acknowledged the need for farm size to increase if they were to survive into the future, although some acknowledged that they would probably not continue into future generations. While this is not specifically a learning issue, it will undoubtedly have some influence on the nature of the farming practices that will be adopted by the farmer in the future.

As observed throughout the literature review there has always been a preference for hands-on or practical learning activities by farmers. Indications for these case studies certainly support this premise overall and this is despite a broad range of educational levels within the group ranging from completion of secondary education through to a postgraduate research qualification. All case study participants acknowledged that the types of learning activities that were most effective for them were the hands-on activities such as field days, workshops, crop trials and demonstrations. Only one participant indicated a slightly different preference for learning in that he preferred doing his own research, but this would reflect the fact that he possessed a postgraduate research qualification and was very familiar with high-level academic research activities. Essentially this reinforces some of the ideas about farmer workplace learning and how it should take place.

It was also possible to provide some answers to the subsidiary research questions.

How do farmers learn that the environment is changing? While a broad range of information sources were highlighted by the participants, the popular media and industry and farming groups and the CSIRO were among the major sources of information. Credibility of those sources indicated by the participants suggested that industry and farming groups were the most trusted by farmers. Both popular media and the CSIRO were criticized for sensational coverage and self-interest.

How do farmers learn about different responses to a changing environment? Attendance at workshops, field days and involvement in crop trials and demonstrations were listed by participants as the most common ways of learning about responses to a changing environment. Participants tended to trust industry and peer farming groups as the best source of information.

How do farmers currently learn about issues relating to farming in a drier environment? Popular media and industry publications were the most quoted sources of information for issues relating to farming in a drier environment. Communication within peer farming groups was also an important source of information for these case study farmers.

What skills do farmers feel they need to acquire to adapt to farming in a drier environment? Participants listed a range of adaptive techniques such as direct drilling, controlled traffic farming, moisture retention techniques such as stubble retention and being prepared to look at new crops and crop varieties for their enterprise. Learning about these practices is essentially a practical activity involving attendance at field days and workshops.

With much of the previous research about farmer learning preferences being done more than a decade ago in the late 1990s, it is clear that there is considerable scope to do more research about the impact of the emergence of the Internet and other digital media services on farmer learning. While these issues were not specifically addressed as part of this research, it was clear that some farmers are making use of digital media as a source of information and self-education. Even during the period of this research new technology such as smart phones and tablets have added a new dimension to the role of computerized information technology for informal farmer learning activities. This would also be a fruitful area for future research on farmer learning preferences and practices.

Finally, it needs to be acknowledged that this is a small qualitative study of just seven mixed or grain farmers in eastern Australian grain belts and hence is a limitation to any conclusions and discussions that can be drawn from the data.

6 Conclusion

In answering the main question posed in this research project; *How do established farmers learn about adapting farming practices in a drier environment most effectively?* The main points that have emerged from this research project are:

- While all participants considered that there was something happening to climatic conditions, it was not necessarily a critical factor for undertaking adaptation. Other issues such as risk management and new farming practices presented by peer groups played a significant role in farming learning in this context.
- Practical learning activities are preferred and appear to be the most effective among grain and mixed enterprise farmers.
- Short, one or two day workshop or field day activities are preferred for learning activities
- Industry and farming groups are considered to be the most credible sources of information for learning about climate change.
- Digital media in various forms has also become a significant vehicle for climate related information for farmers although the credibility of the source was still critical.
- Risk management issues have emerged as a critical factor for farmers in a range of areas including climate change.

The outcomes of this research project for developing programs for effective training of farmers in learning to farm in a drier environment are that hands-on practical, short workshops and field days would be most effective in engaging farmers on this subject. However dealing with climate change does not have to be the focus of the activity but rather learning about some practical farming related techniques that will direct the farmer to reduce their risk in drier or changing conditions would be well received.

It would seem that neither age nor educational level make much of a difference, based on this small sample. Indeed it is not clear that accepting the premise that climate change is happening is even important in getting farmers to participate in any learning activities as the main criteria appears to be the type of learning activity. The implication here is that it is the activity that is important rather than the philosophy behind it. Again the emphasis is on practical, hands-on programs that engage farmers and help them improve their farming activities, thereby creating an environment for behavior change rather than convincing them of human induced climate change.

References

Allan J (2005) Farmers as learners: evolving identity, disposition and mastery through diverse social practices. Rural Soc 15(1):4–18
Aslin H, Giesecke T, Mazur N (2006) Which farmers participate, when? RIRDC publication no 06/106, Barton, ACT
Australian Bureau of Agricultural and Resource Economics (1998) Australian Farm Surveys Report 1998, Canberra
Collis J, Hussey R (2008) Business research a practical guide for undergraduate and postgraduate students, 3rd edn. Palgrave Macmillan, Basingstoke
Commonwealth of Australia (2002) National land and water resources audit. http://lwa.gov.au/files/products/national-land-and-water-resources-audit/pn22042/nlwra-final-report-april-2009-pn22042.pdf. Retrieved 3 Sept 2009
Dunn AM, Wolf EC (2001) Agricultural education: social science in the curricula'. Rural Soc 11(3):271–281
Golding B, Brown M, Foley A (2009) Informal learning: a discussion around defining and researching its breadth and importance. Aust J Adult Lear 49(1):34–56
Halliday-Wynes S, Beddie F (2009) Informal learning: at a glance. National Centre for Vocational Education Research, Australia
Howden M, Stokes C (2010) Adapting agriculture to climate change. CSIRO Publishing, Collingwood
Kilpatrick S (1997) Education and training: impacts on profitability in agriculture. Aust N Z J Vocat Educ Res 5(2):1–36
Kilpatrick S, Rosenblatt T (1998) Information vs training: issues in farmer learning. J Agric Educ Ext 5(1):39–51
Kilpatrick S, Johns S, Murray-Prior R, Hart D (1999) Managing farming, how farmers learn. RIRDC publication 99/74, Barton
Lave J, Wenger E (1991) Situated learning: legitimate peripheral participation. Cambridge University Press, Cambridge
Millar J, Curtis A (1997) Moving farmer knowledge beyond the farm gate: an Australian study of farmer knowledge in group learning. Eur J Agric Educ Ext 4(2):133–142
Reeve IJ, Black AW (1998) Improving farmers' management practices through learning and group participation, RIRDC publication 98/71, Barton, ACT
Remenyi D, Williams B, Money A, Swartz E (1998) Doing research in business and management an introduction to process and methods. Sage, London
Wagner J (1993) Ignorance in educational research: or how can you not know that? Educ Res 22(5):15–23
Yin RK (2003) Case study research – design and methods. Sage, Thousand Oaks

Chapter 5
Economic Impact of Climate Change on Agriculture in SAT India: An Empirical Analysis of Impacts in Andhra Pradesh Using Ricardian Approach

Naveen P. Singh, Cynthia Bantilan, and Kattarkandi Byjesh

Abstract This chapter analyzes the economic impact of climate change on agriculture for the state of Andhra Pradesh, India. The objective of this study is to quantify the impact of climate change on net revenue of paddy crop and 14 other major crops in the 20 districts of the state. The Ricardian approach was used to analyze the effects of climate variables on the net income from the crop. Panel datasets on climatic, agronomic and socio-economic variables were used for this analysis. The results showed that there is significant nonlinear impact of temperature and rainfall on yield over the years on the net income from rice and other crops. On an average in rice; 1 °C rise in temperature will reduce the net income by109 INR (2.42US$) per hectare in these districts and the impact of precipitation are not substantial. Among the analyzed districts, Anantapur face the maximum brunt of the impact of climate change. In case of other crops, there are varied impacts of different climatic variable on net revenue with some district gaining initially and negatively impacting later. As expected, rainfall had positive marginal impacts, however it is very negligible. The socio-economic variable i.e. amount of irrigated area, literacy rate of rural population also showed significant positive effects on the income.

Keywords Climate change • Ricardian analysis • Semi-arid tropics • Agriculture

N.P. Singh (✉) • C. Bantilan • K. Byjesh
Research Program – Markets, Institutions and Policies, International Crops Research Institute for the Semi-Arid Tropics (ICRISAT), Hyderabad, Andhra Pradesh 502324, India
e-mail: np.singh@cgiar.org; c.bantilan@cgiar.org; k.byjesh@cgiar.org

1 Introduction

The semi-arid tropical region of India is home for 45 % of the total population and majority of this population are in rural areas and farming is the supporting means of livelihood for them. Approximately 380 million populations live in the rural area of Indian SAT region and are primarily dependent on farming for their livelihood. As agriculture does have different dimensions including social, economic and environmental, it's good to optimizing these dimensions for the sustainability and development of the masses primarily dependent on farming community for their livelihood. Several studies conducted at ICRISAT emphasize the importance of sustainable farming and income source that determines the present and future socio-economic condition of the smallholder farmers (Bantilan and Anupama 2006; Jodha 1978). However, the frequent occurrence of shocks including changes in the climate, and the risks associated with it have made the farming community in this region vulnerable. Rainfall variability, frequent droughts (inter and intra seasonal), extreme seasonal temperature rise, decreasing soil fertility, diminishing owned assets etc. are the general characteristics of the region. The major crops grown are cereals, pulses and other horticulture crops, which owe their growth to an increased dependency on rain & to suffice with farm water requirements. Having alternate source of irrigation that provides a cushion to ease out an economic incentive for the farmers to grow high value crops. Helping to preserve, conserve and making efficient use of rainfall can be crucial for ensuring & sustaining availability of water requirement for irrigation, drinking and for domestic use.

The increased variability of climate was hastened by the long term effect of bio-geochemical changes happening in the atmosphere by emissions due to anthropogenic activities (IPCC 2007). Increased frequency of drought, decreasing number of rainy days during the south west monsoon, delay of the onset of monsoon, decreasing quantum of rainfall, rising average atmospheric temperature etc. have added woes to the sustenance. In India, most climate related studies confirmed an increasing trend in the surface temperature (Kothawale and Rupa Kumar 2005) and increasing variability in the seasonal precipitation (Sivakumar et al. 2005) in SAT tract of India. The intra seasonal drought at the critical crop developmental stages affect crop and thereby reduce economic yield. Avoiding water stress at the different stages of crop growth can be done by having the capacity to give irrigation through different conserved sources of water. Canals, open dug wells, tanks and exploring ground water are the plausible sources in this region. Population growth together with over exploitation of ground water resources, low adoption of water conservation measures makes the situation even worse. Low soil productivity, rainfall variability, water shortage or scarcity, poor development in rural infrastructure, institutions and markets are major identified characteristics of the SAT India (Bantilan and Keatinge 2007; Shiferaw and Bantilan 2004). Adaptation and copying strategies to climatic and environmental fluctuations are not new to the rural population of the SAT region.

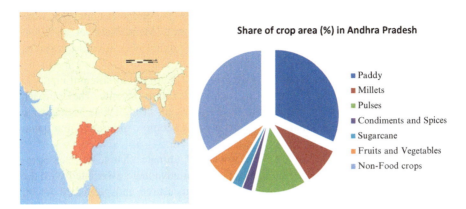

Fig. 5.1 Crop wise distribution of agricultural area in Andhra Pradesh (2008–2009)

Climate Change is a reality now. It has affected many sectors of the human society. It is well known that agriculture is the most important sector that is very much affected by changes in climate as temperature and rainfall are major variables of crops production. Numerous studies have been done to quantify the impact in terms of monetary loss using the Ricardian approach (Khamwong and Praneetvatakul 2011; Mendelsohn et al. 1994; Kumar and Parikh 1998). These studies help the policy makers to formulate suitable adaptation and mitigation mechanisms to minimize the negative effect of climate change. The purpose of the present study is to use the Ricardian modeling approach to study the impacts of climate change on the important crops in districts of Andhra Pradesh, India.

Among all the crops cultivated in Andhra Pradesh, Paddy(Rice) is the most important food crop and it occupies 31.7 % of the cropped area (Season and Crop Report Andhra Pradesh, 2008–2009) (Fig. 5.1).

Hence an attempts are made to quantify climate change impact on rice separately and the other major crops. The present study aims to quantify the impact of climate change on agriculture in Andhra Pradesh districts by employing the Ricardian modeling approach. The study focuses on the changes in the net revenue of (a) paddy crop and (b) 14 major crops due to changes in climate variables.

Three approaches have been widely used in the literature to measure the sensitivity of agricultural production to climate change; agronomic-economic models, cross-sectional models and agro-ecological zone models:

- The agronomic-economic method begins with a crop model that has been calibrated from carefully controlled agronomic experiments (FAO 2000; Kumar and Parikh 1998). Crops are grown in fields or laboratory settings under different possible future climates and carbon dioxide levels keeping all farming methods across experimental conditions fixed so that all differences in outcomes can be attributed to the climate variables, viz., temperature, precipitation, or carbon dioxide.

- In the cross-sectional approach, also known as the Ricardian method, farm performances are examined across climate zones (Mendelsohn et al. 1994; Mendelsohn and Nordhaus 1996; Kumar and Parikh 1998). Ricardo observed that the land values would reflect land productivity at a site (under competition). In this approach the land value is regressed on a set of environmental inputs to measure the marginal contribution of each input to farm income. The approach has been applied to the United States (Mendelsohn et al. 1994; Mendelsohn and Nordhaus 1996). Climate parameters are precipitation, minimum, maximum and diurnal temperature.
- The third approach to measure the impact of climate change utilizes agro-ecological zones (AEZ) (FAO 1996). The main advantage associated with the agro-ecological zones is that they have been measured and published for all developing countries (FAO 1996). Detailed information is available about the climate and soil conditions, crops, and technologies being used throughout the tropical zone. The AEZ model develops a detailed eco-physiological process model. Factors such as length of the growing cycle, yield, leaf area index, and harvest index etc that explain plant growth are inputs to the model. Existing technology, soil, and climate are combined to predict Land Utilization Types (LUT). Combining these variables, the model determines which crops are suitable for each cell. The impact of changes in climate variables on potential agricultural output and cropping patterns are thus simulated.

Agronomic studies of India suggested that extensive warming could cause significant reduction in yield. If temperature rose by 4° centigrade, the yield of major cereal crops would fall by 25–40 %, rice yields by 15–25 % and wheat yields by 30–35 % (Kumar and Parikh 1998). Mendelsohn et al. (1994) indicated that the global warming would decrease the net income by 8 %. In this context, it is important to address the impact of climate change on future crop area, production and productivity of crops using the available data. The agronomic studies indicate that the higher temperatures are likely to be harmful in many developing countries where the climate change is marginal, water is inadequate and temperatures are high. Thus the agronomic studies suggest that the countries of the temperate and polar regions could gain productivity whereas developing countries in the subtropical and tropical zones are likely to lose productivity (IPCC 2001).

There are studies based on Ricardian approach to quantify the impact of global warming on crop production in Tamil Nadu State, India. For example, Palanisami et al. (2009) employed the Ricardian type analysis to study the effects of climate variables on the area and production of three major crops of Tamil Nadu in India. The crops selected for the study were paddy, groundnut and sugarcane that account for a major cultivated area of the state besides being grown in almost all the districts. Dependent variables considered for the analysis were area and the yield of the crops. The results show that there will be a reduction in both the area and yields of major crops by about 3.5–12.5 % due to the impact of climate change. Consequently overall production will decrease by 9–22 % for these crops.

2 Data and Methodology

The Ricardian approach is a cross-sectional model applied to agricultural production. It takes into account how variations in the climate change affects the net revenue or land value. Following Mendelsohn et al. (1994), the approach involves specifying a net productivity function of the form

$$R = \sum p_i q_i(x,f,z,g) - \sum p_x x \qquad (5.1)$$

where R is the net revenue per hectare in the constant rupees, p_i is the market price of crop i, q_i is output of the crop i, x is a vector of purchased inputs (other than land), f is a vector of climate variables, z is a set of soil variables, g is a set of economic variables such as market access, literacy, population density etc, and p_x is a vector of input prices. The farmer is assumed to choose x to maximize the net revenues given the characteristics of the farm and market prices. Assuming a quadratic function for crop output, the standard Ricardian model is specified by the quadratic function

$$R = \beta_0 + \beta_1 f + \beta_2 f^2 + \beta_5 z + \beta_6 g + u \qquad (5.2)$$

where u is an error term and f and f^2 are levels and quadratic terms for temperature and precipitation. The inclusion of quadratic terms for temperature and precipitation ensures non-linear shape of the response function between the net revenues and climate. Normally we expect that farm revenues will have a concave relationship with temperature. When the quadratic term has a positive sign, the net revenue function is U-shaped, but when the quadratic term is negative, the function is hill-shaped. As each crop has an optimal temperature at which it has a maximum growth, the function is expected to have a hill-shape. From the fitted equation, we can find the marginal impact of a climate variable on farm revenue. The marginal impacts are usually found at the mean level of the climate variable. Thus from Eq. (5.2) we have

$$\frac{\partial R}{\partial f} = \beta_1 + 2\beta_2 \bar{f} \qquad (5.3)$$

Where $\bar{f}$ is the mean of the climate variables. This shows that the marginal effect of a particular climate variable is equal to the sum of (i) coefficient of the linear term and (ii) twice the product of the coefficient of the quadratic term multiplied by the mean level of the climate variable. The climate variables included in the model are season temperatures and their squares and season precipitations and their squares. MATLAB[1] software package was used to fit the model.

---

[1] Matrix Laboratory, The Mathworks.

Table 5.1 List of crops grown in this study district of Andhra Pradesh

Sl. No.	Crop	Average crop wise area in '000 ha per year (1970–2008)
1	Rice (paddy)	3,685
2	Sorghum	1,428
3	Pearl millet	305
4	Maize	398
5	Finger millet	177
6	Chickpea	159
7	Pigeon pea	317
8	O. Pulses	1,117
9	Groundnut	1,744
10	Sesamum	166
11	Castor	280
12	O. Oilseeds	2,514
13	Sugarcane	209
14	Cotton	698

The dataset for the present study included panel data on three types of data (i) climatic (ii) crops area and production and (iii) socio-economic. The study included 14 major crops (as given in Table 5.1) grown over 20 districts of Andhra Pradesh for 39 years during the period 1970–2008. The climatic variables included the temperature and precipitation during the four seasons. The crop variables included are area and production for each crop. The socio-economic variables are fertilizer consumption (N, P, K), tractors, pump sets etc. Table 5.2 provides a detailed list of all the variables. These data including the output prices were collected from the various government and other publications. Net revenue per ha was computed as given below

$$Net\,Revenue\,per\,ha = \frac{Gross\,Revenue\,from\,Crops - Fertilizer\,Cost - Estimated\,Labour\,Cost}{Total\,Area\,under\,Crops}$$

Costs attributed to other inputs such as tractors, bullocks, irrigation etc. have not been included as it is difficult to estimate them. However these variables have been used as control variables in the model given in Eq. (5.2). The net incomes were converted to 1981–82 constant prices. As there are 20 districts and the data relates to the 39 years, the sample size is 780.

As suggested by Eq. (5.2), the net revenue per acre was regressed on climate and socio-economic variables. The squares of the climate variables were also included. The list of these variables used in the model is given in the Table 5.2. The methodology followed by Praneetvatakul et al. (2011) was employed in estimating the regression equations.

Table 5.2 List of variables included in the study

Sl. No.	Classification of variables	Variables	Notation used
1	Climatic	South West monsoon temperature (°C)	TSWM
2		North East monsoon temperature (°C)	TNEM
3		Jan-Feb. temperature (°C)	TWP
4		Mar-May temperature (°C)	THWP
5		South West monsoon precipitation (mm)	RSWM
6		North East monsoon precipitation (mm)	RNEM
7		Jan-Feb. precipitation (mm)	RWP
8		Mar-May precipitation (mm)	RHWP
9	Socio-economic	Tractors ('000)	TRACTOR
10		Pump sets ('000)	PUMPSET
11		NPK consumption ('000) tons	NPK
12		Rural literacy (%)	LITPOPRU
13		Population density (%)	POPDEN
14		Percentage of area under high yielding varieties (%)	HYV
15		Percentage of irrigated area to gross crop area (%)	IRR

South-West monsoon: June–August; North-west monsoon season: September–October to November–December

3 Results and Discussions

The Table 5.3 provides the mean values of area and net revenue per ha for rice and all the crops for all the 20 districts.

It can be observed from the Table 5.3 that West Godavari, East Godavari and Krishna districts are predominantly rice growing districts. In these districts the area under rice occupies respectively 80 %, 66.6 % and 58 % of the total area under all major crops. The net revenue per ha for rice ranges between Rs. 1,334 (US$29.6) to Rs.3,631 (US$80.7) across the districts whereas for major crops it ranges from Rs.1,089 (US$24.2) to Rs. 4,712(US$104.7). Kurnool, Anantapur and Guntur are the first three districts with a maximum net revenue per ha for rice. Their net revenues are respectively Rs. 3,631(US$80.7), Rs.3,560(US$79.1) and Rs. 3,542(US$78.7). West Godavari, East Godavari and Krishna districts occupy the first three positions for maximum net revenue per ha for crops with respective net revenues being Rs.4,712(US$104.7), Rs. 3,589(US$79.8) and Rs. 3,438(US$76.4). All these net revenues are weighted by 1981–1982 constant prices. Table 5.4 gives the mean values of all input variables.

From the Table 5.4, it can be seen that the average temperature during the four seasons ranged between 24.5 and 30.8 °C. Average temperature during the summer months (March to May) is 30.8 °C. Across the districts, the average temperature ranges from 26.3(Anantapur) −29.9 (Nellore) °C during South West Monsoon period. During the summer months the average temperature ranges between 27.4 and 32.1 °C. Rainfall has a wide range in the four seasons with South West

Table 5.3 Mean values of area and net revenue of rice and major crops in different districts of Andhra Pradesh

Sl. No.	District	Rice-area (000 ha)	Area under major crops (000 ha)	Rice-net revenue (per ha) in INR	Crops-net revenue (per ha) in INR
1	Adilabad	65.7	593.4	2,288	1,171
2	Anantapur	58.8	1,437	3,560	1,350
3	Chittoor	103.3	675.5	3,252	2,719
4	Cuddapah	62.4	539.2	3,245	1,580
5	East Godavari	378.5	568.4	3,130	3,589
6	Guntur	354.7	922	3,542	2,667
7	Ranga Reddy	43.6	307	1,334	1,094
8	Karimnagar	186.8	545.6	2,950	2,471
9	Khammam	133.1	425.2	2,397	1,999
10	Krishna	366.4	631.6	2,867	3,438
11	Kurnool	102.9	1,102.4	3,631	1,593
12	Mahabubnagar	117.3	1,062.7	2,800	1,089
13	Medak	100.3	460.7	2,433	2,104
14	Nalgonda	234.4	831.7	3,233	1,612
15	Nellore	252.1	520.3	3,125	2,949
16	Nizamabad	135.3	337.5	2,667	3,423
17	Srikakulam	257.8	573.4	1,888	1,956
18	Visakhapatnam	156.7	525.6	1,908	2,296
19	Warangal	147.3	602.1	2,797	1,734
20	West Godavari	426.9	533.3	2,905	4,712
	Average	184.215	659.73	2,797.6	2,277

1US$ = 45 INR (Indian Rupees)

Monsoon (634.3 mm) and North East Monsoon (213 mm) having major contributions. One important feature of the distribution of rainfall is the variability across the districts in the two seasons. The South West Monsoon it ranges between 331 and 926 mm. The North East monsoon has a range of 564 mm across the districts. The contribution of the other two seasons is negligible. All the other input variables also have similar variability. For example, the percentage of irrigated area has a range of 72.6 % with West Godavari having 84.1 % and Adilabad having 11.5 %. Again West Godavari district has the highest percentage of area (67.4 %) under high yielding varieties while Anantapur has 6.2 %.

3.1 The Ricardian Model for Net Revenue per ha for Rice

3.1.1 Model Fit

Considering the importance of rice, separate Ricardian model was fitted for net revenue per ha for rice. Table 5.5 summarizes the model fit.

The results of the Ricardian analysis show that the temperature and its square terms during all the monsoons except during Jan-Feb have significant impact on the

Table 5.4 Mean values of the input variables

District	TSWM	TNEM	TWP	THWP	RSWM	RNEM	RWP	RHWP	TRACTOR	PUMPSET	NPK	LITPOPRU	POPDEN	HYV	IRR
Adilabad	28.6	24.1	24.1	32.1	925.8	98.6	18.1	40.8	0.001	0.032	4.2	19	122.9	18.7	11.5
Anantapur	26.3	23.9	24.4	29.7	331.3	157	4.7	76.3	0.002	0.056	1.1	25	157.7	6.2	16.2
Chittoor	27.5	24	23.7	29.2	428.7	382.3	15.8	93.3	0.007	0.209	8.8	31.4	187.1	17.4	40.2
Cuddapah	27.6	24.3	24.6	30.5	402.5	251.5	6.8	63.2	0.005	0.071	9.1	30.2	163	16.1	34.3
East Godavari	29.4	25.6	24.5	30.2	730.6	309	19.4	101.2	0.008	0.054	84.2	31.5	388.7	57	63
Guntur	29.7	25.6	25.2	31.4	562.4	234.8	17.1	69.8	0.004	0.021	55.7	28.4	262.9	33.4	41
Ranga Reddy	28	24.7	24.9	31.6	619.2	138.1	18.5	68.1	0.008	0.183	26.3	8	684.9	24.6	22.3
Karimnagar	28.1	23.9	24	31.5	773.4	101.5	18.2	47	0.009	0.286	50.8	23.8	241.7	49.8	54.4
Khammam	29.3	25.5	24.9	31.5	880.2	144.2	15.2	75.9	0.004	0.059	37.1	24.9	132	31.4	32.9
Krishna	29.5	25.6	24.9	30.7	662.6	240.6	15.5	65.5	0.007	0.034	94.7	29.7	399.5	48.6	58.8
Kurnool	28.3	25.2	25.6	31.8	478	146	4.6	71.9	0.002	0.025	7.4	24.9	161.7	14.4	18.7
Mahabubnagar	28	25.1	25.4	31.9	501.2	118.4	5.3	61.1	0.005	0.151	6.1	21.7	156.4	18.3	19.2
Medak	27.3	24	24.3	31.4	696.2	108.2	12.8	59.3	0.004	0.154	12.8	24.1	227.3	29.7	29.2
Nalgonda	28	24.3	24.4	31.1	518.9	144.4	10	50.6	0.007	0.16	35.7	27.2	188.3	36.5	40
Nellore	29.9	25.6	25.2	31.2	338.1	662.1	35.7	63.7	0.008	0.098	38.2	30	166	41.9	56.8
Nizamabad	27.4	24.3	24.4	31.7	888.7	111.5	15	43.8	0.005	0.182	32.6	22.7	241.2	39.4	57.9
Srikakulam	26.9	23.6	22.3	27.4	722.9	260.8	23.8	114.7	0.001	0.024	47.7	26.4	352.7	36.9	43.7
Visakhapatnam	28.6	24.8	23.8	29.7	682.2	279.8	25.8	146.8	0.001	0.027	14.0	18.6	283.9	24.5	34.4
Warangal	28.8	24.9	24.7	31.7	793.5	120.5	19.5	58.9	0.004	0.221	33.0	24.5	209.1	33.3	44.9
West Godavari	29.7	25.8	24.8	30.3	749.5	250.3	17	85.7	0.01	0.059	161.4	37.7	419	67.4	84.1
Average	28.3	24.7	24.5	30.8	634.3	213	15.9	72.9	0.005	0.105	38.0	25.5	257.3	32.3	40.2

Table 5.5 Ricardian model for net revenue per ha for rice

Variable	Coefficient	Standard error	t-stat	P-value
Intercept	−33,381.153	13,717.99	−2.433	0.02
TSWM	−1,482.804*	849.94	−1.745	0.08
TNEM	1,657.433*	860.98	1.925	0.05
TWP	265.853	1,033.82	0.257	0.80
THWP	2,066.000**	856.24	2.413	0.02
RSWM	−0.746*	0.40	−1.874	0.06
RNEM	−0.058	0.45	−0.129	0.90
RWP	−0.846	1.96	−0.432	0.67
RHWP	0.432	0.76	0.571	0.57
TSWM-squared	24.236**	11.933	2.031	0.04
TNEM-squared	−35.223**	17.33	−2.032	0.04
TWP-squared	−2.512	21.14	−0.119	0.91
THWP-squared	−31.976**	14.04	−2.278	0.02
RSWM-squared	0.0003	0.00	1.055	0.29
RNEM-squared	−0.0003	0.00	−0.551	0.58
RWP-squared	0.005	0.01	0.361	0.72
RHWP-squared	0.000	0.00	−0.175	0.86
TRACTOR	−22,811.084	8,751.17	−2.607	0.01
PUMPSET	−905.235	378.31	−2.393	0.02
NPK	−1.189	1.20	−0.992	0.32
LITPOPRU	67.723	3.23	20.990	0.00
POPDEN	−1.058	0.24	−4.419	0.00
HYV	−4.246	1.93	−2.206	0.03
IRR	8.687	2.49	3.488	0.00

*Significant at 10 % level; **Significant at 5 % level

net revenue for rice crop. The nonlinear effect of the temperature in these three seasons is implied by the significance of the coefficients of the square terms. Figure 5.2 provides the relationship between the net revenue and temperature in these monsoon seasons when all other input variables are held at their respective mean values.

It shows that during the South West monsoon season the net revenue decreases initially with an increase in temperature and then increases and the net revenue attains a minimum value at about 30.6 °C, while in the North East monsoon season, the net revenue reaches a maximum level at about 23.5 °C. Similarly between March & May, which is a summer period, the temperature reaches a maximum at about 32.3 °C. Thus the relationship is non-linear and it is U (or inverted U) shaped. This finding is consistent with the existing literature (Praneetvatakul 2011; Mendelsohn et al. 1994, Mendelsohn 2003; Kurukulasuriya and Mendelsohn 2008).

Similarly rainfalls during the South West monsoon and its square terms have a significant effect on the net revenue (Fig. 5.3).

It also shows that the relationship is U shaped and confirms to the results of the earlier studies (Praneetvatakul and Khamwong 2011) Thus climate variables have significant non-linear effect on the net revenue per ha for rice crop.

5 Economic Impact of Climate Change on Agriculture in SAT India: An Empirical... 81

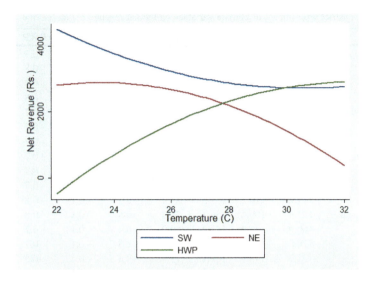

Fig. 5.2 Relationship between net revenue and temperature during the two monsoon seasons

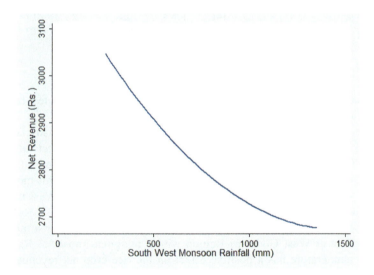

Fig. 5.3 Relationship between South West monsoon rainfall and net revenue for rice

In the case of socio-economic variables, the percentage of irrigated area and literacy of the rural population have positive significant effects. For other variables the coefficients are negative which is difficult to explain but some of the coefficients are significant too. Finally the R-square value for the model fit is 0.563 which shows the adequacy of the fit.

Table 5.6 District wise impact of climate variables on the net revenue per ha for rice

Region	TSWM	TNEM	TWP	THWP	RSWM	RNEM	RWP	RHWP
Adilabad	−96.9	−38.4	144.8	10.1	−0.3	−0.1	−0.7	0.4
Anantapur	−209.9	−25.1	143.3	168.6	−0.6	−0.1	−0.8	0.4
Chittoor	−151.5	−32.5	147.0	201.7	−0.5	−0.3	−0.7	0.4
Cuddapah	−144.0	−51.8	142.4	114.6	−0.5	−0.2	−0.8	0.4
East Godavari	−57.9	−147.9	142.8	132.0	−0.4	−0.2	−0.7	0.4
Guntur	−41.7	−146.1	139.2	56.0	−0.5	−0.2	−0.7	0.4
Hyderabad	−127.8	−82.6	141.0	45.9	−0.4	−0.1	−0.7	0.4
Karimnagar	−122.2	−29.7	145.2	52.5	−0.4	−0.1	−0.7	0.4
Khammam	−62.8	−138.4	140.7	54.1	−0.3	−0.1	−0.7	0.4
Krishna	−50.7	−143.5	140.6	105.5	−0.4	−0.2	−0.7	0.4
Kurnool	−111.2	−116.2	137.2	31.0	−0.5	−0.1	−0.8	0.4
Mahabubnagar	−123.4	−109.6	138.1	27.5	−0.5	−0.1	−0.8	0.4
Medak	−160.9	−34.9	143.7	60.0	−0.4	−0.1	−0.7	0.4
Nalgonda	−126.0	−56.5	143.4	77.0	−0.5	−0.1	−0.7	0.4
Nellore	−33.7	−143.9	139.1	73.9	−0.6	−0.4	−0.5	0.4
Nizamabad	−153.0	−53.4	143.2	40.4	−0.3	−0.1	−0.7	0.4
Srikakulam	−179.4	−3.2	153.8	313.5	−0.4	−0.2	−0.6	0.4
Visakhapatnam	−97.7	−92.2	146.2	165.7	−0.4	−0.2	−0.6	0.4
Warangal	−86.6	−99.1	141.6	36.2	−0.3	−0.1	−0.7	0.4
West Godavari	−43.4	−157.7	141.1	126.4	−0.4	−0.2	−0.7	0.4
Average	−109.0	−85.1	142.7	94.6	−0.4	−0.2	−0.7	0.4

3.1.2 Marginal Impact of Climate Change District Wise

The impact of climate variables on the net revenue were computed based on Eq. (5.3) and the results are presented in Table 5.6.

It is evident from the Table 5.7 that, in general, across the districts, South West and North East monsoon temperatures have a negative effect on the net revenue. The negative effect during the South West monsoon season is at its peak level for Anantapur district with a value of Rs. 209.9 per hectare. This means that when the temperature increases by one degree during the South West monsoon season, the expected net revenue (in 1981–1982 constant prices) is Rs. 209.9. Similarly the temperature rise during north east monsoon will have a maximum adverse effect at West Godavari Region with a marginal impact of Rs. 157.7. However temperature has a positive effect on the rice crop net revenue during the remaining two seasons. In the case of rainfall, South west and North east monsoons have negative effect even though the impacts are not substantial.

3.1.3 Combined Effect of Climate Variables

The Ricardian model assumes that climate variables have an impact on net revenue. However to draw meaningful conclusions, this hypothesis must be tested statistically. With this in view separate regression equations were fitted with and without climate variables and the residual sum of squares were statistically tested. The results (Table 5.7) show that climate variables do have significant contribution.

5 Economic Impact of Climate Change on Agriculture in SAT India: An Empirical... 83

Table 5.7 Testing the combined effect of climate variables on net revenue

Sum of squares with climate variables	476,861,679
Sum of squares without climate variables	418,460,711
Increase in sum of squares	58,400,968
Number of climate variables	16
Increase in mean sum of square	3,650,061
Residual sum of squares with climate variables	369,927,938
Error-- Degrees of freedom with climate variables	756
Residual mean sum of squares with climate variables	489,323
F-Ratio: Numerator	3,650,061
F-Ratio: Denominator	489,323
Calculated-F-Ratio	7.5
F-Ratio-Table value at 5 %	1.7
F-Ratio-Table value at 1 %	2.0

3.2 Ricardian Model for Net Revenue for Major Crops

3.2.1 Model Fit

The Table 5.8 presents the results of Ricardian analysis of other crops.

It is clear from the above table that the South West monsoon temperatures, Jan-Feb. temperature, South West monsoon rainfall and their square terms have strong effect on the net revenue per ha implying that climate variables are important contributing factors for a farmer's income. Further all control variables except the percentage area under the high yielding varieties have significant effect on the net revenue. Further the coefficients of all the control variables except the pump set and NPK consumption have expected signs.

As in the case of rice, the relation between temperature during South West Monsoon season and the Net Revenue per ha from all major crops is also U shaped (Fig. 5.4). It reaches a minimum level of about 28.3°C and the corresponding net revenue is Rs. 2,325 (U$52).

3.2.2 Marginal Effects

The Table 5.9 gives the marginal impact of the climate variables on the net revenue. These values are calculated using Eq. (5.3) when the variables are set at their mean values (Table 5.9).

It can be seen that most of the marginal effects of temperature during the first three seasons are negative implying that one unit increase in the respective climate variables will decrease the net revenues. For example, the South West monsoon temperature and Jan-Feb temperature have a strong negative impact on the many districts. Anantapur district has the most negative marginal net return of Rs. 277.2 during the South West Monsoon season. Rainfall in all seasons has positive marginal impacts in all districts even though they are small.

Table 5.8 Ricardian model fit for major crops

Variable	Coefficient	Standard error	t-stat	P-value
Intercept	242.960	11,498.209	0.021	0.98
TSWM	−3,802.054***	711.800	−5.341	0.00
TNEM	352.065	720.147	0.489	0.63
TWP	2,734.900***	866.038	3.158	0.00
THWP	918.990	716.582	1.282	0.20
RSWM	1.566***	0.332	4.710	0.00
RNEM	0.334	0.377	0.886	0.38
RWP	2.334	1.639	1.424	0.15
RHWP	0.380	0.635	0.599	0.55
TSWM-squared	67.112***	12.535	5.354	0.00
TNEM-squared	−7.277	14.496	−0.502	0.62
TWP-squared	−56.152***	17.712	−3.170	0.00
THWP-squared	−14.149	11.748	−1.204	0.23
RSWM-squared	−0.001***	0.000	−3.443	0.00
RNEM-squared	0.000	0.000	−0.659	0.51
RWP-squared	−0.006	0.011	−0.514	0.61
RHWP-squared	0.000	0.001	−0.409	0.68
TRACTOR	13,388.236	7,098.239	1.886	0.06
PUMPSET	−1,347.551***	284.787	−4.732	0.00
NPK	−2.848***	0.558	−5.105	0.00
LITPOPRU	47.275***	2.829	16.713	0.00
POPDEN	1.492***	0.236	6.322	0.00
HYV	−2.428	1.583	−1.534	0.13
IRR	37.544***	1.892	19.847	0.00

***Significant at 1 % level

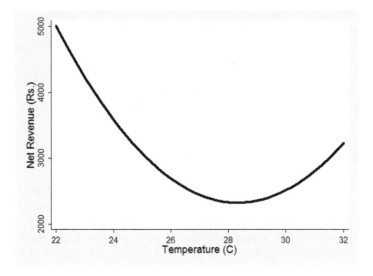

Fig. 5.4 Relationship between South West monsoon temperature and net revenue for major crops

Table 5.9 Marginal impact of climate variables on the net revenue from major crops in the different districts of Andhra Pradesh

Region	TSWM	TNEM	TWP	THWP	RSWM	RNEM	RWP	RHWP
Adilabad	35.7	1.7	29.2	9.3	0.3	0.3	2.1	0.3
Anantapur	−277.2	4.5	−4.0	79.4	1.1	0.2	2.3	0.3
Chittoor	−115.4	2.9	78.4	94.1	1.0	0.1	2.2	0.3
Cuddapah	−94.6	−1.1	−24.4	55.5	1.0	0.2	2.3	0.3
East Godavari	143.8	−20.9	−16.6	63.2	0.5	0.2	2.1	0.3
Guntur	188.6	−20.5	−97.7	29.6	0.8	0.2	2.1	0.3
Hyderabad	−49.9	−7.4	−56.7	25.1	0.7	0.3	2.1	0.3
Karimnagar	−34.4	3.5	37.4	28.0	0.5	0.3	2.1	0.3
Khammam	130.0	−19.0	−62.9	28.7	0.3	0.3	2.2	0.3
Krishna	163.8	−20.0	−64.2	51.5	0.6	0.2	2.2	0.3
Kurnool	−3.9	−14.4	−141.2	18.5	0.9	0.3	2.3	0.3
Mahabubnagar	−37.5	−13.0	−121.6	17.0	0.9	0.3	2.3	0.3
Medak	−141.6	2.4	3.3	31.4	0.6	0.3	2.2	0.3
Nalgonda	−44.9	−2.0	−2.1	38.9	0.8	0.3	2.2	0.3
Nellore	210.8	−20.1	−98.1	37.5	1.1	0.0	1.9	0.3
Nizamabad	−119.7	−1.4	−7.4	22.7	0.3	0.3	2.2	0.3
Srikakulam	−192.8	9.0	229.2	143.5	0.6	0.2	2.1	0.3
Visakhapatnam	33.4	−9.4	60.0	78.1	0.6	0.2	2.0	0.3
Warangal	64.2	−10.8	−42.8	20.8	0.5	0.3	2.1	0.3
West Godavari	184.0	−22.9	−53.8	60.8	0.5	0.2	2.1	0.3
Total	2.1	−7.9	−17.8	46.7	0.7	0.2	2.2	0.3

Table 5.10 Testing the significance of inclusion of climate variables in the Ricardian model

Sum of squares with climate variables	908,627,736.11
Sum of squares without climate variables	874,579,089.06
Increase in the sum of squares	34,048,647.05
Number of climate variables	16.00
Increase in the mean sum of squares	2,128,040.44
Residual sum of squares with climate variables	259,966,593.74
Error – Degrees of freedom with climate variables	756.00
Residual mean sum of squares with climate variables	343,871.16
F-Ratio: Numerator	2,128,040.44
F-Ratio: Denominator	343,871.16
Calculated-F-Ratio	6.19
F-Ratio-Table value at 5 %	1.66
F-Ratio-Table value at 1 %	2.02

3.2.3 Combined Effect of Climate Variables

A statistical test was performed to check whether the climate variables really have an impact on the net revenue. This was performed by running a separate regression (with net revenue as a dependent variable) without climate variables and testing the change in the sum of squares using F-test. The results are presented in Table 5.10.

The result of the test shows that climate variables jointly have a significant effect on the net revenue.

4 Conclusions

The report presented in the study clearly establishes that climate variables do have a significant negative effect on crop revenue for Andhra Pradesh districts. Among the climate variables, South West Monsoon season temperature and rainfall seem to have a significant effect on crops; as it is the major cropping season of this region. The impacts are not uniform across the districts and the impact is very much on Anantapur district for rice and all the major crops. It is hoped that these findings will help policy makers, planners and extension workers to formulate suitable adaptation strategies to nullify the negative effects of climate variables on agricultural production.

References

Bantilan MCS, Anupama KV (2006) Vulnerability and adaptation in dryland agriculture in India's SAT: experiences from ICRISAT's village level studies. E - SAT J 2:1–13

Bantilan MCS, Keatinge JDH (2007) Considerations for determining research priorities: learning cycles and impact pathways. In: Loebenstein G, Thottapilly G (eds) Agricultural research management. Springer, UK, pp 37–64

FAO (1996) Food and agriculture organization. Agro-ecological Zoning Guidelines. FAO Soil Bull 73. Rome. www.fao.org

FAO (2000) Food and Agriculture Organization. In: FAO (ed) The Eco-crop database. Rome, Italy

IPCC (2001) Third assessment report on climate change. Cambridge University Press, Cambridge, UK. www.ipcc.org

IPCC (2007) Climate change 2007. Working group I. The physical science basis. Intergovernmental Panel on Climate Change. Cambridge University Press, Cambridge, UK

Jodha NS (1978) Effectiveness of farmers' adjustments to risk. Econ Polit Wkly 13(25):38–48

Kothawale DR, Rupa Kumar K (2005) On the recent changes in surface temperature trends over India. Geophys Res Lett 32:L18714

Kumar K, Parikh J (1998) Climate change impacts on Indian agriculture: the Ricardian approach. In: Dinar RM, Everson JP, Sanghi A, kumar J, Mckinsey J, Lonergan S (eds) Measuring the impact of climate change on Indian agriculture, World Bank Technical Paper No. 402. Washington, DC: World Bank

Kurukulasuriya P, Mendelsohn R (2008) A Ricardian analysis of the impact of climate change on African Cropland. AFJARE 2(1):23pp

Khamwong C, Praneetvatakul S (2011) Impact of climate change on agricultural revenue and adaptation strategies of farmers in North-eastern Thailand. Kasetsart J (Soc Sci) 32:214–228

Mendelsohn R (2003) Assessing the market damages from climate change. In: Griffin J (ed) Global climate change: the science, economics and politics. Edward Elgar Publishing, Cheltenham/Northampton, pp 92–113

Mendelsohn R, Nordhaus W (1996) The impact of global warming on agriculture: reply. Am Econ Rev Am Econ Assoc 86(5):1312–1315

Mendelsohn R, Nordhaus W, Shaw D (1994) The impact of global warming on agriculture: a Ricardian analysis. Am Econ Rev 84:753–771

Palaniswami K, Paramasivam P, Ranganathan CR, Aggarwal PK, Senthilnathan S (2009) Quantifying vulnerability and impact of climate change on production of major crops in Tamil Nadu, India. In: Headwaters to the Ocean-Hydrological change and watershed Management. Taylor & Francis Group, London

Sivakumar MVK, Brunini O, Das HP (2005) Impacts of present and future climate variability and change on agriculture and forestry in the arid and semi-arid tropics. Clim Change 70:31–72

Shiferaw B, Bantilan C (2004) Agriculture, rural poverty and natural resource management in less favored environments: revisiting challenges and conceptual issues. Food Agric Environ 2(1):328–339

Chapter 6
Vulnerability to Climate Change in Semi-arid Tropics of India: Scouting for Holistic Approach

Naveen P. Singh, Kattarkandi Byjesh, and Cynthia Bantilan

Abstract Characterizing vulnerable regions is a prerequisite in the priority setting for climate change research and allocating resources for the targeted stakeholders such as farmers, researchers, development practitioners and policy makers. In this context, vulnerability profiles are developed at district level for the varied agro-socio-economic environment of Andhra Pradesh and Maharashtra. The districts were indexed, based on the set of indicators representing the three components of climate change vulnerability i.e. exposure, sensitivity and adaptive capacity. The standard and widely accepted IPCC approach was used for the indexing purpose. Results indicated that the majority of the districts of Andhra Pradesh and Maharashtra are vulnerable to climate change. Over the years the analyzed districts experienced a varying level of vulnerability to climate change, with some districts improved and while others slipping into more vulnerable category. This dynamicity existed among the components of exposure, sensitivity and adaptive capacity is reflected in the vulnerability status of the districts. This exercise is a guiding tool for formulating action plans and is way forward in improving adaptive capacity among the rural population.

Keywords Climate change • Semi-arid tropics • Vulnerability index • Exposure • Sensitivity • Adaptive capacity

---

N.P. Singh (✉) • K. Byjesh • C. Bantilan
Research Program – Markets, Institutions and Policies, International Crops Research Institute for the Semi-Arid Tropics (ICRISAT), Hyderabad, Andhra Pradesh 502324, India
e-mail: np.singh@cgiar.org; k.byjesh@cgiar.org; c.bantilan@cgiar.org

1 Introduction

The increasing global relevance of climate change and the attempt to understand its impacts are being comprehensively studied over the years (IPCC 2001, 2007). The impacts are expected to affect livelihood and further aggravate during the years to come. Climate change impact has a multidimensional effect on humanity in terms of several socio-economic parameters like agriculture, human health, sea level rise, scarcity of labor, disease prevalence etc. (Adger 1999). The adversities of climate change induced related events to emphasize the importance of strategies to be adopted to cope against the impacts. Unless proper adaptation strategies are implemented, it can results in a far reaching consequence and certainly contribute to severe impacts on societies and their livelihood especially among the natural resource dependent community (Bantilan and Anupama 2006; Thomas and Twyman 2005). Vulnerability and adaptation to climate change are the major pressing issues among others in many developing countries. Hence, there is a need to recognize, prioritize and channelize the resources to improve the capacity to adapt against these climate induced changes or extremes. However, the impacts, vulnerability and the capacity to adapt to these changes are different with respect to time and space. For the same reason, there exist provisions in the United Nations Framework Convention on Climate Change (UNFCCC) to assist those countries that are thought to be the most vulnerable and the least able to adapt. Furthermore, existing constraints in financing the climate change adaptation applies equally to all the regions. Hence we need to prioritize the regions that need special attention and one should take into account the vulnerabilities and impacts caused by the climate change. For realizing the extent of vulnerability, it is important to understand the climate change impacts at regional/district level, obtaining knowledge of these regions and understanding the direct and indirect consequences that contributes to the vulnerability, and understanding the capacity and ability that these regions possess to cope with it. Maximum disaggregated information on vulnerability would support efficient adaptation planning and aid in targeting regions.

Vulnerability analysis is an attempt to quantify and map vulnerability to climate change for the entity or the target region. Various definitions on climate change 'vulnerability' exist, the Intergovernmental Panel on climate change (IPCC) has defined vulnerability as the degree to which a system is susceptible to, or unable to cope with the adverse effects of climate change, including climate variability and extremes (IPCC 2001; McCarthy et al. 2001) along with others that are usually associated with the natural hazards like floods, droughts, and socio hazards like poverty etc. With the increased importance in climate change research, it has been widely used to compute vulnerability. In the climate change research, vulnerability has three components: exposure, sensitivity and adaptive capacity. (i) Exposure can be interpreted as the direct danger (i.e., the stressor) and the nature and extent of changes to a region's climate variables (e.g. temperature, precipitation, extreme weather events). (ii) Sensitivity describes the human-environmental conditions that can worsen the hazard, ameliorate the hazard, or trigger an impact. (iii) Adaptive capacity represents the potential to implement adaptation measures that help avert potential impacts. An integrated assessment approach in computing vulnerability considers physical impacts of climate change on both socio-economic and bio-physical variables (Fussel and Klein 2006).

Several studies on indexing regions based on climate change vulnerability have proved that vulnerability is closely associated with poverty, as the poor are least capable to respond to these climatic stimuli. Furthermore, regions that are socio-economically under developed are more severely affected by the effects of climate change than others; especially in an economy closely tied to its natural-resource-base and climate-sensitive sectors such as agriculture, water, forestry etc. India where still a large chunk of rural poor and agriculture dependent population lives, faces a major threat because of the current and projected changes in climate. India's large population primarily depends on climate-sensitive sectors like agriculture, livestock and forestry for livelihood. The majority of the vulnerable population of Indian semi-arid tropics are poorly equipped to cope effectively with the adversities of climate change due to low capabilities, weak institutional mechanisms, and lack of access to adequate resources (Ribot et al. 1996). Even in India, studies to profile vulnerable regions have been undertaken, 'in North East India' (Ravindranath et al. 2011), lower Himalaya (Pandey and Jha 2012), and some districts of India (Teri 2003). However, minimum or rather no study was done for characterizing regions in the semi-arid regions of India.

The objective of the paper is to (a) characterize the regions based on the extent of climate change vulnerability in semi-arid tropics of India and (b) Decadal assessment of vulnerability of the region.

2 Methodology

The assessment of vulnerability was done by constructing indices. The three components of vulnerability i.e. exposure, sensitivity and adaptive capacity and their relative contribution determines the level of vulnerability. The first two components together represent the a potential impacts (exposure and sensitivity) and adaptive capacity is the extent and the innate capacity at which these impacts can be averted. Thus, vulnerability is potential impact (I) minus adaptive capacity (AC). This leads to the following mathematical equation for vulnerability, V;

$$V = \int (I - AC)$$

This index is computed based on a set of indicators representing the bio-physical and socio-economic status of the region. It produces an index number which can be used to compare the different regions. All the indicators chosen were assumed to have direct and indirect influence on the vulnerability of the region to climate change.

Table 6.1 List and description of indicators used for computing vulnerability (Exposure and sensitivity)

Components of vulnerability	Description of the component indicators	Functional relationship with climate change
Exposure	Change in rainfall (%) – Percentage change in rainfall from base period to 2007	↑
	Change in the maximum temperature	↑
	Change in minimum temperature	↑
Sensitivity	Percentage of cultivable waste	↓
	Percentage of gross area irrigated to gross area sown	↓
	Fertilizer use (t/ha)	↓
	Population density (persons/km^2)	↑
	Percentage of small farmers	↑
	Percentage of forest area	↓
Adaptation capacity	Livestock nos. (per ha)	↓
	Percentage of rural agricultural laborers	↓
	Percentage of cultivators	↓
	Percentage of rural literates	↓
	Percentage of fodder area	↓
	Cereals production (tons/ha)	↓
	Pulses production (tons/ha)	↓

↑ +^{ve} relationship; ↓ −^{ve} relationship

The semi-arid districts of Andhra Pradesh and Maharashtra state were selected for computing vulnerability. These two states contribute to a major area and population of the semi-arid tropical region of central India. Among the 23 districts in Andhra Pradesh about 19 fall under the semi-arid zone and 25 out of 35 in Maharashtra. For computing periodical vulnerability indices, data sets of indicators from 1971, 1981, 1991 and 2001 were analyzed. The indicators under three groups (a) exposure (b) sensitivity and (c) adaptive capacities were used (Table 6.1). These indicators are normalized based on the methodology used in UNDP's Human Development Index (HDI) (UNDP 2006) and using expert judgement to identify the functional relationship of the indicators with respect to climate change vulnerability. The normalization was done using the formula for indicators having both +ve and −ve functional relationship.

$$Y_i = \frac{(X_i - Min\, X_i)}{(Max\, X_i - Min\, X_i)} \text{ for } + \text{ve functional relationship}$$

$$Y_i = \frac{(Max\, X_i - X_i)}{(Max\, X_i - Min\, X_i)} \text{ for } - \text{ve functional relationship}$$

From the normalized data sets, the indices were generated using Iyengar and Sudarshan (1982) method. The methodology was used for this analysis. The level or stage of vulnerability of the zone/district is assumed to be a linear sum.

$$\overline{Y}_t = \sum_{i=1}^{m} w_i y_i$$

where w's is ($0 < w < 1$ *and* $\sum_{i=1}^{n} w_i = 1$); are the weights. In this method the weights are assumed to vary inversely as the variance over the regions, in the respective indicators of vulnerability. That is, the weight w_i is determined by

$$w_i = \frac{k}{\sqrt{var(y_i)}}$$

where k is a normalizing constant; $k = \left[\sum_{i=1}^{n} \frac{1}{\sqrt{var(y_i)}}\right] - 1$

The vulnerability index so computed lies between 0 and 1 with 1 indicating maximum vulnerability and 0 indicating no vulnerability. A meaningful characterization of the different component wise vulnerability is computed. The summative district total vulnerability from these three components form a probability distribution, Beta distribution, is generally skewed and takes values in the interval (0, 1). This distribution's probability density is given by

$$f(z) = \frac{x^{a-1}(1-x)^{b-1}}{b(a,b)}, 0 \pounds x \pounds 1 \text{ and } a,b > 0$$

Where b (a, b) is the beta function. The beta distribution is skewed and the districts are classified into different intervals. The interval is made with the same probability weight of 20 % (5 classes). These fractile intervals were used to characterize the various stages of vulnerability.

3 Results

3.1 Andhra Pradesh

Climate change vulnerability indices were computed for 19 districts of Andhra Pradesh periodically from 1971 to 2001. The relative contribution of each component to the vulnerability from 1971 at a decadal interval was computed (Table 6.2). The districts were ranked and grouped based on the degree of vulnerability (Fig. 6.1). In 2001, the districts viz., Visakhapatnam, Chittor, Nellore, West Godavari, and Kurnool were identified as very highly vulnerable; Anantpur, Krishna, East Godavari and Nalgonda were highly vulnerable, Cuddapah and Adilabad were grouped under vulnerable districts. The 'moderately vulnerable' districts are

Table 6.2 Periodical indexing of vulnerability to climate change for the state of Andhra Pradesh

Sl. No.	District	1971 E	S	AC	Total	Rank	1981 E	S	AC	Total	Rank	1991 E	S	AC	Total	Rank	2001 E	S	AC	Total	Rank	COV*
1	Kurnool	0.12	0.15	0.38	0.66	1	0.13	0.15	0.35	0.63	1	0.12	0.19	0.26	0.58	4	0.11	0.16	0.3	0.56	5	D
2	Anantapur	0.08	0.21	0.36	0.65	2	0.09	0.17	0.34	0.6	3	0.08	0.2	0.29	0.57	5	0.08	0.16	0.31	0.54	6	D
3	Visakhapatnam	0.12	0.13	0.32	0.57	3	0.08	0.12	0.35	0.55	6	0.12	0.14	0.35	0.61	3	0.12	0.12	0.37	0.6	1	I
4	Nellore	0.11	0.16	0.28	0.55	4	0.12	0.15	0.27	0.53	10	0.16	0.18	0.2	0.54	8	0.21	0.17	0.21	0.59	3	I
5	Nalgonda	0.15	0.15	0.24	0.53	5	0.18	0.15	0.19	0.52	11	0.16	0.18	0.21	0.55	6	0.13	0.15	0.25	0.52	9	D
6	Guntur	0.16	0.14	0.24	0.53	6	0.19	0.13	0.22	0.54	9	0.09	0.19	0.12	0.4	18	0.07	0.15	0.23	0.45	13	D
7	Krishna	0.08	0.16	0.28	0.52	7	0.12	0.18	0.26	0.56	5	0.12	0.22	0.15	0.49	13	0.07	0.16	0.29	0.53	7	I
8	Chittor	0.15	0.13	0.23	0.51	8	0.1	0.12	0.24	0.46	15	0.2	0.17	0.25	0.62	2	0.14	0.14	0.32	0.6	2	I
9	West Godavari	0.09	0.17	0.25	0.51	9	0.11	0.18	0.25	0.55	8	0.13	0.2	0.18	0.51	12	0.09	0.15	0.34	0.58	4	I
10	East Godavari	0.04	0.18	0.29	0.51	10	0.06	0.13	0.24	0.42	17	0.08	0.17	0.28	0.53	10	0.04	0.12	0.36	0.53	8	I
11	Mahabubnagar	0.09	0.13	0.29	0.51	11	0.12	0.11	0.27	0.5	13	0.12	0.14	0.28	0.54	7	0.09	0.12	0.26	0.48	12	D
12	Nizamabad	0.04	0.11	0.36	0.5	12	0.15	0.08	0.32	0.56	4	0.14	0.13	0.25	0.52	11	0.08	0.13	0.23	0.44	16	D
13	Adilabad	0.06	0.08	0.36	0.5	13	0.15	0.09	0.36	0.6	2	0.1	0.11	0.44	0.64	1	0.07	0.08	0.35	0.51	11	I
14	Karimnagar	0.1	0.14	0.25	0.48	14	0.17	0.12	0.21	0.51	12	0.15	0.15	0.17	0.47	14	0.11	0.12	0.2	0.43	18	D
15	Cuddappah	0.04	0.14	0.28	0.46	15	0.06	0.14	0.3	0.49	14	0.07	0.16	0.22	0.45	15	0.14	0.14	0.24	0.52	10	I
16	Srikakulam	0.04	0.19	0.23	0.45	16	0.1	0.16	0.19	0.45	16	0.04	0.19	0.15	0.38	19	0.03	0.15	0.25	0.43	17	D
17	Warangal	0.08	0.11	0.25	0.44	17	0.1	0.09	0.22	0.41	18	0.09	0.13	0.23	0.44	16	0.04	0.12	0.23	0.39	19	D
18	Medak	0.03	0.17	0.2	0.4	18	0.13	0.17	0.24	0.55	7	0.05	0.2	0.29	0.54	9	0.05	0.16	0.24	0.45	14	I
19	Khammam	0.04	0.07	0.27	0.38	19	0.1	0.07	0.23	0.41	19	0.08	0.09	0.25	0.42	17	0.08	0.07	0.29	0.44	15	I

E exposure, S sensitivity, AC Adaptive capacity
*COV change in vulnerability from 1971 to 2001, I Increased, D Decreased

6 Vulnerability to Climate Change in Semi-arid Tropics of India: Scouting... 95

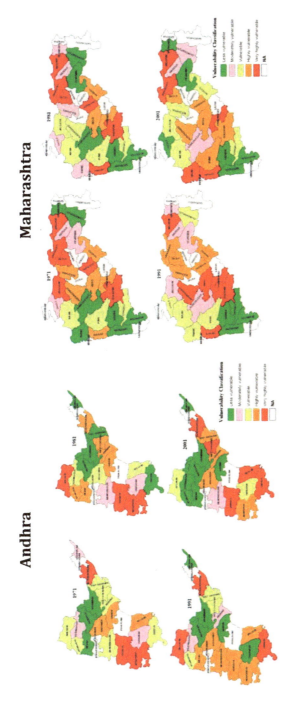

Fig. 6.1 Mapping districts level vulnerability to climate change for Andhra Pradesh and Maharashtra states of India

Mahabubnagar Guntur, Khamam, Nizamabad, Srikakulam, Karimnagar and Warangal were identified as less vulnerable. Over the years the vulnerability of the districts has changed. Medak and Adilabad districts have become comparatively more vulnerable to climate change from 1971 to the present (2001) (Fig. 6.1). Guntur and Nalgonda have improved significantly i.e. were comparatively less vulnerable to climate change.

The districts were ranked according to the ascending order of index value (Table 6.1). There have been positive and negative trends in case of vulnerability status among the districts. Almost half of the analyzed districts vulnerability increased (10 nos.) over the years and for the remaining, there had been changes and vulnerability had fallen back. Southern districts of Andhra Pradesh have mostly became more vulnerable over the years and there has been periodical improvement in the adaptive capacity for the northern districts.

3.2 Maharashtra

For Maharashtra, out of 25 districts analyzed, 18 districts are relatively vulnerable to climate change. The proportion of vulnerable districts hasn't changed over the years (Fig. 6.1). The component wise contribution to climate change vulnerability varied periodically and among the districts. In 2001, Nagpur, Amaravathi, Raigad, Thane and Osmanabad were identified (very highly vulnerable); Pune, parbhani, Solapur, Beed, Aurangabad, Akola (Highly vulnerable), Nasik, Jalgaon, Sangli, Buldhana, Nanded (Vulnerable); Ahmednagar, Wardha, Chandrapur (Moderately Vulnerable); Yeotmal, Bhandara, Dhulia, Satara, Ratnagiri, Kolhapur (Less vulnerable) groups. In 1971, Nagpur, Akola, Amaravathi, Solapur, Parbhani, Jalgaon were identified as very highly vulnerable and Osmanabad, Raigad, Buldhana, Aurangabad, Beed, Nanded were classified as very highly vulnerable and highly vulnerable. The remaining districts i.e., Nasik, Sangli, Pune; Dhulia, Wardha, Yeotmal; and Satara, Thane, Chandrapur, Bhandara, Ratnagiri and Ahmednagar were identified as vulnerable, moderately vulnerable and less vulnerable respectively. Lately, in 1981, Jalagaon was replaced by Raigad in the very highly vulnerable group and in 1981 two districts have improved and advanced from highly vulnerable to moderately vulnerable group. In 1991, districts fall under very highly vulnerable are (Nagpur, Raigad, Osmanabad, Beed, Aurangabad, Solapur); highly vulnerable (Parbani, Akola, Buldhana, Amravathi, Nanded); Vulnerable (Chandrapur, Nasik, Pune, Dhulia, Thane); Moderately Vulnerable (Bhandara, Yeotmal, Jalgaon, Ahmednagar, Wardha) and less vulnerable (Sangli, Ratnagiri, Satara, Kolhapur) (Fig. 6.1 and Table 6.3).

In Maharashtra, Thane and Akola are the districts which have become more vulnerable over the years. Four districts i.e., Aurangabad, Nagpur, Sangli and Solapur held the same position in vulnerability grouping over the years.

6 Vulnerability to Climate Change in Semi-arid Tropics of India: Scouting...

Table 6.3 Periodical indexing of vulnerability to climate change for the state of Maharashtra

District	1971 E	S	AC	Total	Rank	1981 E	S	AC	Total	Rank	1991 E	S	AC	Total	Rank	2001 E	S	AC	Total	Rank	COV*
Nagpur	0.12	0.21	0.45	0.78	1	0.12	0.21	0.45	0.78	1	0.12	0.21	0.44	0.77	1	0.12	0.20	0.40	0.73	1	D
Akola	0.02	0.20	0.40	0.62	2	0.02	0.20	0.36	0.58	3	0.03	0.20	0.38	0.61	8	0.03	0.10	0.40	0.53	11	D
Amravothi	0.02	0.16	0.44	0.62	3	0.02	0.16	0.41	0.59	2	0.04	0.16	0.40	0.59	10	0.04	0.10	0.50	0.62	2	D
Solapur	0.02	0.22	0.37	0.61	4	0.02	0.22	0.34	0.58	4	0.02	0.20	0.41	0.63	6	0.02	0.10	0.40	0.54	8	D
Parbhani	0.03	0.21	0.37	0.61	5	0.02	0.22	0.33	0.57	6	0.02	0.21	0.38	0.61	7	0.02	0.10	0.40	0.54	7	D
Jalgaon	0.02	0.18	0.39	0.60	6	0.02	0.17	0.31	0.50	12	0.06	0.15	0.32	0.53	19	0.06	0.10	0.40	0.52	13	D
Osmanabad	0.03	0.23	0.34	0.60	7	0.03	0.23	0.30	0.56	7	0.03	0.22	0.42	0.67	3	0.03	0.20	0.40	0.57	5	D
Raigad	0.03	0.28	0.27	0.58	8	0.02	0.31	0.25	0.57	5	0.03	0.31	0.34	0.67	2	0.03	0.30	0.30	0.58	4	I
Buldhana	0.01	0.18	0.37	0.56	9	0.04	0.17	0.28	0.49	16	0.04	0.18	0.38	0.60	9	0.04	0.10	0.30	0.51	15	D
Aurangabad	0.02	0.20	0.34	0.56	10	0.02	0.20	0.30	0.51	10	0.03	0.20	0.42	0.64	5	0.03	0.10	0.40	0.54	10	D
Beed	0.02	0.21	0.33	0.56	11	0.02	0.22	0.31	0.54	8	0.02	0.21	0.42	0.65	4	0.03	0.20	0.40	0.54	9	D
Nanded	0.01	0.20	0.34	0.56	12	0.01	0.21	0.30	0.52	9	0.03	0.18	0.37	0.59	11	0.05	0.10	0.30	0.51	16	D
Nasik	0.03	0.19	0.34	0.56	13	0.03	0.18	0.29	0.49	14	0.04	0.18	0.34	0.56	13	0.04	0.10	0.40	0.53	12	D
Sangli	0.03	0.19	0.32	0.54	14	0.02	0.17	0.31	0.51	11	0.03	0.16	0.29	0.49	22	0.04	0.10	0.30	0.51	14	D
Pune	0.02	0.22	0.27	0.52	15	0.02	0.22	0.26	0.50	13	0.03	0.22	0.30	0.55	14	0.03	0.20	0.40	0.56	6	I
Dhulia	0.02	0.13	0.34	0.50	16	0.04	0.13	0.31	0.47	18	0.04	0.14	0.38	0.55	15	0.04	0.10	0.30	0.43	22	D
Wardha	0.02	0.16	0.30	0.48	17	0.03	0.18	0.28	0.49	15	0.03	0.18	0.29	0.50	21	0.03	0.10	0.30	0.47	18	D
Yeotmal	0.03	0.17	0.28	0.48	18	0.06	0.18	0.23	0.46	20	0.06	0.17	0.31	0.54	18	0.06	0.10	0.30	0.46	20	D
Satara	0.03	0.19	0.25	0.47	19	0.05	0.20	0.17	0.42	23	0.05	0.18	0.23	0.46	24	0.05	0.10	0.20	0.43	23	D
Thane	0.03	0.22	0.22	0.47	20	0.04	0.24	0.21	0.48	17	0.04	0.25	0.26	0.55	16	0.04	0.20	0.30	0.59	3	I
Chandrapur	0.05	0.12	0.29	0.46	21	0.05	0.11	0.29	0.45	22	0.05	0.11	0.42	0.58	12	0.05	0.10	0.30	0.47	19	D
Bhandara	0.05	0.18	0.23	0.46	22	0.04	0.19	0.24	0.47	19	0.06	0.18	0.30	0.54	17	0.06	0.10	0.30	0.45	21	D
Ratnagiri	0.05	0.24	0.17	0.46	23	0.05	0.22	0.14	0.42	24	0.05	0.22	0.21	0.48	23	0.05	0.20	0.20	0.40	24	D
Ahmednagar	0.06	0.18	0.22	0.46	24	0.06	0.18	0.22	0.46	21	0.06	0.17	0.31	0.53	20	0.04	0.10	0.30	0.49	17	I
Kolhapur	0.05	0.17	0.21	0.44	25	0.05	0.17	0.18	0.40	25	0.05	0.15	0.13	0.33	25	0.05	0.10	0.20	0.35	25	D

E exposure, *S* sensitivity, *AC* Adaptive capacity
*COV change in vulnerability from 1971 to 2001, *I* Increased, *D* Decreased

4 Discussion

Spatial and temporal changes in weather pattern, socio-economic indicators, effective rural development programs, improved rural institutions and efficient social capital over the years have influenced climate change vulnerability (Heltberg et al. 2008). Drastic changes in the climate, especially rainfall have multi-dimensional effect on the cropping pattern, cropped area, irrigated area, farmers land holding, farm and non-farm income etc. This successively effect people's earning and livelihood in the semi-arid tropics. This could be very severe for those who are directly and indirectly linked to farming. There exists a support system from the government and non-government agencies to identify constraints and to uplift and address food security issues, socio-economic concerns etc. Government supported developmental programs are meant to improve the living conditions of the people during extreme events like drought or other climate hazards. The idea of improving the adaptive capacity calls for an overall improvement in the several sectors, strata and communities (Berkhout et al. 2006). Hence, plans, action and financing for transformational change are important to transform and improve resiliency to climate change. So, this exercise is an initial step is used as a guide and this outcome as an input in developing climate change adaptation policies for the targeted regions.

This indexing of vulnerability to climate change is a comprehensive way of profiling based on the vulnerability to climate change. The set of indicators that are used for the analysis are largely determined by the availability rather than the need. The indicators should capture the major issues of food security, income security, environmental capacity, economic resources, social condition, and institutional capacity (Eriksen and Kelly 2007). Most cases, the lack of availability of desired datasets end up in measuring what they can rather than what they must do (Nelson et al. 2010). However, to start these approaches with definite functional framework is useful. There have been numerous research articles on approaches and methods to assess vulnerability of socio-economic and linking climate into its exposure component along with factors contributing sensitivity and adaptive capacity is always a challenge (Aggarwal et al. 2010). The factors contributing to the adaptive capacity of a region differs widely with respect to space and time. The types of indicator we should use that capture the climate change vulnerability of the region in depth and breadth is still not well addressed. Furthermore, the processes that shape up vulnerability is inadequately captured by using national level indicators which usually neglect dynamic indicators representing local conditions. However, the major limitations lie in the unavailability of datasets that could cover a minimum set of indicators representing vulnerability.

Even though the IPCC approach tries to address the weaknesses of the other approaches, it also has its limitations. The main one is that there is no standard method for combining the biophysical and socioeconomic indicators. The relative importance of different variables used in this approach has not been taken into account and thus need much care in using this approach. The other weakness of this approach is that it does not account for the dynamism in vulnerability. The dynamism underlying in the process of adaptation involves a continual change of strategies to take advantage of opportunities, which are missing in this approach (Deressa et al. 2008).

5 Conclusion

The vulnerability assessment is important in identifying target regions as an initial step to guide in the process of developing climate change adaptation policies in order to guide resources allocation and enable effective management of climatic extremes. The semi-arid tropical regions of India are considered as vulnerable hot spot to current and future climatic impacts. The majority of districts of semi-arid tropics of India are vulnerable to climate change with varied intensity. Identification of the prone area will help the governmental and non-governmental agencies to prioritize and to re-direct their focus for the development of these regions. Unless these prioritized regions are focused in improving the adaptive capacity against climatic shocks and variability, this could hamper the innate capacity to cope with these changes.

References

Adger WN (1999) Social vulnerability to climate change and extremes in coastal Vietnam. World Dev 2:249–269

Aggarwal PK, Baethegan WE, Cooper P, Gommes R, Lee B, Meinke H, Rathore LS, Sivakumar MVK (2010) Managing climatic risks to combat land degradation and enhance food security: key information needs. Procedia Environ Sci 1(2010):305–312

Bantilan MCS, Anupama KV (2006) Vulnerability and adaptation in dryland agriculture in India's SAT: experiences from ICRISAT's village level studies. J SAT Agric Res 2(1):1–13

Berkhout F, Hertin J, Gann DM (2006) Learning to adapt: organisation adaption to climate change impacts. Clim Change 78:135–156

Deressa T, Hassan RT, Ringler C (2008) Measuring Ethiopian farmers' vulnerability to climate change across regional states. International Food Policy Research Institute. Discussion paper 00806. Environmen and Production Technology Division, Washington, DC

Eriksen SH, Kelly PM (2007) Developing credible vulnerability indicators for climate adaptation policy assessment. Mitig Adapt Strateg Glob Change 12:495–524

Fussel HM, Klein RJT (2006) Climate change vulnerability assessments: an evolution of conceptual thinking. Clim Change 75(3):301–329

Heltberg R, Jorgensen SL, Siegel PB (2008) Climate change, human vulnerability and social risk management. Workshop paper on social aspects of climate change, World Bank, Washington, DC, 5–6 March

IPCC (2001) Climate change: the scientific basis. http://www.ipcc.ch/

IPCC (2007) Climate change impacts, adaptation and vulnerability. Summary for policy makers. Inter-Governmental Panel on Climate Change. Cambridge University Press, Cambridge, UK, p 22

Iyengar SS, Sudarshan P (1982) A method of classifying regions from multivariate data. Econ Pol Wkly 17:2043–2052 (Special Article)

McCarthy J, Caniziani O, Leary N, Dokken D, White K (eds) (2001) Climate change 2001: impacts, adaptation, and vulnerability. Cambridge University Press, Cambridge

Nelson R, Kokic P, Crimp S, Meinke H, Howden SM (2010) The vulnerability of Australian rural communities to climate variability and change: Part I – conceptualising and measuring vulnerability. Environ Sci Policy 13:8

Pandey R, Jha SK (2012) Climate vulnerability index-measure of climate change vulnerability to communities: a case of rural lower Himalaya, India. Mitig Adapt Strat Glob Chang 17:487–506

Ravindranath NH, Rao S, Sharma N, Nair M, Goplakrishanan R, Rao AS, Malaviya S, Tiwari R, Sagadevan A, Munsi M, Krishna N, Bala G (2011) Climate change vulnerability profiles for North East India. Curr Sci 101(3):384–394

Ribot JC, Magalhaes AR, Panagides SS (1996) Climate variability, climate change and social vulnerability in the semi-arid tropics. Cambridge University Publication/Cambridge University Press, New York, p 170

TERI (2003) The energy and resources institute coping with global change: vulnerability and adaptation in Indian agriculture. The Energy and Resources Institute, New Delhi, 26 pp

Thomas DSG, Twyman C (2005) Equity and justice in climate change adaptation amongst natural resource-dependent societies. Glob Environ Change 15(2):115–124

UNDP (2006) Human development report, United Nations Development Program. Available at: http://hdr.undp.org/hdr2006/statistics/

Chapter 7
Sustainable Agriculture and Rural Development in the Kingdom of Saudi Arabia: Implications for Agricultural Extension and Education

Mirza B. Baig and Gary S. Straquadine

Abstract Saudi Arabia is a country consisting largely of desert terrain, with limited naturally occurring ground water and is subject to extremes of temperature typical of an arid climate. The country is scarcely ideal for agricultural development. However, recognizing the importance of food security, the kingdom initiated several programs to advance facilities, services and farm inputs, including water, for the farming communities to increase food production and feed more of its citizens. With less than favorable circumstances, the kingdom successfully attained self-sufficiency in wheat, dates, eggs, fresh milk, and some vegetables. Although its share of the GDP is quite low, the kingdom's fertile regions make agriculture and agri-business the third largest sector in GDP revenues. The agriculture sector has also helped improve the livelihoods of rural population. Yet, producing crops with the heavy application of irrigation from the limited and meager water resources posed serious threats to the sustainability of natural resources of the kingdom. The situation required major changes in the present farming practices and the adoption of sustainable agriculture concepts and practices in the Saudi Arabia. Increasingly, it has become important to educate the farming community to use the natural resources, including water, with sustainable mindfulness. This article presents the current scenario of the agricultural sector, the potential of sustainable agriculture, and the possible role of extension and education in converting agricultural

---

M.B. Baig (✉)
Department of Agricultural Extension and Rural Society, Faculty of Food and Agricultural Science, College of Food and Agricultural Sciences, King Saud University, P.O. Box 2460, Riyadh 11451, Kingdom of Saudi Arabia
e-mail: mbbaig@ksu.edu.sa

G.S. Straquadine
Chair, Department of Agricultural Communication, Education, and Leadership,
Ohio State University, 208 Agricultural Administration Building,
2120 Fyffe Road, Columbus, OH 43210-1067, USA
e-mail: straquadine.5@osu.edu

production to sustainable practices in the kingdom. The purpose of this article is to establish the importance of agricultural extension in developing the relationship between sustainable agriculture and rural development initiatives.

Keywords Sustainable agriculture • Rural development • Natural resource management • Environment • Agricultural extension • Capacity building

1 Introduction

With a population of 28.7 million (FAO 2009), the kingdom of Saudi Arabia comprises about 80 % of the Arabian Peninsula (Wikipedia 2010) and occupies approximately an area of 2.25 million km^2 (UNDP 2010). As a landmass, it constitutes a distinct geographical entity. It is bounded on the north by Jordan, Iraq and Kuwait; on the east by the Gulf, Bahrain, Qatar and the United Arab Emirates; on the south by the Sultanate of Oman and Yemen and in the west by the Red Sea with a coastline of 1,750 km (FAO 2009). Its skies are cloudless, climatic conditions are quite harsh – arid and dry, and landscape consists of dry desert with extreme temperatures. During the 1970s and 1980s, the kingdom initiated many extensive agricultural programs with the objectives to gain food security through self-sufficiency while at the same time improving rural incomes (Wikipedia 2010). The kingdom has successfully increased yields of several important crops and foodstuffs through the introduction of modern agricultural technologies.

Yet, the objectives of the agricultural development program have been only partially met. While the kingdom has not only gained self- sufficiency, with a sufficient surplus for export (Country Studies 2010), the kingdom realized that surpluses were at the cost of overexploitation of water, causing severe and serious damage to its water resources (Al-Subaiee et al. 2005).

The agriculture sector in the kingdom faces many constraints, the prominent are shortage of water (Al-Ibrahim 1990), lack of experienced technical personnel; soil and water salinity, and the cost of desalination; marketing problems, low prices for products, pests and diseases (Al-Zeir 2009). However, among these, scarcity of fresh water resources remains the most serious challenge in the kingdom (Al-Zahrani and Baig 2011). Because the kingdom lacks perennial rivers or permanent reservoirs, water poses a continual challenge, as does the depletion of underground water resources (Al-Alawi and Abdul Razzak 1994; Al-Tukhais 1997; U.S. Department of Energy 2002; Ray 2003). Ray (2003) also noted that the underground aquifers were being drawn down faster than the recharge rate. He warned that the rapidly growing population could end up competing with agriculture for the scarce water resources.

On the other hand, the climate is harsh, hot and arid in most parts of the kingdom and any minor change in climate could cause a pronounced impact. Keeping in view the importance of this potential threat, a study was conducted to assess the potential impact of climate change on the kingdom's agriculture and water supplies. Data collected over 4 decades indicate that an increase in temperature and decrease in precipitation (climate change) could cause severe negative impacts on agriculture and water resources (Alkolibi 2002).

However, the author suggests the decision-makers of the kingdom to adopt a "strong policy" to focus on the ever increasing population and recent harsh climatic conditions. Further, he urges for the implementation of monitoring systems in the policy to avoid future environmental and socio-economic issues that may emerge due to any climate change (Alkolibi 2002). Finally the kingdom faces numerous environmental challenges such as 5 % arable land (US Department of Energy 2002), desertification and creeping sands (UNCCD 2000).

The need for agriculture prompts scientists to evolve sustainable farming practices; genuinely they are advocating for the adoption of innovative environmental friendly technologies to sustain the agriculture sector. This article proclaims that sustainable agriculture must become a viable vehicle for rural development. At this critical juncture, the role of agricultural extension could be very curative and constructive in addressing constraints, enhancing agriculture and improving the rural livelihood. In this article, an effort has been made to examine the role of sustainable agriculture in achieving rural development. Further it is argued that agricultural extension can promote sustainable agriculture to seek sustainable rural development in the kingdom.

2 An Overview of Agriculture in the Kingdom

Historically, small scale agriculture was practiced by the rural population with the help of nomads (Beduin) in the rural areas. They used the limited arable lands with little vegetation and were forced to raise their livestock into a nomadic pattern. With the development of oil industry in the kingdom, particularly between the periods of 1970–1980, the kingdom began to recognize the importance of food security. However, with significant efforts to initiate agricultural development in the 1970s under the first development plan of the kingdom (Royal Embassy of Saudi Arabia, USA 2010) and to move further in this direction, emphasis was placed on the development of infrastructure. The important components of agricultural sector are discussed further as:

2.1 The Initiatives

Among the numerous initiatives taken by the kingdom to promote contemporary, sustainable agriculture, the most important ones include establishment of rural roads, irrigation networks and storage, export facilities and agricultural research and training institutions (Royal Embassy of Saudi Arabia, USA 2010). Appreciable progress made in the last four decades can be attributed to an array of programs, launched by the kingdom. They include: the provision of soft, long term interest-free loans, technical and support services, free seeds and access to fertilizers (Saudi Arabia Magazine 2001), low cost water, fuel and electricity, duty free imports of raw materials and machinery, and installation of drainage and irrigation networks. In addition, through a land reclamation program, free land has been allocated to farmers to increase area under cultivation and to encourage crop and livestock production. The farmers are also encouraged and assisted on the diversification and efficient farming systems.

2.2 The Progress

Growth in production of all basic foods has been noticed across the kingdom reaching self-sufficiency in a number of food items. Such growth has been reported by many sources (Al-Hazmi 1997; FAO 2009; Royal Embassy of Saudi Arabia, Ottawa, Canada 2010; Oxford Business Press 2010). According to Al-Hazmi (1997), the successful farming sector of the kingdom was able to produce cereals like wheat, barley, sorghum and millets. Among the vegetables, tomatoes, potato, watermelon, eggplant, cucumber and onions are worth to mention. Fruits like date-palm, citrus and grapes were grown in abundance and alfalfa remained the prominent fodder crop grown on huge areas. The kingdom's most remarkable agricultural accomplishment was its rapid transformation from importer to exporter of wheat (Royal Embassy of Saudi Arabia, Ottawa, Canada 2010). In 1978, the kingdom also built its first grain silos to store extra grains (Saudi Arabia Magazine 2001).

2.3 Agriculture Production

An account of production of various agricultural and animal commodities is presented in Table 7.1. Grains are cultivated on 699,221 ha, while the area allocated for production of grains and fodders amounted to about 700,000 ha. Vegetables are grown over an area of 114,000 ha. About 2.2 million tons of fresh vegetables, notably 424,000 tons of tomatoes; 318,000 tons of potatoes; 233,000 tons of muskmelon and 283,000 tons of watermelon are produced annually in the kingdom. The fruits are cultivated on an area of about 198,000 ha and their production exceeds 1.33 million tons. Similarly, dates are cultivated on an area of about 142,000 ha while the production of dates amounts to 884,000 tons. The kingdom produces 468,000 tons of chickens and 2,498 million eggs. The livestock industry is also flourishing at a rapid pace as cattle have increased to 333,000 heads, sheep to 16.124 million heads, and camels to 824,000 heads. The kingdom produces more than 1.2 million ton of milk and some 165,000 tons of red meat (SAMIRAD 2005).

Table 7.1 Agricultural production for the years 1999, 2003, 2004 and 2008 in the Kingdom of Saudi Arabia (KSA)

Produce	Year 1999	Year 2003	Year 2004	Year 2008
	Production in tons			
Eggs	129,000	137,000	145,000	188,000
Fish	55,000	67,300	67,000	93,495
Fruits	1,190,000	1,333,000	1,450,000	1,616,000
Milk	1,039,000	1,200,000	1,230,000	1,370,000
Poultry	418,000	4,68000	522,000	427,000
Red meat	160,000	165,000	167,000	–
Vegetables	1,930,000	2,200,000	2,480,000	2,696,000
Wheat	1,804,000	2,552,000	2,952,000	1,986,000

Source: SAMIRAD (2005). *The Saudi Arabian Market Information Resource*. Accessed on March 03, 2010; Data for the year 2008 are drawn from (1) Annual Agricultural Statistical Year Book (ASYB 2009) and (2) the website of the Ministry of Agriculture, KSA

2.4 Self-Sufficiency

By 1984, Saudi Arabia became self-sufficient in wheat and shortly thereafter, the kingdom started exporting wheat to thirty countries, including China. In addition, substantial amounts of other grains such as barley, sorghum and millets are also grown in the country (Royal Embassy of Saudi Arabia, USA 2010). Intensive programs on dairy, meat, poultry and egg farming were introduced and by 1985 the domestic requirements for these food items are met by the local farms. The kingdom exports dates, dairy products, eggs, fish, poultry, fruits, vegetables and flowers to markets around the world (Royal Embassy of Saudi Arabia, Ottawa, Canada 2010). Self-sufficiency and export of some products, such as dates and watermelons, has been achieved (Al-Hazmi 1997).

Locally produced vegetables satisfy 85 % of domestic consumption (Strategic Media 2009). Saudi farmers export potatoes, tomatoes, carrots, and pumpkins, and 60,000 tons of vegetables are shipped to the neighboring countries. Among fruits, strawberries are important to mention (Saudi Arabia Magazine 2001). Fresh cut flowers are exported to Netherlands (Strategic Media 2009). The kingdom achieved self-sufficiency for some important food crops by adopting new innovative and modern technologies (Al-Subaiee et al. 2005; FAO 2009).

The kingdom also witnessed great success in wheat production for many years and has recently been exporting rather than importing wheat. However, currently exports have been completely disallowed to conserve water resources (Strategic Media 2009). Gradually, less area was brought under wheat cultivation. Consequently from 1994 to 2004, the production of grains also declined from 4.86 million tons to 2.95 million tons while the production of fruits and vegetables increased (SAMIRAD 2005). The decline is the result of a policy to discourage farmers to grow less wheat and to popularize the idea of diversification of crops in order to save water resources and achieve their sustainability. According to Strategic Media

(2009), the kingdom now produces wheat for local consumption only, which is roughly 2.5–2.6 million tons per year. It does not allow exporting wheat and the government no longer buys barley from the farmers to save its water resources.

2.5 Maximizing the Agricultural Production

The kingdom attaches great importance and top priority to its scarce fresh water resources, therefore, has been continuously looking for ways and means to produce more with less drops of water, overcome water deficit and remain sustainable in water use (Al-Shayaa et al. 2012). An example is the use of treated urban wastewater to irrigate agricultural crops (Saudi Arabia Magazine 2001). By the year 2002, almost 548 million m^3 treated waste water was available, of which 123 million m^3 were re-used. There were 70 sewage plants operating in the kingdom in 2003. In year 2006, about 166 million m^3 treated waste water was re-used and remains a potentially important source of irrigation and for other uses as well. The kingdom is promoting the use of drip irrigation system to enhance water use efficiency (FAO 2009).

2.6 Quest for Sustainability

The kingdom has attained a high self-sufficiency in wheat, fresh milk, eggs and potatoes. These products now meet 100 % of the kingdom's needs whereas poultry production satisfies 50 % of national needs and general vegetables and fruits provide 85 and 65 % respectively (Strategic Media 2009). However, this over-production has put the water resources of the kingdom under stress and is absolutely the result of the over-exploitation and over-utilization of resources and high inputs. This boom and boost in agriculture is viewed to be only short-term, short lived and unsustainable. Therefore, the country needs to focus on long term sustainability and must adopt productive, yet preventive farming practices.

3 The Scope and Potential of Sustainable Agriculture in the Kingdom

In the kingdom, the natural resources are faced with various types of stresses and are considered at risk today more than ever, therefore, the scientists struggle to devise the ways to use the natural resources wisely and sustainably. Al-Subaiee and co-authors (2005) examined the future and potential of sustainable agriculture in the kingdom of Saudi Arabia and reported partial degradation of natural resources in the kingdom based on conflict between agriculture production and the environment.

For the past 40 years, the economic development of Saudi Arabia has been broadly governed by the series of ongoing five-year economic plans. In 1970, Saudi Arabia introduced the first five-year development plan to build a modern economy capable of producing consumer goods that previously had been imported and emphasized the development of the kingdom's infrastructure. The Sixth Plan, which began in 1995, called for broadening the technical skills of the Saudi population, and an even stronger emphasis was placed on economic diversification of industrial and agricultural sectors by increasing the role of the private sector in the economy (The Saudi Network 2010). The Saudi Development Plans ensured the rational use and conservation of natural resources, raising income levels and improving the living standards of rural people. The Ministry of Agriculture (MOA) is the primary agency responsible for the execution and implementation of agricultural policy and activities related to agriculture. The MOA also provides extension and research based information to farmers in the kingdom (Royal Embassy of Saudi Arabia, Ottawa, Canada 2010). The intensification of agricultural extension programs to raise awareness among farmers regarding the significance of water conservation and development of manpower in the agriculture sector have received due attention in the plans (Ministry of Planning 2000; Al-Subaiee et al. 2005). The plans include initiatives that focus on the practice and promotion of sustainable agriculture in the kingdom.

In simple terms, sustainable agriculture means practicing farming by using the natural resources and other agricultural inputs within safe limits for optimum benefits without harming the resources. Sustainable practices ensure leaving the remaining resources in safe conditions for the use of next generations. Sustainable agriculture is a system of farming, consisting of production systems where natural resources are being used but without causing any environmental damage. Such systems also aim at producing food that is nutritious, and uncontaminated with products that might harm human health (MacRae 1990). According to Sullivan (2003), sustainable agriculture aims at producing healthy and ample food for both the present and the future through the wise and judicious use of natural resources. However, sustainable agriculture can be viewed as the ecosystem management of complex interactions among soil, water, plants, animals, climate, and people. The object is to integrate all these factors into a production system that is suitable and appropriate for the environment, the economic conditions, and the people of farming areas. Sustainable agriculture has the potential to meet all the three objectives i.e. economic, environmental, and social simultaneously. However, these three objectives overlap in providing benefits and are managed together.

3.1 Economic Sustainability

Sustainability contains a few other forms, such as economic sustainability which requires selecting of profitable enterprises by implementing comprehensive financial planning. Another type, social sustainability, involves keeping money circulating in the local economy, and maintaining or enhancing the quality of life of the farmers (Sullivan 2003). The information presented in the above paragraphs leads to establish that the sustainable agriculture has the great potential to address issues like water shortage and its over-utilization. Sustainable agriculture is the main focus and significant component of Saudi strategic plans and recently has become a serious concern among Saudi scientists and professionals (Al-Mogel 1999).

The kingdom achieved self-sufficiency in wheat but this was realized with the huge consumption of water. The kingdom now has launched a new plan that stresses the farmers to produce more crops with less water. The basis of the sustainable agriculture strategy focuses to scale down water consumption to 50 % of the present volume. The plan suggests not cultivating the green fodder that consumes some six billion cubic meters (BCM) of water in a year. In the strategy, the government plans to provide subsidies for the least water requiring crops. The sustainable agriculture plan has been scheduled for a period of five years and that would also focus on the development of the distribution chain and the promotion of organic farming in the kingdom (Abou-Hadid 2010). On the other hand to conserve water, farmers are being advised by the kingdom to use greenhouses and drip irrigation instead of ground irrigation to conserve water (Strategic Media 2009). According to Oxford Business Group (2010), a number of segments in the agriculture sector have been identified that can be sustainable. However, five segments deserve prime attention which includes: poultry, aquaculture, greenhouse agriculture, technology and dairy production. To enter into the business of sustainable agriculture, Saudi agriculture is passing through a period of great transition engaged in evolving new farming techniques and technologies suitable for arid climates, as reported at its (SADRC) Sustainable Agriculture Development Research Centre- (Oxford Business Group 2010).

Currently, there is a dire need to create an awareness of sustainable agriculture practices among the farmers and extension. However, launching of successful sustainable agriculture programs would require the support of Extension Service and its workers (Al-Shayaa et al. 2012). However, before initiating any extension program, it is important to evaluate the current perceptions and knowledge of the extension agents toward that particular project. Bearing in mind this important aspect, Al-Subaiee et al. (2005) conducted a study to determine the attitude and perceptions of Saudi extension agents toward sustainable agriculture. The study indicated that the sustainable agriculture and its embedded practices are capable of addressing many problems faced by the Saudi agriculture such as (low soil fertility; conservation of natural resources including water, variety selection, environmental protection etc). Their study further revealed that extension agents in Saudi Arabia generally had a positive perception toward sustainable agriculture, both practices and concepts. However, extension agents should be provided in-service training programs and encourage them to attain higher education with an emphasis on extension education and/or technical agricultural fields as well as environmental concepts. Their study also established the need to develop sustainable agriculture programs to educate farmers on sustainable agriculture technologies (Al-Subaiee et al. 2005).

3.2 Environmental Sustainability

Most often, sustainable agriculture is viewed as the set of ecologically sound practices that cause little or no negative impact on the natural ecosystems, rather they enhance environmental quality and strengthen the natural resource base on which the agricultural economy depends. Usually, it involves the features like protecting, recycling, replacing and maintaining the natural resources base such as land, water wildlife that contribute towards conservation of natural capital (Kassie and Zikhali 2009). The kingdom of Saudi Arabia, under the umbrella of sustainable agriculture, has adopted measures like improving water use efficiency, reducing the irrigation water losses and applying good agronomic practices like precision laser leveling, soil amelioration, short-duration varieties, etc. In fact these adopted practices have been able to promote efficient water use and resultantly ensured environmental sustainability (Al-Zahrani et al. 2011). In fact, the adoption of farm-level modernization technologies like ditch lining or tubing and the associated improvement of water-lifting point, buried pipes, and similar activities certainly improved agricultural water-use that ultimately helped achieving environmental sustainability. Al-Zahrani et al. (2012) revealed that the existing cropping patterns are not quite suitable for different regions. They believe that there is a great potential of enhancing and ensuring food and water security in the Kingdom by simply modifying the existing cropping pattern. There is need to encourage crop production in the regions where that are suitable to grow. To save and conserve fresh water resources, the use of gray water has come up as a viable option when the demand for water on the increase. Outdated infrastructures have either been replaced or renovated so as to ensure fair distribution, especially to landholdings on the tail-end of a command area (Abou-Hadid 2010).

3.3 Social Sustainability

Sustainable agriculture being the production system has the potential of addressing many issues and constraints faced by the resource-poor farmers and at the same time it is quite socially acceptable. It refers to the capacity of agriculture over time to contribute to the overall welfare of the communities by providing them sufficient food and other goods and services in the ways that are socially acceptable, economically viable and environmentally sound. Whereas social sustainability refers to the quality of life of those involved in the farming, as well those in the surrounding communities (Kassie and Zikhali 2009). The strategies and practices reported in the previous paragraphs were put into practice under sustainable agriculture initiatives, proved quite capable to ensure social sustainability and at the same time were well-received in the Saudi farming communities (Al-Zahrani and Baig 2011).

4 An Overview of Infrastructure That Directly or Indirectly Promotes Agriculture and Rural Development

According to the Embassy of Saudi Arabia (2009), the present kingdom was established in 1932. Before then, its agricultural exports and trade were limited to producing and selling dates to the pilgrims visiting the holy cities of Makkah and Madina. The weak infrastructure of the kingdom was insufficient and incapable to support an appreciable economic growth. Right after the emergence of the kingdom as an economic power, the importance of the infrastructure in achieving the sustainable development received due attention. Consequently, the kingdom's successfully development is worth mentioning as its world-class infrastructure as second to none (Long Term Strategy 2002). One of the objectives of creating infrastructure, and the expanding of the agricultural base, was to encourage social development (Long Term Strategy 2002).

In fact, the implementation of an improved infrastructure is viewed as the first step in the developmental process that can offer a boost to teach sector of the economy, including agriculture. The agricultural development has been realized due to the application of modern agricultural technology and the roads that link farmers with urban consumers and facilitate the transport of agricultural commodities from the farms-to-the-forks (Royal Embassy of Saudi Arabia, UK 2009). Additional growth and socio-economic development can be realized in the kingdom through the better infrastructure and its overview is presented in the Table 7.2. According to WHO (2007) the country has witnessed an improvement in socio-economic development in the past forty years, and considerable progress has been made in the areas of health, education, housing, and the environment. The country enjoys a network of modern roads, highways, airports, seaports, and power desalination plants (UNDP 2010).

Discussing the basic facilities and service available in the kingdom, UNDP (2010) reported that all the residents had access to safe water resources in both rural and urban area, and 90 % of households had sustainable access to sanitation services in 2000, with 100 % service coverage in urban areas and 72 % in rural areas. The kingdom believes that real development can only be achieved by equipping the nation with quality education. The kingdom has successfully established a network of educational institutions that includes 24 public universities, eight private universities, more than 100 colleges and more than 26,000 schools in order to realize an educated society. To promote education, numerous more educational institutions are planned. At present some five million students have been enrolled in the existing educational facilities. An appreciable student to teacher ratio i.e. 11-to-1 exists in the kingdom and seems quite low as compared to the world standards (Royal Embassy of Saudi Arabia, USA 2009).

Table 7.2 An overview of infrastructure and services available to the society in the Kingdom of Saudi Arabia

Population in 1,000 (2009 est.)[a]	25,721
Population of Saudi Arabia[b]	28,287,855
Percent population average annual growth rate (2005–2010)[c]	2.1
Percent population in rural areas (2005)[d]	15
Percent urban population (2007)[c]	81.4
Percent contribution of agriculture to GDP (2010)[e]	5
Agriculture, value added (% of GDP in 2008)[d]	2
GDP per capita – PPP (2008 est.)[f]	US$ 24,208
Gross national income per capita (PPP International $) (2006)[g]	US$ 22,300
Percent arable land (2010)[e]	5
Percent labor force in agriculture (2009)[h]	6.7
Percent of the annual kingdom budget for education including vocational training[i]	25
Percent of GDP spent on Education (2004)[d]	5.7
Percent combined primary, secondary and tertiary gross enrolment ratio (2004)[f]	98.6
Percent population with access to modern infrastructure, utilities and services (2010)[e]	80
Percent population with access to safe water supplies (2010)[e]	100
Total per capita water consumption (cubic meters) (2010)[e]	1,010
Area of the kingdom (million square kilometers) (2010)[e]	2.25
Undernourished as percent of total population (2009)[f]	<5.0
Total consumption of water in billion cubic meters 2000 (2010)[e]	21.1
Percent water supply coming from non-renewable deep aquifers (2010)[e]	57.4
Percent fresh water withdrawal by agriculture (2010)[e]	89.2
Total renewable surface water sources in cu km (1997)[j]	2.4

Sources:
[a] FAOSTAT
[b] The World Bank 2012. Data available at: http://data.worldbank.org
[c] United Nations Statistics Division 2009
[d] World Development Indicators Database, September 2009
[e] MDGs – Arab States – Saudi Arabia UNDP 2010
[f] Human Development Report 2009, New York, UNDP 2010
[g] World Health Statistics 2008
[h] FAO Country Brief
[i] Millennium Development Goals Report for the Kingdom of Saudi Arabia 2002
[j] FAO 2009, FAOAQUASTAT

5 Implications for Extension Education

In the kingdom, extension education services are provided by the Ministry of Agriculture (MOA). Extension agents, typically with the B.Sc. Agriculture degrees, work in the different agricultural regions of the country. They help farmers by responding to their enquiries and interacting with researchers at the ministry to solve local problems. The MOA aims to realize sustainable agriculture and extends collaboration with scientists and researchers at the three colleges of agriculture in the kingdom.

Initiatives of the Ministry towards sustainable agriculture include many examples. Recently, the Ministry of Agriculture (MOA) signed an agreement with the King Saud University (KSU) to establish a Sustainable Agriculture Development and Research Centre (SADRC) in Riyadh. The centre's goal is to explore new methods of conserving water in agriculture. Staffed with researchers and scientists, KSU provides research facilities to develop the professional cooperation with the international organizations to undertake research initiatives in the kingdom (Arab News 2010).

Another unique example on the issue of sustainability and extension work on the part of MOA is a Date Palm project. Date Palm is viewed as one of most important tree crops of the kingdom. But it has been severely infested with the Red Palm Weevil. The Ministry of Agriculture (MOA) established the Red Palm Weevil Research Chair (RPWRC) at the KSU. Since the emergence of the problem of the Red Palm Weevil, the Ministry of Agriculture has made dedicated efforts to control this harmful pest in the kingdom. The MOA has chalked out many programs to address agricultural issues by focusing on a strong network of extensionists and agricultural personnel in all the provinces to promote sustainable agriculture.

The King Saud University (KSU), in collaboration with the MOA, will launch the joint extension ventures (RPWRC 2010). The MOA contributes to promote the extension activities through different channels such as extension leaflets and bulletins, diagnostic services, recommendations for pests and diseases treatment, conducting training programs and extension workshops. King Saud University, established in 1957 is the oldest and the largest university in the kingdom. King Saud University (KSU) was recognized as the top university in the Arab World, and one of the best 300 in the world, according to the Times Higher Education – QS World University Rankings (The World University Ranking 2009). KSU has established an Agricultural Extension Centre in the Department of Agricultural Extension and Rural Society, College of Food and Agriculture Sciences, Riyadh. This is the only department in the entire Gulf Region to offer academic degree program in the subject of Agricultural Extension and Rural Development. The department is delivering quality education, offering modern agricultural knowledge and upgrading the skills of the extensionists to better communicate with the farming communities in the rural areas. The mandate and activities of the department are in line with the mission and objectives of both the college and the university. Like all other developed countries around the globe, the kingdom, also attaches great importance to Extension Education and is consistently viewed as an important subject and essential service for many reasons, for example, protecting the environment and conserving the natural resources have been the major objectives of Saudi Arabia's national development strategy.

The kingdom also places great emphasis on its relatively limited fresh water (Al-Ibrahim 1990) and arable land resources. About 89.2 % of fresh water is consumed in agriculture, while 57.4 % comes from non-renewable fresh aquifers (UNDP 2010). Water being a non-renewable commodity in Saudi Arabia, made planners to seriously re-think about its wise and judicious use. Recently, a draft water code was developed for the kingdom (Glennon 2009). However, according to Al-Shayaa et al. (2012) Extension Education can have the following implications and can help in:

- Protecting the environment and conserving the natural resources of Saudi Arabia and it also happens to be the prime objective of the kingdom policy. The kingdom attaches a significant importance to its relatively limited fresh water and arable land resources. Extension education can help creating awareness on their wise use among the rural dwellers.
- Guiding farmers on the judicious and moderate use of farm inputs to operate in safe limits to conserve natural resources and protect environment in the era of high input agriculture.
- Addressing the concerns in rural communities about implementing sustainable agriculture such as lower yields, and limited profit margins.
- Educating the farmers to adopt modern scientific farming practices to elevate their crop yields, minimize losses, increase incomes and enhance livelihood strictly only through sustainable agriculture.

While carrying out extension activities in the rural areas, extension professionals must focus on these points to help farmers towards sustainable agriculture – for it all means sustainable rural development. There is a great need for a strong and efficient extension service staffed with well-trained extension professionals in the country. To refresh their knowledge and upgrade their skills, Extensionists must regularly receive in-service training.

6 Conclusions and Recommendations

By employing adequate irrigation volumes, the kingdom was able to produce surplus wheat and attained the level of exporter. Not only has it caused a serious drain on the kingdom's water resources, drawn mainly from non-renewable aquifers, but it also required the use of huge quantities of chemical fertilizers to boost yields which was not sustainable (Country Studies 2010). This sort of agricultural production does not fall under the definition and limits of sustainable. The kingdom should only focus on the segments capable of resulting sustainability and improving crop yields rural livelihood of the farming communities. Such focus must include aquaculture, poultry, greenhouse agriculture, dairy production and technology to produce more diversified crop with less water. The following steps are of importance to achieve sustainability:

- The initiatives taken by the kingdom for improving the livelihood and elevating the living standards of the rural population are quite appreciable. However, real rural development in the country will definitely depend upon its agriculture and its diversification.
- The role of agricultural extension has not been fully exploited and explored. It is proclaimed that awareness campaigns, technology transfer initiatives and

agricultural extension can produce encouraging and tangible results in educating the farming communities on the wise, appropriate and judicious use of agricultural inputs including water. There is an increasing demand to educate our farmers in this respect so they would view natural resources, such as land and water, as the non-renewable resources. Water and land use should be always treated as the most precious and valuable resources.
- All the three disciplines, sustainable agriculture, rural development and agricultural extension go hand-in-hand and are inter-related. The kingdom must focus and attach equal importance to these three essential ingredients of production and development in the national policy and agendas.
- There is a need to initiate in-service training programs for the extension agents. Extension agents should also be encouraged and provided with opportunities to attain higher education with an emphasis in extension education and/or technical agricultural fields as well as environmental concepts. Extension educators need to develop sustainable agriculture programs to educate farmers in sustainable agriculture and farming technologies to achieve sustainable rural development.

Acknowledgements The authors are extremely thankful to the Saudi Society for Agricultural Sciences, Saudi Arabia for extending the possible financial assistance for the completion of the studies. The valuable support extended for the publishing of this research is thankfully acknowledged.

References

Abou-Hadid AM (2010) Chapter 4: agricultural water management. In: El-Ashry M, Saab N, Zeitoon B (eds) Arab environment –water sustainable management of a scarce resource. Arab Forum for Environment and Development, Beirut, pp 56–70

Al-Alawi M, Abdulrazzak M (1994) Water in the Arabian Peninsula: problems and perspectives. In: Rogers P, Lydon P (eds) Water in the Arab world: perspectives and progress, division of applied sciences. Harvard University, Cambridge, MA

Al-Hazmi AS (1997) Plant nematode problems and their control in the near east region: Saudi Arabia. Food and Agriculture Organization of the United Nations, Rome

Al-Ibrahim AA (1990) Water use in Saudi Arabia: problems and policy implications. J Water Res Plann Manage 116:375–388

Alkolibi FA (2002) Possible effect of global warming on agriculture and water resources in Saudi Arabia: impacts and responses. Clim Change 54:225–245

Al-Mogel A (1999) The higher agricultural education in Saudi Arabia and its future roles within the changes of the new centenary. In: The large university symposium: agricultural aspect. King Saud University, Riyadh: University Print

Al-Shayaa MS, Baig MB, Straquadine GS (2012) Agricultural extension in the Kingdom of Saudi Arabia: difficult present and demanding future. J Anim Plant Sci 22(1):239–246

Al-Subaiee FS, Yoder EP, Thomson JS (2005) Extension agents' perceptions of sustainable agriculture in the Riyadh region of Saudi Arabia. J Int Agric Ext Educ 12:5–14

Al-Tukhais AS (1997) Water resources and agricultural production in Saudi Arabia: present and future. Water resources and its utilization in Saudi Arabia. In: Proceedings of the first Saudi conference on agricultural sciences, College of Food and Agricultural Sciences, King Saud University, Riyadh, 25–27 Mar 1997

Al-Zahrani KH, Baig MB (2011) Water in the Kingdom of Saudi Arabia: sustainable management options. J Anim Plant Sci 21(3):601–604

Al-Zahrani KH, Al-Shayaa MS, Baig MB (2011) Water conservation in the Kingdom of Saudi Arabia for better environment: implications for extension and education. Bulgarian J Agric Sci 17(3):389–395

Al-Zahrani KH, Muneer SE, Taha AS, Baig MB (2012) Appropriate cropping pattern as an approach to enhancing irrigation water efficiency in the Kingdom of Saudi Arabia. J Anim Plant Sci 22(1):224–232

Al-Zeir K (2009) Protected agriculture in the Kingdom of Saudi Arabia

Arab News (2010) Saudi KSU to have SR187m Agriculture Centre. February 11, 2010. http://www.meed.com/sectors/economy/education/saudi-ksu-to-have-sr187m-agriculture-centre/3061197.article. Accessed 14 May 2010

ASYB (2009) Agricultural statistical year book (ASYB). Issue # 22. Agricultural Research and Development Affairs, Department of Studies Planning & Statistics. www.moa.gov.sa

Country Studies (2010) Agriculture in the Kingdom of Saudi Arabia. Accessed 3 Mar 2010 from http://www.country-studies.com/saudi-arabia/agriculture.html

FAO (2009) Irrigation in the middle East region in figures AQUASTAT survey 2008. In: Frenken K (ed) FAO water report 34, country report Saudi Arabia, FAO Land and Water Division, Food and Agriculture Organization of the United Nations, Rome 2009, pp 325–337. Available at: ftp://ftp.fao.org/docrep/fao/012/i0936e/i0936e00.pdf

Glennon R (2009) Saudi Arabia's water woes. The Huffington post. December 8, 2009. Robert Glennon posted January 14, 2009. http://www.huffingtonpost.com/robert-glennon/saudi-arabias-waterwoes_b_157817.html. Accessed 2 Mar 2010

Human development Report (2009) Saudi Arabia – the human development index – going beyond income. Available at http://hdrstats.undp.org/en/countries/country_fact_sheets/cty_fs_SAU.html

Kassie M, Zikhali P (2009) Brief on sustainable agriculture. Prepared for the expert Group meeting on "Sustainable Land Management and agricultural practices in Africa: bridging the gap between research and farmers" Gothenburg, Sweden, 16–17 Apr 2009

Long Term Strategy (2002) Long term strategy 2025- Saudi Arabia. Ministry of Economy and Planning, Government of Saudi Arabia, Riyadh

MacRae R (1990) Strategies for overcoming the barriers to the transition to sustainable agriculture. PhD thesis, McGill University (Macdonald Campus) Ste-Anne-de-Bellevue, QC, Canada, H9X 3V9. Definition of the term "Sustainable Agriculture". Accessed 30 Mar 2009 from http://eap.mcgill.ca/sustain.htm

Millennium Development Goals (MDGs) (2002) Report for the Kingdom of Saudi Arabia 2002, UN, Riyadh, KSA

Ministry of Planning (2000) The seventh Saudi development plan. Ministry of Economy and Planning, Government of Saudi Arabia, Riyadh, The Kingdom of Saudi Arabia

Oxford Business Group (2010) Great transition – Saudi Arabia planting new seeds. Published by Economics & Development, Global Arab Network. Monday, 26 April 2010. Available at http://farmlandgrab.org/post/view/12434-great-transition-saudi-arabia-planting-new-seeds

Ray DE (2003) Income-rich Saudi Arabia prefers grow-their-own food security. MidAmerica Farmer Grower 20(12)

Royal Embassy of Saudi Arabia, Ottawa, Canada (2010) Agriculture and water. Accessed 1 Jan 2010 from http://www.mofa.gov.sa/detail.asp?InNewsItemID=103852&InTemplateKey=print

Royal Embassy of Saudi Arabia, USA (2009) The Kingdom of Saudi Arabia – Report on political, economic and development initiatives, Information Office, Washington, DC, USA, November 2009 Accessed 30 Mar 2010 from http://www.saudiembassy.net/affairs/reports/

Royal Embassy of Saudi Arabia, USA (2010) Agriculture and water. Royal embassy of Saudi Arabia, Washington, DC USA. http://www.saudiembassy.net/about/country-information/agriculture_water/. Accessed 12 May 2010

RPWRC (2010) Red palm weevil research chair. http://www.rpwrc-ksu.org/index.php?page_id=93. Accessed 27 May 2010

SAMIRAD (2005) The Saudi Arabian market information resource. Agricultural developments in Saudi Arabia. 21 September 2005. Accessed 30 Mar 2009 from http://www.saudinf.com/main/f41.htm

Saudi Arabia Magazine (2001) Making the desert come alive. Petrol Miner. http://www.saudiembassy.net/files/PDF/Publications/Magazine/2001-Spring/Agriculture.htm. Accessed 24 May 2010

Strategic Media (2009) Cultivating sustainable agriculture. The Kingdom of Saudi Arabia. Strategic powerhouse, global strength, Part II, special report prepared by Media Report.

Sullivan P (2003) Applying the principles of sustainable farming. Fundamentals of sustainable agriculture. The electronic version of applying the principles of sustainable farming. Accessed 30 Mar 2009 from http://www.attra.ncat.org/attra-pub/PDF/Transition.pdf

The Saudi Network (2010) Agricultural development in Saudi Arabia. http://www.the-saudi.net/saudi-arabia/agriculture.htm. Accessed 12 May 2010

The World Bank (2012) Data Available at: http://data.worldbank.org

The World University Ranking (2009) http://ksu.edu.sa/AboutKSU/Pages/qs.aspx. Accessed 24 May 2010

U.S. Department of Energy (2002) Saudi Arabia: environmental issues. Energy information administration. Accessed 30 Mar 2009 from http://www.earthscape.org/r1/ES15071/doe_saudi

United Nations Convention to Combat Desertification (UNCCD) (2000) Executive summary. The third national report for the kingdom of Saudi Arabia about the implementation of the United Nations Convention to combat desertification. UNCCD program, Switzerland

United Nations Development Program (UNDP) (2010) MDGs -Arab Kingdoms -Saudi Arabia progress towards environmental sustainability. Accessed 30 Mar 2009 from http://www.undp.org/energyandenvironment/sustainabledifference/PDFs/ArabKingdomKingdoms/SaudiArabia.pdf

United Nations Statistics Division (2009) MDG progress chart 2009. http://unstats.un.org/unsd/mdg/Resources/Static/Products/Progress2009/MDG_Report_2009_Progress_Chart_En.pdf

WHO (2008) World health statistics (2008) http://www.who.int/gho/publications/world_health_statistics/EN_WHS08_Full.pdf

Wikipedia (2010) Saudi Arabia. Accessed 3 Mar 2010 from http://en.wikipedia.org/wiki/Saudi_Arabia

World Development Indicators Database (2009) World Resources Institute, Washington, DC

World Health Organization (WHO) (2007) Country cooperation strategy at a glance. Accessed 30 Mar 2009 from http://www.who.int/countryfocus

Part II
Adapting Agriculture to Climate Risks: Selected Successful Practices

Chapter 8
Economic Impact of Climate Change on Tunisian Agriculture: The Case of Wheat

Ali Chebil, Brian H. Hurd, Nadhem Mtimet,
Boubaker Dhehibi, and Weslati Bilel

Abstract This paper measures the potential economic impact of climate change on durum wheat in Tunisia using the Ricardian approach. A model using panel data was estimated for the period 1990–2010 over the main cultivation regions. Gross margin of the durum wheat under rainfed conditions was used as the dependent variable while the explanatory variables were mainly related to climate such as precipitation and temperature, technological progress, and type of soil. Empirical findings show that precipitations during different stages of the growing season affect positively net-income. In addition, the interactions variables between temperature and precipitation in different growth stages are negative. The assessment impact of technology shows a positive coefficient of trend parameter but not statistically significant. Finally, the soil quality index parameter is positively correlated with the net revenue per hectare. This finding indicates that good quality of soil may improve the net income of farmers by increasing wheat yields. Climate change impact was simulated using scenarios from the HadCM3 global circulation. Empirical results indicate that

---

A. Chebil (✉)
Department of Agricultural Economics, National Institute of Research in Rural Engineering, Water and Forestry (INRGREF), P. O. Box 10, 2080 Ariana, Tunisia
e-mail: chebila@yahoo.es

B.H. Hurd
Department of Agricultural Economics and Agricultural Business,
New Mexico State University, Gerald Thomas Hall Rm. 374, MSC3169,
P. O. Box 30003, Las Cruces, NM 88003-8003, USA
e-mail: bhurd@nmsu.edu

N. Mtimet • W. Bilel
Department of Agricultural Economics, Mograne High School of Agriculture (ESA Mograne),
1121 Mograne-Zaghouan, Tunisia
e-mail: mnadhem@hotmail.com; weslati.bilel@hotmail.fr

B. Dhehibi
Social, Economic and Policy Research Program (SEPRP), International Center for Agricultural Research in the Dry Areas (ICARDA), P. O. Box 5466, Aleppo, Syria
e-mail: B.Dhehibi@cgiar.org

economic impacts are not uniformly distributed across the different regions of Tunisia. These impacts are likely to be more accentuated in the arid regions. A rise in temperature and a reduction in rainfall would cause reductions in gross margin by 4 % in sub-humid areas and 24 % in arid zones. The results further suggest the necessity for wider diffusion of drought-tolerant varieties among farmers and the identification of new agricultural practices as advisable adaptation strategies in order to alleviate the effects of climate change on farmer's income.

Keywords Climate change • Wheat • Gross margin • Panel data • Tunisia

1 Introduction

The Intergovernmental Panel on Climate Change (IPCC) reports that the climate could warm by as much as 5 °C over the next 100 years, and estimates that we have already seen a warming of about 0.7 °C since 1900. Agricultural activity which is very sensitive to climate change (CC) will be among the most affected sectors from this phenomenon (IPCC 2007).

Most studies on the economic impact of CC on agriculture have been mainly conducted in developed countries (Adams 1989; Adams et al. 1999; Kaiser et al. 1993; Mendelsohn et al. 1994, 1999; Mendelsohn 2001; Adams and McCarl 2001; Reinsborough 2003; Deschenes and Greenstone 2007; Mendelsohn and Reinsborough 2007). In developing countries, most analysts agree that agriculture is more vulnerable to CC, especially in Africa where the region already endures high heat and low precipitation, in addition to the fact that African farmers rely basically on low technology (Rosenzweig and Parry 1994; Mendelsohn et al. 2000, 2006; Deressa et al. 2005; Gbetibouo and Hassan 2005; Kurukulasuriya et al. 2006; Seo and Mendelsohn 2008). While some efforts have been conducted to quantify the economic impact of CC on agriculture in Africa (Mendelsohn et al. 2000; Deressa et al. 2005; Eid et al. 2007; Gbetibouo and Hassan 2005; Molua and Lambi 2007; Kurukulasuriya and Mendelsohn 2008; Seo and Mendelsohn 2008), little effort has been devoted to the case of Tunisian agricultural sector (MAE 2005), despite its importance for the country's economy.

In Tunisia, agriculture accounts for around 12 % of the national GDP. It is also an economically important sector in terms of employment with 23 % of the total human population. Like other agricultural sectors, wheat cultivation in Tunisia is expected to be significantly influenced by CC (Mougou et al. 2008; Lhomme et al. 2009). The latter studies adopted agronomic crop models with climatic data projected by General Circulation Models (GCM). However, to the author's knowledge, this is the first attempts to estimate the economic impacts of CC on durum wheat in Tunisia.

Our study aims to measure the potential economic impact of CC on wheat production in three different production areas of Tunisia: Béja, Siliana and Kairouan. Wheat is a major cereal in Tunisia in terms of its output and cultivated land area. It occupies about 50 % of all cereals area and represents almost 55 % of the total cereals production. Accordingly, and given the social and economic importance of the wheat sub-sector, the potential impacts of CC should be a major concern for the Tunisian agricultural policy makers because of food security issues.

The rest of this paper is structured as follows. In the next section, the methodological framework is outlined. In Sect. 3, we describe the study areas, and data used in the analysis. Section 4 discusses the main empirical results obtained after applying the Ricardian approach to Tunisian wheat sector using panel data, and predictions for climate change on net-income per hectare for HAdCM3 climate model. Finally, the main conclusions are summarized in Sect. 5.

2 Methodological Framework

2.1 Approaches to Measuring the Economic Impacts of Climate Change

Agro-economic and Ricardian approaches are the two major methods that have been adopted to quantify the economic impact of CC on agriculture (Mendelsohn and Dinar 2003). The agro-economic model uses a combination of controlled experiments on specific crops, agronomic modelling, and economic modelling to predict climate impacts (Adams and McCarl 2001; Adams et al. 1998). The presumed changes in yields from the agronomic model are fed into the economic model, which determines crop choice, production, and market prices. However, it is difficult to apply this model to developing countries, because it has been criticized for underestimating the adaptive responses to CC. Secondly, there have not been sufficient experiments to determine the nature of agronomic responses in most developing countries. Finally, agro-economic models for developing countries are usually poorly calibrated (Mendelsohn et al. 1999).

The Ricardian approach is a cross-sectional model applied to agricultural production. It explains how variations in CC affect net revenue of land value. The method was named after Ricardo because of his original observation that land rents would reflect the net productivity of farmland. Farmland value reflects the present value of future rents, which are tied to future net productivity (see Mendelsohn et al. 1994).

The advantage of Ricardian approach is its ability to capture the adaptation farmers might make in response to local environmental conditions. In addition, the model is cost-effective, since secondary data on cross-sectional sites can be relatively easy to collect on climatic, production, and socio-economic factors. In contrast, the major weakness of this approach is that it does not include carbon fertilization and price effects (Cline 1996).

In this research, the Ricardian approach is used for two main reasons: the method has been successfully applied and empirically validated by several authors (Mendelsohn et al. 1994, 1999: Dinar et al. 1998; Molua and Lambi 2007; Seo and Mendelsohn 2008). The second potential reason is the availability of panel data where using this kind of data can generate different types of information helping government to adopt, deliver and maintain the adequate policies (Baltagi 2001).

2.2 Ricardian Model

The structural Ricardian Model is a micro econometric model in which an agent makes a choice from multiple alternatives in a first stage, and maximizes net revenues in the second stage conditional on the choices (Seo and Mendelsohn 2008).

The original Ricardian studies used land value as a dependent variable. The Ricardian approach, instead of looking at the yields of specific crops, examines how climate in different places affects the *net rent* or *value* of farmland (Mendelsohn et al. 1994). However, in many developing countries data on land values are not available. Annual net revenue per hectare is used as a proxy variable since conceptually land value should be directly related to the present value of a future stream of net revenue (Dinar et al. 1998).

The Ricardian model is specified as follows:

$$NR_i = a_0 + \sum \left(a_s P_s + b_s P_s^2 + d_s T + c_s T_s^2 + f_s P_s T_s\right) + \sum e_c Z_c + u \quad (8.1)$$

where *NR* represents net revenue per hectare, Ts and Ps represent respectively normal temperature and precipitation variables in each season, Z represents relevant socio-economic variables, and u is the error term. The quadratic term reflects the nonlinear shape of the climate impact response function.

Using parameters from the fitted net revenue model, the impact of changing climatic variables on the net revenue per hectare has the following expression:

$$CNR = (NRFS - NRA)/NRA \quad (8.2)$$

Where NRFS is the predicted net revenue per hectare from the estimated net revenue model under the future climate scenario, NRA is the predicted value of the net revenue per hectare from the estimation model under the current climate and CNR is the percentage of the change in the net revenue per hectare (i.e. difference between actual trend and scenario levels). If the value is negative (positive), then the climate change has caused damage (benefits).

3 Data and Empirical Model Specification

3.1 Study Areas and Data

In this study, time series data for the period 1990–2010 were pooled over three main wheat production regions in Tunisia. These regions were chosen from different bioclimatic zones (see Map1). Béja (latitude 36°73′N, longitude 9°18′E, altitude 222 m) is located in the sub-humid zone: its mean annual rainfall is 575 mm. Siliana (latitude 36°08′N, longitude 9°36′E, altitude 420 m) is located in the semi-arid zone and its mean annual rainfall is 403 mm. Kairouan (latitude 35°40′N, longitude

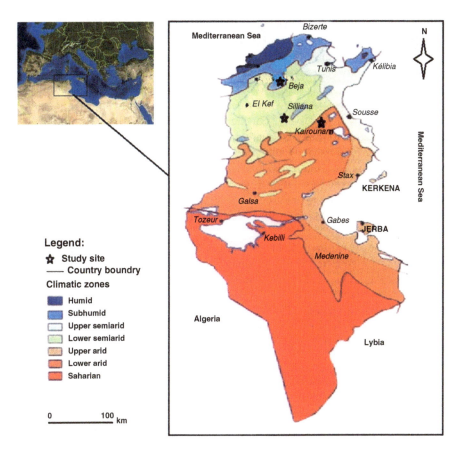

Fig. 8.1 Map of Tunisia showing the regions considered in the study

Table 8.1 Cultivated land in the study areas (1,000 ha)

	Cereals	Forages	Vegetables	Fruit trees	Others	Total
Béja	145.170	47.360	6.700	34.600	22.460	256.290
Siliana	171.210	53.990	2.500	72.920	8.990	309.610
Kairouan	127.120	34.430	5.310	106.500	0.550	273.910

Source: MAE 2010

10°06′E, altitude 68 m) is located in the arid zone and its mean annual rainfall is 297 mm (Fig. 8.1).

The main crops in the Target regions are cereals, fruits trees and forages. Horticultural and others crops (food legumes and industrial crops) occupy a reduced surface. The major crop in the target regions is cereal. This latter is grown on 145.170 ha, 171.210 ha and 127.120 ha in Béja, Siliana and Kairouan, respectively (Table 8.1).

The cereal crops species used are predominantly wheat. According to the Tunisian Ministry of Agriculture, these regions cover about 45 % of all the country's wheat cultivated areas. More than 90 % of wheat crops are grown under rainfed conditions and are consequently very sensitive to climactic fluctuations.

Usually there is a delay of approximately two weeks in the occurrence of the all crop growth stages between the southernmost and northernmost regions of Kairouan and Béja, respectively. For the central region of Siliana, timing of the different wheat growth stages is usually intermediate. Sowing is most often carried out in November, immediately after receiving the first autumn rainfall. Harvesting is usually made in end June-early July with a two weeks delay as we move from south to north between the studied regions. Similar geographic delays also affect the operations of tillering and stem elongation with the former occurring in February and latter in March-April. Heading and ripening occur respectively in May and June.

The data on wheat crop yields and prices (fixed every year by the Tunisian government) are collected from the annual agriculture statistics provided by the Ministry of Agriculture and Environment (MAE 2010). The climatic data were obtained from the National Meteorological Institute (INM: Institut National de Météorologie in French), and average monthly data were used. Temperatures are measured in degrees Celsius and precipitations are measured in millimeters. Soil data are obtained from the Food and Agricultural Organization (FAO 2003).

3.2 Empirical Model Specification

As discussed earlier, the main objective of this research is to assess the economic impact of CC on Tunisian wheat farms using the analytical specification of the Ricardian approach expressed in Eq. (8.1). In the empirical estimation, the following variables were used:

Dependent variable: Gross margin (revenue less variable costs) under rainfed conditions, per hectare per year, in 1990 Tunisian National Dinar (1TND ≈ 0.70 $).

y_{it}: Gross margin of the durum wheat per hectare in region i and at year t.

Explanatory variables: climate (precipitation and temperature), technological progress, and type of soil.

Climate variables: the effect of climatic variables was captured by using average temperatures and total rainfall in critical stages of wheat growing seasons. Wheat cycle generally extends from November to June in Tunisia.

P1: Months precipitation (November–February)
P2: Months precipitation (March- April)
P3: Months precipitation (May-June)
$P1^2$: Months precipitation squared (November –February)
$P2^2$: Months precipitation squared (March- April)
$P3^2$: Months precipitation squared (May-June)

Table 8.2 Description of variables used in the analysis (Period: 1990–2010)

Region	Béja	Siliana	Kairouan	All regions
NR (TND/ha)	433.3113	220.29	345.08	332.89
Mean P1 (mm)	410.95	257.03	203.60	290.53
Mean P2 (mm)	114.5476	90.37	59.48	88.13
Mean P3 (mm)	49.94	56.55	33.48	46.61
Mean T1 (°C)	11.70	10.75	13.94	12.13
Mean T2 (°C)	13.84	13.13	16.79	14.59
Mean T3 (°C)	22.05	21.75	24.87	22.89
Soil quality index	20.62	18.65	22.90	

Source: Authors' analysis based on data from MAE, INM and FAO

T1: Mean temperature for the months November-February
T2: Mean temperature for the months March-April
T3: Mean temperature for the months May-June
T1^2: Mean Temperature squared (November-February)
T2^2: Mean Temperature squared (March-April)
T3^2: Mean Temperature squared (May-June)
P1*T1: Interaction between precipitation and temperature (November-February)
P2*T2: Interaction between precipitation and temperature (March-April)
P3*T3: Interaction between precipitation and temperature (May-Jun)

Trend (T): included as a proxy variable to assess technological progress, such as new crop varieties and improved cropping practices.

Soil quality index (Soil): six different main types of soils are encountered in wheat growing regions of Tunisia (Ben Hassine 2002). Soils are assigned a score from 1 to 6 according to their suitability for wheat cropping: 1 for poorest quality soils and 6 for best ones. The regional soil quality index is constructed by weighting each type of soil according to its presence in the agricultural region, and then summing these weights.

The Ricardian model used for the empirical analysis is as follows:

$$Y_{it} = a_0 + a_1 P_1 + a_2 P_1^2 + a_3 P_2 + a_4 P_2^2 + a_5 P_3 + a_6 P_3^2 + a_7 T_1 + a_8 T_1^2 + a_9 T_2 + a_{10} T_2^2 + a_{11} T_3 \\ + a_{12} T_3^2 + a_{13} P_1 T_1 + a_{14} P_2 T_2 + a_{15} P_3 T_3 + a_{16} trend + a_{17} soil_i + u_{it}$$

(8.3)

Table 8.2 shows the averages of net revenue (NR) per hectare, temperature, rainfall, and soil quality index for the three studied regions. On average, the NR for these regions is about 332.89 TND per hectare. In addition, Béja offers the highest NR per hectare of farmland. The soil quality index for wheat cropping is higher in Kairouan compared to the remaining two regions.

The mean rainfall and temperatures, during the growing stages of wheat cultivation, varied across the regions and over time. Béja receives the highest rainfall among the selected regions. Kairouan presents the highest temperatures at all different stages of wheat growth. As expected, the mean monthly temperature of May-June (T3) is the highest.

4 Results and Discussion

4.1 Empirical Results

The software package Stata was used to estimate the Ricardian model Eq. (8.3). Fixed versus random effects specification of panel data is tested using Hausman test where the null hypothesis is that the preferred model is random effects vs. the alternative the fixed effects (Baltagi 2001). The test results indicate that Hausman statistic is equal to 4, which is not significant at the 5 % level. Hence, we accept the null hypothesis implying that random effects model is preferred (Baltagi 2001).

Parameter estimates, along with the t-ratios of the random effects estimators of the Tunisian wheat farms are presented in Table 8.3. The overall regression showing net-revenue per hectare is significant at the 1 % level (Likelihood ratio 41.92 is higher than χ^2_{17} table).[1] The sign of the estimated climate coefficients are as expected and statistically significant. Precipitations during different stages of the growing season affect positively net-income. The interactions variables between

Table 8.3 Parameters estimates of the Ricardian model

Variable	Coefficient	t-value
Intercept	−3,175.26	−1.11
P₁	2.659	1.52*
P₁* P₁	9.630E-04	1.23
P2	4.966	1.11**
P2*P2	−3.300E-03	−1.22
P3	22.546	2.10*
P3*P3	2.552E-02	−1.19
T1	68.440	0.34
T1*T1	−1.813	−0.23
T2	−185.42	−1.08
T2*T2	6.615	1.20
T3	204.827	0.93
T3*T3	−4.162	−0.88
P1*T1	−0.235	−1.63
P2*T2	−0.264	−0.90
P3*T3	−0.916	−2.29**
Trend	5.661	1.10
Soil	75.369	2.55**
No observations		63
LR¹		41.92**

Note:
*Significant at 10 %; **Significant at 5 %

---

[1] *Likelihood Ratio(LR)* = − 2(*l*(*res*) − *l*(*unres*)).

Where l(res) denotes the restricted maximum likelihood value (under the null hypothesis), while l(unres) denotes the unrestricted maximum likelihood value.

Table 8.4 Future projection of climate change under HadCM3 model by 2030

Region	Temperature change (°C)	Precipitation change (%)
Béja	+0.8	−5
Siliana	+1.0	−5
Kairouan	+1.0	−7

Source: IPCC 2000

Table 8.5 Expected effects of climate change on net revenue in three regions of Tunisia under scenario A2 of HadCM3 model

Region	Percentage of change in net income per hectare
Béja	−4.01
Siliana	−11.92
Kairouan	−23.58

temperature and precipitation in different growth stages are negative and significant at the 5 % level for the period May-June. This could be explained by the fact that an increase in temperature and precipitation during the same period of the growing season could favor disease and consequently cause a loss in wheat productivity.

Assessing the impact of technological variable, empirical findings show a positive coefficient of trend parameter but not statistically significant. This could be attributed to a number of factors including low adoption rate of new technology by farmers and problem of varieties adaptation.

Finally, as expected, the soil quality index parameter is positively correlated with the net revenue per hectare. This finding indicates that good quality of soil may improve the net income of farmers by increasing wheat yields and quality.

4.2 Estimation of Climate Change Impacts

The last step of the analysis explains the CC impacts on Tunisian wheat farmers' net revenue. Table 8.4 presents predictions of future CC from HadCM3 model under scenario A2. From this table, it appears that the model forecasts increasing temperature levels and decreasing precipitations by 2030.

To assess the consequences of CC scenario for the year 2030, predicted values of temperature and rainfall from HadCM3 model were applied to better understand the impact of CC on wheat production in Tunisia and farmers net-revenue. The change in net-revenue per hectare was calculated using Eq. (8.2). The main empirical results from this estimation are presented in Table 8.5 where regional impacts of future CC scenario on net-revenue per hectare are described. These findings indicate that only climate variables are subject to change. All other factors remained unchanged. The main findings indicate that CC has negative impacts on the net-revenue in all three regions. As expected this impact is not equally distributed between these regions. The loss ranges from −4 % of current net-income in Béja to −24 % in Kairouan. These results are considered consistently with the reality of arid zones where are more vulnerable to CC than sub-humid areas. This negative impact of CC on wheat production is in line with the results obtained in previous studies in Mediterranean regions (Quiroga 2005; Mougou et al. 2008; Gasmi et al. 2011).

Finally, the estimated annual damages per region amount to 2.523 million TND, 4.522 million TND and 10.309 million TND in Béja, Siliana and Kairouan, respectively.

5 Conclusions

This study aims to assess the potential economic impact of climate change on rainfed wheat production in Tunisia using the Ricardian approach. Panel data model was estimated by Maximum likelihood method using annual data for the period ranging between 1990 and 2010 over three cropping regions.

Empirical results indicated that net revenue per hectare was sensitive to climate variables. Rainfall appears to be the most influencing factor. Indeed, the climate change has significant nonlinear impacts on net revenue per hectare of wheat production in Tunisia. The interactions variables between temperature and precipitation in different periods of growing season affect negatively the net income. This could be explained by the fact that an increase in temperature and precipitation during the same stage of growth could favor disease and consequently the loss of wheat productivity.

The coefficient of technological variable was positive but not statistically significant. This could be attributed to a number of factors including low adoption rate of new technology by farmers and problem of varieties adaptation. The soil quality index parameter was positively related to the net revenue per hectare which indicated that good quality of soil could improve the net income of farmers.

The assessment of climate change impact indicates that economic impact is not uniformly distributed across the different regions of Tunisia. Thus, a rise in temperature and a reduction in rainfall predicted by A2 scenarios of HadCM3 would cause reductions in gross margin by 4 % in sub-humid areas and 24 % in arid regions.

The sub-humid regions of Tunisia are likely to be less vulnerable to climate change than the arid ones. Given the sensitivity of the Tunisian wheat income to climate change, the government must encourage farmers to use adaptive farm management strategies in order to respond to CC. Widespread diffusion of drought tolerant varieties and the identification of new agricultural practices would be appropriate adaptation strategies that could efficiently alleviate climate change effects on farmer's income.

Acknowledgment The first author acknowledges financial support for this research from the Fulbright foundation.

References

Adams RM (1989) Global climate change and agriculture: an economic perspective. Am J Agric Econ 71:1272–1279

Adams RM, McCarl BA (2001) Agricultural: agronomic-economic analysis. In: Mendelson R (ed) Global warming and the American economy: a regional assessment of climate change. Edward Elgar Publishing, Cheltenham, pp 18–31

Adams RM, Hurd BH, Lenhart S, Leary N (1998) Effects of global climate change on agriculture: an interpretative review. Clim Change 11:19–30

Adams RM, McCarl BA, Segerson K, Rosenzweig C, Bryant KJ, Dixon BL, Conner R, Evenson RE, Ojima D (1999) The economic effect of climate change on US agriculture. In: Mendelsohn R, Neumann J (eds) The economic impact of climate change on the economy of the United States. Cambridge University Press, Cambridge/London, pp 18–54

Baltagi BH (2001) Econometric analysis of panel data, 2nd edn. Wiley, New York

Ben Hassine H (2002) Etude de l'évolution des propriétés chimiques et physico-hydriques des principaux types de sols céréaliers du Nord-Ouest tunisien. Effets sur les productions céréalières. Thèse, Université de Provence, Aix-Marseille I, 299 pages

Cline WR (1996) The impact of global warming on agriculture: comment. Am Econ Rev 86(5):1309–1311

Deressa TR, Hassan R, Poonyth D (2005) Measuring the economic impact of climate change on South Africa's sugarcane growing regions. Agrekon 44(4):524–542

Deschenes O, Greenstone M (2007) The economic impacts of climate change: evidence from agricultural output and random fluctuations in weather. Am Econ Rev 97(1):354–385

Dinar A, Mendelsohn R, Evenson R, Parikh J, Sanghi A, Kumar K, McKinsey J, Lonergan S (1998) Measuring the impact of climate change on Indian agriculture, vol 402, Technical paper. World Bank, Washington, DC

Eid HM, El-Marsafawy SM, Ouda SA (2007) Assessing the economic impacts of climate change on agriculture in Egypt. A Ricardian approach, vol 4293, Policy research working paper. The World Bank, Washington, DC

FAO (Food and Agriculture Organization) (2003) The digital soil map of the world: version 3.6. Rome, Italy

Gasmi F, Belloumi M, Matoussi MS (2011) Climate change impacts on wheat yields in Tunisia: an econometric analysis. Paper presented at the annual conference of the economic research forum, Antalya, Turkey

Gbetibouo G, Hassan R (2005) Economic impact of climate change on major South African field crops: a Ricardian approach. Global Planet Change 47:143–152

Institut National de Météorologie (INM). www.tunisiemeteo.com

Intergovernmental Panel on Climate Change (IPCC) (2000) Special report on emissions scenarios. Cambridge University Press, Cambridge

Intergovernmental Panel on Climate Change (IPCC) (2007) Impacts, adaptations, and vulnerability. Fourth assessment report, Cambridge University Press, Cambridge, UK

Kaiser HM, Riha SJ, Wilks DS, Rossiter DG, Sampath R (1993) A farm-level analysis of economic and agronomic impacts of gradual warming. Am J Agric Econ 75:387–398

Kurukulasuriya P, Mendelsohn R (2008) A Ricardian analysis of the impact of climate change on African cropland. Afr J Agric Resour Econ 2:1–23

Kurukulasuriya P, Mendelsohn R, Hassan R, Benhin J, Diop M, Eid HM, Fosu KY, Gbetibouo G, Jain S, Mahamadou A, El-Marsafawy S, Ouda S, Ouedraogo M, Sène I, Maddision D, Seo N, Dinar A (2006) Will African agriculture survive climate change? World Bank Econ Rev 20:367–388

Lhomme JP, Mougou R, Mansour M (2009) Potential impact of climate change on durum wheat cropping in Tunisia. Clim Change 96:549–564

Mendelshn R (ed) (2001) Global warming and the American economy: a regional assessment of climate change. Edward Elgar Publishing, Cheltenham

Mendelsohn R, Dinar A (2003) Climate, water, and agriculture. Land Econ 79(3):328–341
Mendelsohn R, Reinsborough M (2007) A Ricardian analysis of US and Canadian farmland. Clim Change 81:9–17
Mendelsohn R, Nordhaus W, Shaw D (1994) The impact of global warming on agriculture: a Ricardian analysis. Am Econ Rev 84:753–771
Mendelsohn R, Nordhaus W, Shaw D (1999) The impact of climate variation on US agriculture. In: Mendelsohn R, Neumann J (eds) The impacts of climate change on the U.S. economy. Cambridge University Press, Cambridge, UK, pp 55–74
Mendelsohn R, Dinar A, Williams L (2006) The distributional impact of climate change on rich and poor countries. Environ Dev Econ 11:1–20
Mendelson R, Dinar A, Dalfelt A (2000) Climate change impacts on African agriculture. http://www.google.com/url?sa=t&rct=j&q=&esrc=s&source=web&cd=5&cad=rja&uact=8&ved=0CEgQFjAE&url=http%3A%2F%2Fwww.researchgate.net%2Fpublication%2F46471979_Climate_Change_Impacts_On_Us_Agriculture%2Ffile%2F9fcfd5076a0a06236e.pdf&ei=K_KmU7jHLqTK0QWP-oCwAQ&usg=AFQjCNF8Sqp2rChJheJp91Tx6bpWKwtjsg&sig2=jrGMBkOQteBii53VnQqX4w&bvm=bv.69411363,d.d2k
Ministère de l'Agriculture et de l'environnement (MAE) (2005) Changement climatique : Effets sur l'économie tunisienne et stratégie d'adaptation pour le secteur agricole et les ressources naturelles. Tunisie, 296 p
Ministère de l'Agriculture et de l'environnement (MAE) (2010) Annuaire des statistiques Agricoles, Tunisie
Molua EL, Lambi CM (2007) The economic impact of climate change on agriculture in Cameroon, vol 4364, Policy research working paper. World Bank, Washington, DC
Mougou R, Abou-Hadid A, Iglesias A, Medany M, Nafti A, Chetali R, Mansour M, Eid H (2008) Adapting dryland and irrigated cereal farming to climate change in Tunisia and Egypt. In: Leary N, Adejuwon J, Barros V, Burton I, Kulkarni J, Lasco R (eds) Climate change and adaptation book. Earthscan, London, pp 181–195
Quiroga S (2005) Modelos de decisión y análisis empírico de las relaciones entre el clima y la productividad agraria. PhD thesis, Complutense University of Madrid, Spain
Reinsborough MJ (2003) A Ricardian model of climate change in Canada. Can J Econ 36:21–40
Rosenzweig C, Parry M (1994) Potential impact of climate change on world food supply. Nature 367:133–138
Seo SN, Mendelsohn R (2008) Measuring impacts and adaptation to climate change: a structural Ricardian model of African livestock management. Agric Econ 38(2):150–165

Chapter 9
Greenhouse Gas Mitigation Options in Greek Dairy Sheep Farming: A Multi-objective Programming Approach

Alexandra Sintori

Abstract Dairy sheep farming is an important agricultural activity in the Mediterranean region. In Greece, sheep farming offers employment and income to thousands of families. On the other hand, ruminant livestock farming has been identified as a considerable source of Greenhouse Gases (GHGs). In this analysis, multiple objectives of policy makers are incorporated into a decision making model that yields a number of alternative mitigation strategies, for Greek dairy sheep farming. Each policy alternative achieves the environmental and socio-economic objectives at certain levels. The policy maker can then select the preferred alternative. The model utilizes detailed farm level data, which increases the accuracy of the results. The analysis is undertaken on two different farming systems identified in Continental Greece and indicates that there is a considerable degree of conflict among the GHGs minimization objective and the gross margin and labor maximization objectives. The results also indicate that the mitigation options for sheep farming involve the reduction or/and the intensification of the activity and also changes in the production orientation and feeding practices. The model, can, therefore, be a useful tool for policy makers, since it allows them to design appropriate measures, according to the mitigation option that best meets their preferences.

Keywords Sheep farming • Multiple objectives • Compromise programming • Greenhouse gas emissions • Mitigation

A. Sintori (✉)
Department of Agricultural Economics and Rural Development,
Agricultural University of Athens, Iera Odos 75, Athens 118 55, Greece
e-mail: al_sintori@yahoo.gr

1 Introduction

Dairy sheep farming is an important agricultural activity in the Mediterranean region. In Greece, the activity is mainly located in less favored areas of the country and utilizes less fertile and abundant pastureland. The number of sheep bred in Greece is approximately 9,000,000, held in about 128,000 farms (NSSG[1] 2000). These farms are dairy farms, since they aim primarily at the production of sheep milk that is responsible for over 60 % of their gross revenue and secondarily at the production of meat (Kitsopanides 2006). It is estimated that almost 40 % of the total milk produced in Greece is sheep milk (NSSG 2006). Furthermore, the activity contributes highly in regional development and helps maintain the population in the rather depressed areas, where it is located. It is apparent that the preservation of the dairy sheep farming activity is important not only for farmers but also for policy makers.

Most decision making models used in agricultural planning and policy reflect the basic socioeconomic criterion of gross margin maximization. But the rising public concern on the adverse effects of agriculture to the environment has encouraged the development of models that incorporate not only socioeconomic but also environmental criteria. A number of studies employ multi-objective programming techniques to address complex issues in environmental and agricultural economics. These studies mainly focus on irrigated agriculture and the optimal allocation of water and other resources (see for example: Zekri and Romero 1993; Romero 1996; Tiwari et al. 1999; Latinopoulos 2007; Ragkos and Psychoudakis 2009).

One major environmental concern associated with ruminant livestock farming is climate change and Greenhouse Gas (GHG) emissions. Livestock production and livestock systems undergo a number of changes as a result of climate change. The impact of climate change on livestock production involves quantity and quality of feeds, heat stress and production loss, water supply, animal health and reproduction, biodiversity, land use and livestock systems and other indirect impacts (Thornton et al. 2009; Nardone et al. 2010). On the other hand agriculture has been identified as a significant source of GHGs, and therefore not only adaptation but also mitigation options have to be examined.

GHG emissions are particularly high in the case of ruminant livestock farming because of methane produced through enteric fermentation (Pitesky et al. 2009). Therefore, farmers are, nowadays, urged to adopt not only economically viable but also environmentally sound farming practices, while agricultural policy making should acknowledge the significance of this factor. The issue of GHGs abatement and the evaluation of alternative mitigation policy measures have been addressed in a number of studies that focus mainly in the estimation of GHGs in livestock farms and the impact of the abatement on the farm income. Furthermore, the majority of these studies refer to dairy cow and cattle farming (Olesen et al. 2006; Weiske et al. 2006; Briner et al. 2012) or to meat and wool sheep farming (Petersen et al. 2002; Benoit and Laignel 2008).

---

[1] National Statistical Service of Greece.

Table 9.1 Population of livestock in Greece and their contribution to CH_4 production (%)

	Population size (1,000)	CH_4 production (%)
Dairy cattle	135.26	10.54
Non-dairy cattle	514.66	18.65
Buffalo	1.91	0.07
Sheep	8,831.59	52.12
Goats	5,154.93	16.79
Swine	875.10	0.85
Poultry	29,079,22	0.36
Horses	27.39	0.30
Mules and asses	45.65	0.32
Total		**100.00**

Source: M.E.E.C.C. 2012

It should be noted that though dairy sheep farming is an important and common activity in the Mediterranean region, there is little evidence on its contribution to GHGs and mitigation options have not yet been explored. As can be seen in Table 9.1, the 9,000,000 sheep that are bred in Greece are responsible for over 50 % of the emitted methane from enteric fermentation, which is the main GHG associated with livestock production (MEECC[2] 2012).

Following the aforementioned studies that focus on the use of multi-objective techniques to address environmental issues, this study aims at the construction of a decision making model that is used to explore GHG mitigation options in Greek sheep farming. Moreover, the compromise programming technique is employed to derive a number of alternative options for policy makers, that each achieves the conflicting environmental (minimization of GHGs) and socio-economic (maximization of gross margin and employment) objectives at certain levels. The multi-objective model is based on a whole-farm, mixed-integer programming model that utilizes detailed farm level data and is build to reflect the complexity of the resource allocation problem in livestock farms.

It should be noted, that Greek sheep farms are characterized by a high degree of diversification in terms of invested capital, production orientation, breeding system, herd size, milk yield and other technicoeconomic characteristics, indicating heterogeneity in economic performance and GHG emissions. In the extensive breeding system, feed requirements are met mainly through grazing, while supplementary feed is used only a few months of the year. These farms are characterized by low invested capital and low productivity (HMRDF[3] 2007).

More modern and intensive farms are also present in lowland areas of the country. These farms have a higher invested capital and aim to increase their productivity through supplementary feeding, mainly from on-produced fodder. To take into account this high degree of diversification of the sheep farming activity in Greece, the alternative mitigation options are presented separately for the

---

[2] Ministry of Environment, Energy and Climate Change.
[3] Hellenic Ministry of Rural Development and Food.

semi-intensive and the extensive sheep farming system, which are the prevailing farming systems in the country.

In the next section the multi-objective technique used in this analysis is presented. Next, the mathematical programming model and the data used are described in more detail. Section four contains the results of the analysis and the final section includes some concluding remarks.

2 Methodology

2.1 The Compromise Programming Technique

As mentioned above, multi-objective programming (MOP) is commonly used to assist in the decision making process of policy makers, when the simultaneous optimization of multiple objectives is involved. MOP seeks to identify a set of efficient solutions (Pareto optimal solutions), among the feasible set, since optimal solutions cannot be defined in the case of several, conflicting objectives (Romero and Rehman 2003). The most common techniques of MOP are the constraint method and the weighted method. Both of the above methods are used to derive the efficient set, by transforming the multi-objective problem to a single objective one. The efficient set is then presented to the decision maker, who selects the optimal solution. The drawback of the above methodologies is the computational effort they require. Furthermore, they often yield a large number of alternatives and therefore other techniques are also implemented to reduce the number of alternatives presented to the decision maker.

A technique commonly used to accommodate these problems of MOP is Compromise Programming (CP) (Zekri and Romero 1993). The main assumption of CP is that the decision maker seeks a solution as close as possible to the ideal point, since the ideal point cannot be reached (utopian point) (Romero and Rehman 2003). In other words, the method aims to identify the best compromise solutions, through the use of distance functions.

To implement this method, all objectives are first express mathematically as a function of the decision variables. Thus,

$$\begin{aligned}
\text{Max (or Min)} f_1 &= a_{11}x_1 + a_{12}x_2 = + \ldots + a_{1n}x_n \\
\text{Max (or Min)} f_2 &= a_{21}x_1 + a_{22}x_2 = + \ldots + a_{2n}x_n \\
&\ldots \\
\text{Max (or Min)} f_q &= a_{q1}x_1 + a_{q2}x_2 = + \ldots + a_{qn}x_n
\end{aligned} \quad (9.1)$$

where, q is the total number of criteria (objectives), n is the number of the decision variables x_j and a_{ij} the coefficients of the decision variables. Then, the pay-off matrix is estimated to obtain the ideal and anti-ideal values of the objectives and to observe the degree of conflict among them. To obtain the pay-off matrix, each objective is optimized separately over the feasible set and the values of the rest of the objectives are estimated at the optimal solution. The values of the diagonal of the pay-off matrix are the ideal values of the objectives.

To measure the distance from the ideal point, the family of L_p metrics can be used:

$$L_p(w) = \left(\sum_{i=1}^{q} w_i^p \left| \frac{f_i^* - f_i(x)}{f_i^* - f_{i*}} \right|^p \right)^{1/2} \quad (9.2)$$

where f_i^* and f_{*i} are the ideal and anti-ideal values of the i-th objective, identified by the pay-off matrix, w_i represents the weights attached to the deviations according to the relative importance of each objective and p is the parameter that acts as a weight of the magnitude of the deviations. For each set of p and w_i a best compromise solution is derived.

When the L_1 metric is used ($p=1$) the best compromise solution can be identified by solving the following linear programming model:

$$MinL_1(w) = \sum_{i=1}^{q} w_i \frac{f_i^* - f_i(x)}{f_i^* - f_{*i}} \quad (9.3)$$

subject to:

$$x \in F$$

where, F is determined by the constraints of the model. The objectives used in the above mathematical formation are normalized (they are divided by their range) (Zekri and Romero 1993). When the L_∞ metric is used ($p=\infty$) the maximum deviation (D) is minimized (when $p=\infty$ is used emphasis is given on the largest deviation). The minimization of the largest deviation corresponds to the following linear programming model:

$$MinL_\infty = D$$

subject to:

$$w_1 \frac{f_1^* - f_1(x)}{f_1^* - f_{*1}} \leq D \quad (9.4)$$

$$\cdot$$
$$\cdot$$
$$\cdot$$

$$w_q \frac{f_q^* - f_q(x)}{f_q^* - f_{*q}} \leq D$$

$$x \in F$$

Using other metrics to derive best compromise solutions, results in non linear algorithms (Romero and Rehman 2003). But the two metrics presented above, define the compromise set, therefore all other best compromise solutions fall between the bounds provided by the L_1 and the L_∞ metrics. Thus, to derive the compromise set, which is a subset of the efficient set, only the above two linear programming models for each set of weights need to be solved.

2.2 Modeling GHGs in Livestock Farms

The multi-objective technique described above is based on a whole-farm, mixed-integer programming model. Whole-farm models are commonly used to represent farming systems because of their ability to capture the complexity of the farm operation and the interrelationships of the alternative activities, as well as the substitution possibilities between them. This is particularly important when issues of GHG emissions in livestock and crop-livestock systems are addressed, because of the diversity of the emission sources. Attempts to mitigate a particular greenhouse gas can cause the increase of gases from other sources and therefore, only when the whole-farm approach is adopted can these effects be predicted.

Most models used to estimate GHG emissions in livestock farms are either Life Cycle Assessment (LCA) models or Simulation models (see Table 9.2). Whole-farm optimization models are also commonly used for the estimation of GHG emissions in farming, because they can incorporate all possible sources of emissions and capture all trade-offs among them.

An extended review of whole-farm models used to estimate GHG emissions in beef and dairy cattle farms can be found in Crosson et al. (2011). Schils et al. (2007b) also present four models used to predict mitigation options in ruminant livestock systems, namely DairySim, FarmGHG, SIMS$_{DAITY}$ and FarmSim. Table 9.2 includes a presentation of selected models used to estimate GHG emissions in livestock and crop-livestock systems and summarizes some of their basic characteristics. Specifically, the nature of each model and its methodological basis is presented. Furthermore, the table includes the main sources of emissions incorporated in each of the selected models.

3 Case Study

3.1 Data

The decision making model used in the analysis utilizes detailed farm level data to capture the complexity of the livestock activity and GHG emissions. The farm level data was gathered from two sheep farms located in Continental Greece and refer to

Table 9.2 Models used to predict GHG in livestock farms

	Method and objective	Emission sources
De Cara and Jayet (2000)	Farm level, linear programming models representing the French agricultural sector were used to estimate GHG emissions and abatement costs	Enteric fermentation and fertilizer use. Carbon sequestration was also taken into account
Petersen et al. (2002)	Emission sources are incorporated in the MIDAS linear programming model to assess the impact of GHG abatement policies on sheep farming systems in Australia	Enteric fermentation, manure management and application, fertilizer use, fuel use, stubble burning
Gibbons et al. (2003)	Farm-adapt a mixed-integer programming model was used to determine the most cost-effective adaptations at the farm-level for reducing GHG emissions in England and Wales	Enteric fermentation, manure management, N_2O from grasslands, pre-chain emissions from manufacture of fertilizers and pesticides
Olesen et al. (2004)	The FarmGHG, a C and N flow-based simulation model was developed to estimate emissions in livestock farms	Direct, indirect and leaching emissions: Enteric fermentation, manure management and application, fertilizer use, diesel and electricity. Pre-chain emissions are also included
Casey and Holden (2005, 2006)	The LCA method is used to asses emissions from the average Irish milk production system (2005) and from sucker-beef production in Ireland (2006)	Enteric fermentation, manure management and spreading, concentrate feed production transport and processing, N fertilizer production, transport and application, diesel and electricity
Smith and Upadhyay (2005)	A linear programming model is developed to explore mitigation options in representative crop-livestock farming in Canada	Enteric fermentation, manure management and application, fertilizer use, energy use, crop residues. Pre-chain emissions and carbon sequestration are also included in the analysis
Schils et al. (2007a)	The DairyWise empirical model was developed and evaluated using data from 29 dairy farms in the Netherlands	Enteric fermentation, manure management and application, grazing, fertilizer use, crop residues, mineralization from peat soils, grassland renewal, biological N fixation, diesel and electricity. Pre-chain emissions were also taken into account
Fiorelli et al. (2008)	The FarmSim simulation model was coupled with PASIM and CERES-EGC to predict greenhouse gas emissions in crop-ruminant farms	Enteric fermentation, respiration, manure management and use, fertilizer use, fuel and electricity. Pre-chain emissions were included in the analysis

(continued)

Table 9.2 (continued)

	Method and objective	Emission sources
Veysset et al. (2010)	Two models were coupled, Opt'INRA, a linear programming bio-economic model and PLANETE an environmental assessment model, to study the performance of French Charolais Suckler Farms	Enteric fermentation, manure management, N fertilization management, energy consumption, pre-chain emissions (including buildings and machinery). Indirect N_2O emissions were not taken into account
Foley et al. (2011)	The BEEFGEM simulation model was designed to quantify GHG emissions from pastoral suckler beef cow production systems	Enteric fermentation, slurry storage and spreading, deposition of excreta by grazing animals, fertilizer use, silage effluent, diesel use. Indirect N_2O emissions and other pre-chain emissions were included
Lesschen et al. (2011)	The MITERRA-Europe environmental assessment model was used to estimate emissions from livestock sectors of Europe. The model is based on the CAPRI and GAINS models and is supplemented with N, soil C and mitigation modules	Enteric fermentation, manure management and application, fertilizer use, crop residues, deposition of excreta by grazing animals, organic soils and liming, fossil fuel, electricity and fertilizer production emissions were included
Sise et al. (2011)	A software model was developed to estimate greenhouse gas emissions of sheep-beef farming systems	Enteric fermentation, urine and dung, fertilizer use, fuel and electricity. Carbon sequestration is also estimated
Briner et al. (2012)	The INTSCOPT linear programming model was designed to evaluate mitigation options in Swiss suckler cow farm	Enteric fermentation, manure management, fertilizer and manure use, diesel use and pre-chain emissions associated with production and transport of feedstuff and fertilizers. Emissions from electricity use were not included
Weiss and Leip (2012)	The LCA methodology is used, to assess greenhouse gas emissions of the EU livestock sector, with the help of the Capri simulation model	Enteric fermentation, manure management, and application, excreta from grazing animals, fertilizer use, crop residues, on-farm energy use. Pre-chain emissions from production and transport of inputs and emissions (removals) of land use changes were also considered

the agricultural year 2006–2007. The first farm represents the semi-intensive farming system, which is the prevailing farming system in lowland areas of the country and the second farm represents the extensive farming system, which is the prevailing farming system in mountainous and semi-mountainous areas. It is estimated that

33,452 and 24,445 sheep farms are represented by the semi-intensive and the extensive farm, respectively. The two farms are chosen to have a similar size, close to the average in Continental Greece. Semi-intensive farms are characterized by higher milk yield and gross margin per ewe. Also, the feeding of the livestock depends more on fodder, compared to the extensive system. The latter depends more on grazing and therefore, it is associated with higher GHG emissions. On the other hand, intensive farms are more efficient not only in economic but also in environmental terms since they yield low emissions per kilogram of milk. The results of the analysis are presented separately for the two farming systems so that the decision maker can select different optimal mitigation options for upland and lowland areas of Greece.

3.2 Model Specifications

The decision variables of the model used in the analysis, the constraint matrix and the GHG emission sources that were taken into account are presented in this section.

3.2.1 Crop and Livestock Activities

Crop activities of the sheep farms involve, mainly, forage and grain production for livestock feeding. In the model, farmers can produce feed either for consumption in the farm or for sale. Livestock activities incorporated in the model refer to sheep milk and meat production per month and alternative lambing periods.

3.2.2 Feeding Variables

The produced forage and grains are used for the feeding of the livestock. A set of variables is used to approximate monthly distribution of the produced feed. Additionally, monthly consumption of purchased feed corresponds to another set of model variables. Finally, the model includes decision variables that reflect the monthly use of pastureland and the consumption of grass.

3.2.3 Labor Variables

The final set of variables incorporated in the model involves the monthly labor inputs. The model distinguishes between monthly family and hired labor used in crop and livestock activities.

3.2.4 Feed Requirements

The main component of the model reflects the balance of the monthly feed requirements of the livestock. Minimum intake of dry matter, net energy of lactation, digestible nitrogen and fiber matter is ensured through monthly constraints. The feed requirements of the livestock are estimated according to Zervas et al. (2000). For the productive ewes these feed requirements include requirements for preservation, activity and pregnancy. Extra requirements for lactation are estimated per kilogram of produced milk. For the rams, the requirements refer to their preservation, activity and extra requirements during the reproduction period. For the replacement animals, the feed requirements are estimated every month taking into account the live-weight increase. The weight increase is also taken into account in the case of the lambs, for which feed requirements are estimated for the period that they remain in the farm. The lambs are allowed to consume hay after the first 2 weeks and grains after the first 4 weeks. It should be noted that lambing usually occurs in late autumn or early spring, or in both periods.

On-produced feed crops, external feed inputs and available pastureland are used for the balance of the feed requirements of the flock. The nutritional value per kilogram of feedstuff and grass are taken from Kalaisakis (1965), Jarrige (1980) and Zervas et al. (2000). Additional monthly constraints are incorporated in the model to ensure minimum and realistic intake of concentrate feed, according to the feeding practices of the farms.

3.2.5 Additional Constraints

Another component of the model ensures that monthly labor requirements of all production activities are balanced, mainly with the family labor inputs. Additional hired labor can be used, if necessary, in both livestock and crop activities. Land constraints are also incorporated in the model to ensure that the total area utilized by the various crop activities and pastureland is smaller than the available land of the farm. Moreover, land constraints refer to the total irrigated land of the farms. A final set of constraints reflects the demography of the livestock and the maximum milk and meat production per ewe.

3.2.6 GHG Emissions

In order to accurately derive mitigation options for the sheep farms, it is important to identify all potential sources of GHGs. Otherwise the model will substitute the activities with acknowledged sources of emissions with activities for which the emissions have not been included. The main GHGs, in livestock farms are methane (CH_4) from enteric fermentation and excreta and nitrous oxide (N_2O) from excreta. In addition, in a crop-livestock farm, nitrous oxide emissions (N_2O) from nitrogen fertilizers should also be accounted for (see for example Petersen et al. 2002; Schils et al. 2007a). Carbon dioxide emissions (CO_2) from the use of machinery are an additional source of GHGs. In our analysis, all the potential sources of GHGs have been taken into account. It should be noted that CH_4 and N_2O have been converted to CO_2-equivalents using the conversion factors proposed by the IPCC (2006). The method used to estimate emissions from various sources in the sheep farms is described in more detail in the following paragraphs. Emissions from all sources estimated as CO_2-equivalents are added together to estimate total GHG emissions of the sheep farms. Carbon sequestration has also been taken under consideration. Specifically, a carbon sequestration of 0.3 t C/ha for irrigated crops, 0.2 t C/ha for non irrigated crops and 0.1 t C/ha for pasture is assumed (see also Pretty and Ball 2001). Net emissions are estimated after subtracting carbon sequestration from total emissions.

3.2.7 CH₄ from Enteric Fermentation

Methane production from enteric fermentation is the most important source of GHGs in livestock farms and it is associated with the feeding practices of each farm. Farmers choose to feed their livestock with on-produced feed and purchased feed taking into account their cost and their nutritional value. Mathematical programming models select the optimal combination of feedstuff and suggest the least cost ration. For this reason, the ration used in this analysis is not fixed and methane emissions are predicted from intake, taking into account the requirements of the livestock (see also Petersen et al. 2002). Following the work of De Cara and Jayet (2000), methane emissions from livestock are estimated according to the following equations, for simple and compound feedstuff respectively:

$$\text{E-CH}_4/\text{EB} = -1.73 + 13.91 \text{ dE} \qquad (9.5)$$

$$\text{E-CH}_4/\text{EB} = 5.62 + 4.54 \text{ dE} \qquad (9.6)$$

Where, E-CH₄/EB is the percentage share of gross energy of each feedstuff loss in methane and dE is a digestibility index. The digestibility index for each feedstuff is taken from Kalaisakis (1965).

3.2.8 N₂O from Manure

Methane produced from livestock excreta is considered negligible, since the conditions that exist during the management of manure or grazing of livestock are mainly aerobic (Petersen et al. 2002; IPCC 2006). On the other hand direct and indirect N_2O emissions from livestock excreta during manure management and grazing are included in the analysis. Direct and indirect N_2O emissions from manure management and pastureland are estimated according to the Tier 1 methodology proposed by the IPCC (2006). Emissions from leaching occurring in pastureland have also been taken into account but were considered negligible for manure management.

3.2.9 N$_2$O from Fertilizer Use

In our analysis, we have included direct and indirect N$_2$O emissions from the use of nitrogen fertilizers. First, the total amount of nitrogen applied in fields has been calculated using the amount and the type of fertilizer (De Cara and Jayet 2000; Petersen et al. 2002). Then direct, indirect and leaching emissions from the applied N have been estimated according to the Tier 1 methodology and the emission factors proposed by the IPCC (2006).

3.2.10 CO$_2$ from Energy Use

CO$_2$ from energy use is another source of GHG emissions in crop-livestock farms. The main sources of energy in these farms are fuel (mainly diesel) and electricity (see also Olesen et al. 2006). To estimate the emissions from energy use, fuel or electricity, requirements for every operation and type of machinery is estimated and multiplied by emission factors (Petersen et al. 2002).

In our study, pre-chain emissions have also been estimated and included in the analysis, following the work of Olesen et al. (2006). As mentioned above, farmers choose whether to feed their flock with on-produced or purchased feed. Therefore, N$_2$O emissions from nitrogen fertilizers and CO$_2$ emissions from energy requirements have also been estimated per kilogram of purchased feed. Other inputs, like fertilizers and pesticides have also caused GHG emissions when they were manufactured. These emissions have been taken into account as well, using farm level data to estimate the amount of inputs used and related literature to estimate the emissions caused by the manufacture of these inputs. CO$_2$ emissions from the manufacture of fertilizers are assumed 1.2 kg of CO$_2$ eq/kg of fertilizer (see also Wood and Cowie 2004). Energy requirements for the manufacture of herbicides are assumed 287 MJ/kg, for insecticides 263 MJ/kg and for fungicides 195 MJ/kg (see also Helsel 2006). Emissions are then calculated by multiplying the total energy requirements with 0.069 kg of CO$_2$.

3.3 Objectives of the Policy Maker

In our analysis, the main objectives concerning the sheep farming activity, from the perspective of the policy maker, are:

- *The maximization of gross margin.* In this analysis, the maximization of gross margin corresponds to the economic criteria of the decision maker.
- *The maximization of total labor.* Total labor is used in this analysis as a measure of the level of employment in sheep farms, which is considered an important social objective.
- *The minimization of GHG emissions.* This is considered as a major environmental objective in ruminant livestock farms.

4 Results

First, the pay-off matrix of the objectives for the two farming systems is obtained (Tables 9.3 and 9.4). The first part of each table represents the solution for the representative farm and the second part contains the aggregate results, for the total number of farms that each farm represents. To obtain the pay-off matrix, each objective is optimized separately, over the feasible set. The first row of the pay-off matrix contains the results of the model when the gross margin objective is optimized. The second and third row, contain the results of the model, when labor is maximized and when GHGs are minimized, respectively. The values of the diagonal represent the optimal (ideal) values of the objectives, but because the objectives are conflicting, this optimal point is infeasible. This means that when all objectives are included in the decision making process, and there is some degree of conflict among them, then the objectives cannot be optimized simultaneously.

The results of both farms indicate that there is some degree of complementarity among the gross margin maximization and the labor maximization objectives. This means that the optimum value of labor is very close to the value of labor when the gross margin objective is optimized. On the other hand, there is a high degree of conflict between the gross margin and labor maximization objectives and the GHGs minimization objective. In the case of the semi-intensive farming system the optimal value for the GHGs minimization objective leads to about 46 % lower emissions compared to the gross margin maximization objective, but also to a 48 % lower employment level and 30 % lower gross margin.

Table 9.3 Pay-off matrix for the semi-intensive farming system

	Representative farm			Total number of farms		
	Gross margin (€)	Labor (Hours)	GHGs (kg-CO_2 Eq)	Gross margin (1,000 €)	Labor (Annual work units)	GHGs (1,000 tonnes-CO_2 Eq)
Max gross margin	**14,543**	2,282	66,801	**486,477**	43,629	2,235
Max labor	11,787	**2,487**	69,802	394,287	**47,540**	2,235
Min GHGs	10,180	1,172	**35,964**	340,534	22,407	**1,203**

Table 9.4 Pay-off matrix for the extensive farming system

	Representative farm			Total number of farms		
	Gross margin (€)	Labor (Hours)	GHGs (kg-CO_2 Eq)	Gross margin (1,000 €)	Labor (Annual work units)	GHGs (1,000 tonnes-CO_2 Eq)
Max gross margin	**14,341**	1,720	54,671	**350,560**	24,026	1,336
Max labor	11,681	**1,753**	56,469	285,539	**24,485**	1,380
Min GHGs	10,039	1,014	**29,872**	245,392	14,170	**730**

In the case of the extensive farming system, similar results are obtained. GHGs are particularly high, when gross margin and labor are optimized. The optimal level of GHGs is over 45 % lower compared to the value of GHGs when gross margin and labor are maximized. Reduced GHGs lead to a 30 % reduction of gross margin and a to 41 % reduction in total labor. This reveals the degree of conflict among the two objectives.

It should be noted that the solution the model yields, when gross margin is maximized, though utopian, is relatively closer to the existing situation in livestock farming, compared to the optimal solutions when the other two objectives are optimized. This is because, from the farmers' point of view, the maximization of gross margin is the main objective, and therefore, when gross margin is maximized the farm level model simulates the operation of the farms and represents their actual performance, more accurately than when the other objectives are optimized. From the policy maker's point of view, it is assumed that decision making is currently based mainly on the economic objective of gross margin maximization, since, so far, no GHGs mitigation measures have been introduced in Greece. This indicates that this environmental objective is not yet included in policy making.

After the pay-off matrix is obtained the compromise programming technique is used to derive the alternative best compromise solutions for the decision maker. The two linear programming models described in Sect. 2, are solved for different sets of weights to reflect differences in the environment in which the decision making process takes place. First, equal weights are attached to the deviations of the objectives, assuming that the decision maker gives equal importance to all of them (Scenario 1). The second set of weights, reflects the decision making process for a policy maker that is mainly interested in the mitigation of GHGs. In this case, the weight attached next to the deviation of the environmental objective is two times the weight attached to the deviations of the socioeconomic objectives (Scenario 2).

The results for the L_1 and the L_∞ for each set of weights are presented in Tables 9.5 and 9.7 for the semi-intensive and the extensive farming system respectively. The tables also contain the results of the linear programming model when each objective is optimized. The values of some important variables, like the number of productive ewes and economic indicators, like the production value of milk, are also presented in the two tables. The model allows for the precise estimation of the changes the sheep farming activity undergoes under each scenario, which is important information for the selection of the optimal mitigation alternative, by the decision maker.

Finally, the aggregate results for the semi-intensive farming system and the extensive farming system are presented in Tables 9.6 and 9.8, respectively, so that the derived alternatives can be compared, according to their impact on the sheep farming sector. This way, the policy maker can conceptualize the impact various levels of mitigation may have on the sheep farming activity.

The analysis indicates that in both production systems, over 60 % of the total emitted GHGs come from methane produced through enteric fermentation. The results also indicate that emissions (mainly methane) per kilogram of milk are higher in the extensive farming system. In general, the analysis indicates that the semi-intensive farming system is more efficient in socioeconomic and environmental terms. The following paragraphs contain a more detailed presentation of the results of the analysis, for each farming system.

Table 9.5 Results of the compromise programming method for the semi-intensive farm

	Max gross margin (f_1)	Max labor (f_2)	Min GHGs (f_3)	Scenario 1 L_1	Scenario 1 L_∞	Scenario 2 L_1	Scenario 2 L_∞
Objectives							
Gross margin (€)	14,543	11,787	10,180	13,617	12,392	10,372	11,575
Total labor (hr)	2,282	2,487	1,172	2,228	1,839	1,173	1,593
Net emissions (Kg-CO_2 Eq)	66,801	69,802	35,964	66,387	52,645	36,652	47,473
Variables and economic indicators							
Number of ewes	102	115	50	100	82	50	70
Meat production value (€)	9,773	9,357	4,615	9,596	7,866	4,798	6,717
Milk production value (€)	13,770	14,563	6,750	13,500	11,070	6,750	9,450
Value of purchased feed (€)	3,447	5,684	8	3,903	2,226	90	1,538
Value of produced feed (€)	6,121	5,688	3,468	5,938	4,973	3,501	4,461
Gross margin from cash crops (€)	989	0	2,840	838	1,137	2,960	1,997
Greenhouse gases							
CH_4 emissions (Kg-CO_2 Eq)	50,595	55,567	24,162	51,469	40,294	23,567	34,396
N_2O emissions (Kg-CO_2 Eq)	15,437	15,877	10,304	14,906	12,653	10,445	12,153
CO_2 Emissions (Kg CO_2)	17,293	13,871	17,814	16,364	14,973	18,279	17,062
Carbon sequestration (Kg CO_2)	16,524	15,512	16,317	16,351	15,274	15,639	16,138

4.1 Semi-intensive Farming System

As far as the semi-intensive farming system is concerned, the results of the analysis indicate that the emissions per kilogram of produced milk are 2.4 Kg-CO_2 Eq. This is considered low since, all emission sources are taken into account. The main element of the emissions is methane (61 %). Nitrous oxide emissions account for 19 % of the total emissions and carbon dioxide emissions account for the remaining 20 %. Methane emissions are lower in the semi-intensive system, mainly due to low grass consumption (0.6 tonnes/ewe/year). The proportion of carbon dioxide is significant because of the crop production in semi-intensive farms. This proportion may vary, according to the mitigation alternatives as indicated in Table 9.5. It is also different among the optimal solutions of the objectives. It should also be noted that the main production orientation of the semi-intensive farm is milk production, since in all alternatives milk accounts for over 59 % of the total production value of the sheep farming activity.

Table 9.6 Aggregate results of mitigation alternatives for the semi-intensive farming system

	Scenario 1				Scenario 2			
	L_1		L_∞		L_1		L_∞	
	Absolute value	% Deviation from f_1	Absolute value	% Deviation from f_1	Absolute value	% Deviation from f_1	Absolute value	% Deviation from f_1
Socioeconomic indicators								
Gross margin (1,000 €)	455,520	−6.36	414,531	−14.79	346,967	−28.68	387,196	−20.41
Total labor (AWU)	42,593	−2.38	35,150	−19.43	22,376	−48.71	30,443	−30.22
Number of ewes (millions)	3,345	−1.96	2,743	−19.61	1,673	−50.98	2,342	−31.37
Milk production (tonnes)	501,780	−1.96	411,460	−19.61	250,890	−50.98	351,246	−31.37
Environmental indicators								
Net emissions (1,000 tonnes-CO_2 Eq)	2,221	−0.62	1,761	−21.19	1,222	−45.13	1,588	−28.93
Methane emissions (1,000 tonnes-CO_2 Eq)	1,722	1.73	1,348	−20.36	788	−53.42	1,151	−32.02
N_2O emissions (1,000 tonnes-CO_2 Eq)	499	−3.44	423	−18.04	349	−32.34	407	−21.28
CO_2 emissions (1,000 tonnes-CO_2 Eq)	547	−5.37	501	−13.42	611	5.70	571	−1.34

When the optimal solutions for the three objectives are examined, it can be seen that in the labor maximization objective the optimal solution indicates a higher degree of specialization in sheep production. This is the result of the high labor requirements of the livestock activities compared to crop farming. Specifically, the number of ewes is increased and cash crop production is absent. Under the hypothesis of labor maximization, lambs are sold several months after lambing (up to six), compared to the case of gross margin maximization and GHGs minimization objectives. Also in this optimal solution, carbon sequestration is lower because of the small crop production. In the gross margin maximization solution homegrown feed is high, which indicates that the use of on-produced feed lowers cost and increases gross margin. Finally, it should be mentioned that there is a significant conflict among the environmental and labor objectives, since in the optimal solution of the environmental objective; the value of labor is 53 % lower than its optimal value. Finally, in the GHGs minimization objective, the carbon dioxide emissions are higher, due to crop production.

4.1.1 Scenario 1: Objectives of Equal Importance

As indicated in Table 9.5, assigning equal weights to the deviations of the objectives results in a relatively small compromise set for the socioeconomic objectives but the set is larger for the environmental objective. In the L_1 solution, values are closer to the ideal, when the socio-economic objectives are examined, but the distance from the ideal value of the environmental objective is large. Livestock size is almost the same compared to the gross margin maximization objective, which results in only minor deviations in emissions compared to the gross margin solution (1 % reduction). Mitigation is achieved in the L_1 bound through a small reduction of livestock size (less than 2 %) and homegrown feed, which leads to a small reduction in carbon dioxide and nitrous oxide emissions (5 % and 3 %, respectively). But the reduction in homegrown feed leads to a higher grass consumption per ewe and a 2 % increase in methane emissions. This solution can be considered as a low mitigation alternative for the semi-intensive farming system.

The mitigation level is higher in the case of the L_∞ solution (21 % compared to the gross margin maximization solution) which corresponds to a 15 % deviation from the optimal gross margin and a 26 % deviation from the optimal labor. This solution resembles the optimal solution of gross margin maximization, in terms of farm structure. In specific, there is no change in production orientation, since milk and meat yields per ewe remain constant and there is also no significant change in livestock feeding, though the livestock size is 20 % smaller. The reduction of the livestock size leads to a 20 % reduction in methane emissions. In the L_∞ solution, cash crops are cultivated, and mitigation is achieved mainly by the restriction of livestock size.

4.1.2 Scenario 2: Emphasis on the Environmental Objective

In this scenario, and in both the L_1 and L_∞ bounds, mitigation is achieved with no adjustment on milk and meat yield per ewe. Like in the previous scenario, the milk and meat yield per ewe take their maximum values. But in this scenario there is a more evident change in the production orientation. Livestock size is reduced in both solutions, but this reduction is more significant in the L_1 bound.

In the L_1 compromise solution, mitigation is achieved by a significant shift towards crop production, which causes higher carbon dioxide emissions. Mitigation is also achieved by a significant reduction in livestock size (51 %), which reduces nitrous oxide and methane emissions. The methane emissions are reduced by 53 %, which indicates a modification of the feeding practices of the sheep farm. Indeed, a shift to more concentrate homegrown feed and particularly barley and wheat is also responsible for the total 45 % reduction in net emissions.

Thus, the GHGs minimisation objective receives, under this solution, a value almost equal to the ideal value (less than 2 % deviation), which is the lowest value of emissions, that can be achieved. On the other hand, the underachievement of the gross margin objective in this compromise solution reaches 29 %, and the underachievement of the labor objective is 53 %. Compared to the gross margin maximization solution, labor is reduced by 49 %, which corresponds to 21,253 AWU (see also Table 9.6). What should also be noted is that in this solution, the 51 % decrease in livestock size causes analogous decrease in milk production which corresponds to over 260,000 tonnes.

In the case of the L_∞ bound, there is also a significant shift towards crop production and a reduction of livestock size, but to a smaller extent compared to the previous compromise solution. The GHGs minimization objective is underachieved by 32 %, but the other two objectives are 20 % and 36 % smaller than their ideal value, for gross margin and labor, respectively. Compared to the gross margin maximization solution, the value of GHGs is 29 % lower. This alternative, should also be chosen, when high levels of mitigation need to be achieved.

The alternative mitigation options of the semi-intensive farm are summarized in Fig. 9.1. As can be seen, a switch to crop production is an alternative mitigation options for this farm type. However, in order to achieve the maximum mitigation level a combination of changes has to take place in the farm, namely livestock restriction, switch to cash crops and changes in feeding practices.

4.2 Extensive Farming System

In the case of the extensive farming system, methane represents 70 % of the total GHGs (see Table 9.7). Another 21 % refers to nitrous oxide emissions and only 9 % of total emissions come from carbon dioxide. Carbon dioxide emissions are related mainly to crop production, which is less developed in extensive farms. On the other hand, in the extensive farming system, grazing livestock, consumes a lot of grass

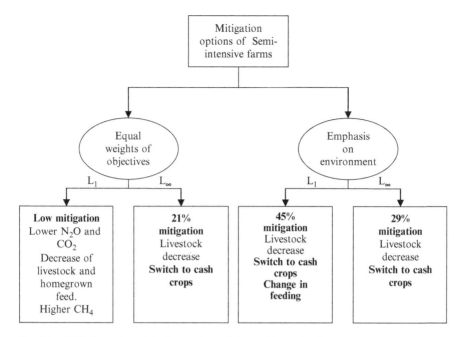

Fig. 9.1 GHG alternative mitigation options of the semi-intensive farm

(1.6 tonnes/ewe/year), which is linked to higher methane and therefore net emissions compared to the semi-intensive system. Methane emissions are estimated at 2.3 Kg-CO_2 Eq/kg of milk in the extensive system and are significantly higher than the methane emissions of the semi-intensive system (1.9 Kg-CO_2 Eq/kg of milk). The analysis also indicates that meat production is very important in extensive breeding farms, since milk production accounts for only 52 % of the total production value of sheep farming (Table 9.7). It should also be noted that the gross margin per ewe is smaller in extensive farms. As indicated in Table 9.7, the objectives of gross margin and labor maximization lead to similar farm structure. When GHGs are minimized, the livestock size decreases, milk yield per ewe increases and there is also a shift towards crop farming.

4.2.1 Scenario 1: Objectives of Equal Importance

The assumption of equal weights attached to the deviations of the objectives generates a relatively small compromise set in the case of the socioeconomic objectives. Mitigation is achieved differently in the L_1 and the L_∞ solutions. In the L_1 solution the value of GHGs is very distant from the ideal point, since less than 1 % mitigation is achieved, compared to the gross margin maximization objective. This hardly affects gross margin and labor since there is a small increase in the cost of purchased feed but there is also a slight increase in milk yield. The feeding

Table 9.7 Results of the compromise programming method for the extensive farm

	Max gross margin (f_1)	Max labor (f_2)	Min GHGs (f_3)	Scenario 1 L_1	Scenario 1 L_∞	Scenario 2 L_1	Scenario 2 L_∞
Objectives							
Gross margin (€)	14,341	11,681	10,039	14,312	12,325	13,426	11,631
Total labor (hr)	1,720	1,753	1,014	1,734	1,407	1,567	1,290
Net emissions (Kg-CO_2 Eq)	54,671	56,469	29,872	54,419	42,365	47,875	38,249
Variables and economic indicators							
Number of ewes	116	117	62	116	89	100	80
Meat production value (€)	9,519	6,993	5,087	9,516	7,301	8,211	6,541
Milk production value (€)	10,288	10,429	5,797	10,500	8,731	9,810	7,831
Value of purchased feed (€)	188	458	0	392	290	0	207
Value of produced feed (€)	6,525	6,507	2,614	6,515	4,781	5,863	4,078
Gross margin from cash crops (€)	0	0	1,447	0	524	233	851
Greenhouse gases							
CH_4 emissions (Kg-CO_2 Eq)	49,914	51,946	27,970	49,506	38,690	43,285	35,256
N_2O emissions (Kg-CO_2 Eq)	15,315	15,789	9,079	15,353	12,215	13,412	11,207
CO_2 emissions (Kg CO_2)	7,229	7,349	6,637	7,321	6,676	6,975	6,670
Carbon sequestration (Kg CO_2 Eq)	17,787	18,613	13,814	17,761	15,215	15,797	14,885

practices, and therefore methane emissions per ewe, remain almost constant. The L_1 solution corresponds to a *very low mitigation alternative* for the decision maker.

In the case of the L_∞ solution, the value of GHGs is underachieved since there is a 42 % deviation from the optimal value. 23 % mitigation is achieved by a significant reduction of the livestock activity, since the number of ewes is 23 % smaller compared to the gross margin maximization objective. This leads to mitigation of both methane and nitrous oxide emissions. In the L_∞ solution labor and gross margin are closer to their ideal points, since they are 20 % and 14 % smaller than their ideal values, respectively. In this solution milk yield per ewe is significantly increased by 11 %. Farms become more labor intensive, since labor per ewe increases by 1 h compared to the gross margin maximization objective. Also, in this compromise solution, there is a shift towards crop production. This alternative is characterized *as a moderate mitigation alternative* for the extensive farming activity and is achieved mainly through reduction of the livestock size, significant increase of milk yield and a small shift towards crop production.

4.2.2 Scenario 2: Emphasis on the Environmental Objective

In the L_1 solution of the second scenario, the values of gross margin and labor are very close to the ideal (6 % and 11 % deviation, respectively) while the GHGs minimization objective is underachieved. In this solution, methane is decreased due to the reduction of the size of the livestock. Also, *changes in feeding practices* take place since homegrown feed per ewe is increased, and grass consumption and purchased feed decrease. Finally, milk yield also increases, which indicates that the farm tends to intensify. But in this scenario, intensification occurs not only in terms of milk yield but also in terms of feeding practices.

The L_∞ solution leads, as in the previous scenario, to a higher mitigation level, compared to the L_1 solution, since emissions are reduced by 30 % compared to the gross margin maximization objective. This mitigation leads to a 19 % deviation from the ideal point of the gross margin maximization objective and to a 26 % deviation from the ideal point of the labor maximization objective. It is also significant to note that the number of ewes is decreased by 31 %, compared to the gross margin maximization objective, which leads to a 24 % reduction in milk production (Table 9.8). In the extensive farming system, the percent reduction of milk production is smaller than the percent reduction of the number of ewes, in all compromise solutions. This means that milk yield per ewe increases, compared to the gross margin maximization solution.

The L_∞ solution of the second scenario corresponds to the intensification in terms of production alternative, since milk production and gross margin per ewe increase. But in terms of feeding practices, there is no shift towards supplementary feeding and concentrates, and therefore emissions per ewe remain high. Finally, the L_∞ solution indicates a shift towards crop production. This shift however is small, compared to the semi-intensive system. This solution achieves the highest mitigation level for the extensive farm as the result of a combination of mitigation options, namely livestock size decrease, intensive milk production, and shift towards crop production.

Table 9.8 summarizes the alternatives for the extensive farming system. It can be seen that in all alternatives, the percent reduction in GHGs is always higher than the percent reduction in gross margin and labor, which indicates that mitigation can be achieved at lower cost in extensive than in intensive systems (see also Table 9.6). In L_∞ solutions, though, the reduction in gross margin and labor can be significant and important for the mountainous and semi-mountainous areas, where this production system is commonly found.

Figure 9.2 presents the mitigation options of the extensive farm. Apart from livestock size restriction, increase in milk yield and intensification of the sheep farming activity are the main mitigation alternatives for this farm type, as opposed to the semi-intensive farms that can switch to cash crop production.

Table 9.8 Aggregate results of mitigation alternatives for the extensive farming system

	Scenario 1				Scenario 2			
	L_1		L_∞		L_1		L_∞	
	Absolute value	% Deviation from f_1	Absolute value	% Deviation from f_1	Absolute value	% Deviation from f_1	Absolute value	% Deviation from f_1
Socioeconomic indicators								
Gross margin (1,000 €)	350,560	−0.20	301,285	−14.06	328,190	−6.38	284,318	−18.90
Total labor (AWU)	24,026	−0.82	19,652	−18.21	21,886	−8.91	18,017	−25.01
Number of ewes (millions)	2.835	0.00	2.176	−23.28	2.445	−13.79	1,955	−31.03
Milk (tonnes)	279,446	2.06	237,141	−15.14	266,450	−4.65	212,706	−23.88
Environmental indicators								
Net emissions (1,000 tonnes-CO_2 Eq)	1,336	−0.46	1,036	−22.51	1,170	−12.43	935	−30.04
Methane emissions (1,000 tonnes-CO_2 Eq)	1,220	−0.82	946	−22.49	1,058	−13.28	862	−29.37
N_2O emissions (1,000 tonnes-CO_2 Eq)	375	0.25	299	−20.24	328	−12.42	274	−26.82
CO_2 emissions (1,000 tonnes-CO_2 Eq)	179	1.28	163	−7.65	170	−3.52	163	−7.73

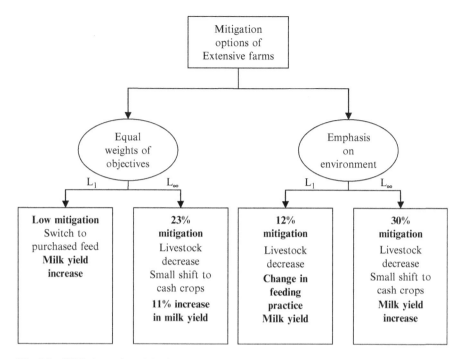

Fig. 9.2 GHG alternative mitigation options of the extensive farm

5 Conclusions

Sheep farming is an important agricultural activity in Greece, since it offers income and employment to a large number of families. On the other hand small ruminant livestock farming is responsible for a considerable amount of GHGs, mainly methane emissions. Although, most decision making models take into account the welfare of the farmers, in terms of income maximization, the adverse effects of the sheep farming activity, should also be considered in policy making. In this study, the environmental and socioeconomic objectives of the decision maker- in this case policy maker- are incorporated in a multi-objective model that yields a number of policy alternatives. Each of the alternatives achieves the conflicting objectives at certain levels, and the policy maker can then select the optimal one, according to specific preferences. Compromise programming is implemented, in order to identify the best compromise solutions.

The environmental objective in our analysis is the GHGs minimization, while the socio-economic objectives are the gross margin and the labor maximization. By giving alternative weights to the deviations of the values of the objectives from their ideal values, alternative mitigation strategies in the sheep farming activity can be explored. The model is built using farm level data, which allows for a precise estimation of the structural changes of the farms, associated with each best compromise solution. The data was collected from two sheep farms, selected to represent the extensive and the semi-intensive farming systems of Continental Greece. It should also be noted that the model used includes all main sources of GHG emissions associated with the sheep farming activity (CH_4, N_2O and CO_2).

The results of the analysis denote the high degree of conflict among the GHGs minimization objective and the maximization of gross margin and labor objectives. The results also indicate that the semi-intensive farming system is more efficient in economic and environmental terms. It can achieve mitigation mainly through shift towards crop production and livestock size decrease, but this mitigation has a significant impact on gross margin and especially labor and milk production. The extensive farming system, on the other hand, has smaller crop production abilities. In this case, mitigation is achieved by a decrease in the livestock size and intensification, mainly in terms of milk production, which leads to increase of the gross margin per ewe. In both cases the analysis indicates that mitigation is possible by a combination of actions that includes the reduction of livestock size, intensification and shift towards crop production.

Policy makers should consider the above findings when policy measures are designed. They should also acknowledge that the solution that best achieves the environmental objective, may have a significant impact not only on the income generated from the sheep farming activity and the amount of labor it utilizes, but also on the produced milk and meat. The high dependency of farms on the sheep farming activity should also be taken under consideration. The above findings emphasize the need for the incorporation of several socioeconomic and environmental criteria in the decision making models used in agricultural planning and policy.

Acknowledgments This research has been co-financed by the European Union (European Social Fund – ESF) and Greek national funds through the Operational Program "Education and Lifelong Learning" of the National Strategic Reference Framework (NSRF) – Research Funding Program: Heracleitus II. Investing in knowledge society through the European Social Fund.

References

Benoit M, Laignel G (2008) Sheep for meat farming systems in French semi-upland area. Adapting to new context: increased concentrates and energy prices, and new agricultural policy. In: Proceedings of the 8th symposium of the International Farming Systems, Clermond-Ferrand, France, 6–10 July 2008

Briner S, Hartmann M, Finger R, Lehmann B (2012) Greenhouse gas mitigation and offset options for suckler cow farms: an economic comparison for the Swiss case. Mitig Adaptation Strateg Glob Change 17:337–355

Casey JW, Holden NM (2005) Analysis of greenhouse gas emissions from the average Irish milk production system. Agric Syst 86(1):97–114

Casey JW, Holden NM (2006) Quantification of GHG emissions from sucker-beef production in Ireland. Agric Syst 90:79–98

Crosson P, Shalloo L, O'Brien D, Lanigan GJ, Foley PA, Boland TM, Kenny DA (2011) A review of whole farm systems models of greenhouse gas emissions from beef and dairy cattle production systems. Anim Feed Sci Technol 166–167:29–45

De Cara S, Jayet PA (2000) Emissions of greenhouse gases from agriculture: the heterogeneity of abatement costs in France. Eur Rev Agric Econ 27(3):281–303

Fiorelli JL, Drouet JL, Duretz S, Gabrielle B, Graux AI, Blanfort V, Capitaine M, Cellier P, Soussana JF (2008) Evaluation of greenhouse gas emissions and design of mitigation options: a whole farm approach based on farm management data and mechanistic models. In: Proceedings of the 8th European IFSA symposium, Clermont-Ferrand, France, 6–10 July 2008

Foley PA, Crosson P, Lovett DK, Boland TM, O'Mara FP, Kenny DA (2011) Whole-farm systems modeling of greenhouse gas emissions from pastoral suckler beef cow production systems. Agric Ecosyst Environ 142:222–230

Gibbons JM, Ramsden SJ, Blake A (2003) Mitigation of greenhouse gas emissions from agriculture: socio-economic costs and impacts. DEFRA project CC0262, Final report, 22 pp

H.M.R.D.F. (2007) Sheep and Goat Sector Development. [Online]. Available at: http://www.minagric.gr/images/stories/docs/ypoyrgeio/dimosieyseis-Arthra/meleti_gia_Nea_KAP/filadia_zoikis/aigoproboatotrofias.pdf. Accessed 2 Feb 2010 (in Greek)

Helsel ZR (2006) Energy in pesticide production and use. Encyclopedia Pest Manage 1(1):1–4

IPCC (2006) Guidelines for national greenhouse gas inventories, vol 4: Agriculture, forestry and other land use [Online]. Available at: http://www.ipcc-nggip.iges.or.jp/public/2006gl/vol4.html. Accessed 9 Oct 2009

Jarrige R (1980) Alimentation des Ruminants, 2nd edn. INRA (Institut National de la Recherche Agronomique), Paris

Kalaisakis P (1965) Applied animal nutrition. Giourdas M. Publishing, Athens (in Greek)

Kitsopanides G (2006) Economics of animal production. ZITI Publishing, Thessaloniki (in Greek)

Latinopoulos D (2007) Multicriteria decision making for efficient water and land resources allocation in irrigated agriculture. Environ Dev Sustain 11:329–343

Lesschen JP, van der Berg M, Westhoek HJ, Witzke HP, Oenema O (2011) Greenhouse gas emission profiles of European livestock sectors. Anim Feed Sci Technol 166–167:16–28

M.E.E.C.C. (2012) Annual inventory submission under the convention and the Kyoto Protocol for greenhouse and other gases for the years 1990–2020. [Online]. Available at: http://www.ypeka.gr/Default.aspx?tabid=470&locale=el-GR&language=en-US. Accessed 2 Apr 2012

Nardone A, Ronchi B, Lacetera N, Ranieri MS, Bernabucci U (2010) Effects of climate changes on animal production and sustainability of livestock systems. Livest Sci 130:57–69

NSSG (2000) Agriculture and livestock census. [Online]. Available at: http://www.statistics.gr/portal/page/portal/ESYE/PAGE-themes?p_param=A1008&r_param=SPK61&y_param=2000_00&mytabs=0 (in Greek). Accessed 24 Feb 2010

NSSG (2006) Annual agricultural survey. [Online]. Available at: http://www.statistics.gr/portal/page/portal/ESYE/BUCKET/A1008/Other/A1008_SPK63_TB_AN_00_2006_07_F_GR.pdf (in Greek). Accessed 24 Feb 2010

Olesen JE, Weiske A, Asman WA, Weisbjerg MR, Djurhuus J, Schelde K (2004) FarmGHG. A model for estimating greenhouse gas emissions from livestock farms. Documentation. DJF Internal report no. 202. Danish Institute of Agricultural Sciences, Tjele, Denmark, 54 pp

Olesen JE, Schelde K, Weiske A, Weisbjerg MR, Asman WAH, Djurhuus J (2006) Modelling greenhouse gas emissions from European conventional and organic dairy farms. Agric Ecosyst Environ 112:207–220

Petersen EH, Schilizzi S, Bennet D (2002) The impacts of greenhouse gas abatement policies on the predominantly grazing systems of south-western Australia. International and development economics, working paper 02–9, Asia Pacific School of Economics and Government, Australian National University, Australia

Pitesky M, Stachhouse KR, Mitloehner FM (2009) Clearing the air: livestock's contribution to climate change. Adv Agron 103:1–40

Pretty J, Ball A (2001) Agricultural influences on carbon sequestration: a review of evidence and the emerging trading options. Center for Environment and Society Occasional Paper 2001–03, Center for Environment and Society and Department of Biological Studies, UK, University of Essex

Ragkos A, Psychoudakis A (2009) Minimizing adverse effects of agriculture: a multi-objective programming approach. Oper Res 9:267–280

Romero C (1996) Multicriteria decision analysis and environmental economics: an approximation. Eur J Oper Res 96:81–89

Romero C, Rehman T (2003) Multiple criteria analysis for agricultural decision making, 2nd edn. Elsevier, Amsterdam

Schils RLM, de Haan MHA, Hemmer JGA, van den Pol-van Dasselaar A, de Boer JA, Evers AG, Holshof G, van Middelkoop JC, Zom RLG (2007a) DairyWise, a whole-farm dairy model. Am Dairy Sci Assoc 90:5334–5346

Schils RLM, Olesen JE, del Prado A, Soussana JF (2007b) A review of farm level modelling approaches for mitigating greenhouse gas emissions from ruminant livestock systems. Livest Sci 112:240–251

Sise JA, Kerslake JI, Oliver MJ, Glennie S, Butler D, Behrent M, Fennessy PF, Campbell AW (2011) Development of a software model to estimate daily greenhouse gas emissions of pasture-fed ruminant farming systems. Anim Prod Sci 51:60–70

Smith EG, Upadhyay M (2005) Greenhouse gas mitigation on diversified farms. Western Agricultural economics Association-Western Economics Association International Joint Annual Meeting, San Francisco, CA, USA, 6–8 July 2005

Thornton PK, van de Steeg J, Notenbaert A, Herrero M (2009) The impacts of climate change on livestock and livestock systems in developing countries: a review of what we know and what we need to know. Agric Syst 101:113–127

Tiwari DN, Loof R, Paudyal GN (1999) Environmental-economic decision making in lowland irrigated agriculture using multi-criteria analysis techniques. Agric Syst 60:99–112

Veysset P, Lherm M, Bédin D (2010) Energy consumption, greenhouse gas emissions and economic performance assessments in French Charolais suckler cattle farms: model-based analysis and forecasts. Agric Syst 103:41–50

Weiske A, Vabitsch A, Olesen JE, Schelde K, Michel J, Friedrich R, Kaltschmitt M (2006) Mitigation of greenhouse gas emissions in European conventional and organic dairy farming. Agric Ecosyst Environ 112:221–232

Weiss F, Leip A (2012) Greenhouse gas emissions from the EU livestock sector: a life cycle assessment carried out with the CAPRI model. Agric Ecosyst Environ 149:124–134

Wood S, Cowie A (2004, June) A review of greenhouse gas emission factors for fertilizer production. IEA Bioenergy Task 38, Research and Development Division, State Forests of New South Wales, Cooperative Research Centre for Greenhouse Accounting, Uk, South Wales.

Zekri S, Romero C (1993) Public and private compromise in agricultural water management. J Environ Manage 37:281–290

Zervas G, Kalaisakis P, Feggeros K (2000) Farm animal nutrition. Stamoulis Publishing, Athens (in Greek)

Chapter 10
Adaptation Strategies of Citrus and Tomato Farmers Towards the Effect of Climate Change in Nigeria

O. Adebisi-Adelani and O.B. Oyesola

Abstract Climate change is a major challenge to agricultural development in Africa and the world at large. Agriculture, being one of the most weather-dependent of all human activities is highly vulnerable to climate change. Horticulture which is therefore an aspect of Agriculture is not an exemption. Farmers had to live with the realities of climate change and be able to manage the situation to maintain their enterprise. Agricultural Development Programme (ADP) structure was adopted to draw a sample of 441 horticultural farmers using multi-stage sampling technique. Four agricultural zones (Southeast, Southwest, Northeast and North central) were purposively selected due to prevalence in horticultural production. From each of these zones, one state each was randomly selected (Oyo, Imo, Gombe and Benue). ADP zones with high production of fruits and vegetables were purposively selected, 25 % of the blocks within the selected zones was randomly selected, 50 % cells within the selected blocks were randomly sampled and 10 % of the respondents from the generated list of fruits and vegetable producers, farmers group were randomly selected. Both quantitative and qualitative methods of data collection were used. Data for the study were analyzed using descriptive and statistics, and T-test at $p = 0.05$, Mean age of horticultural farmers was 46.0 years ±9.3, 83.9 % were male, and mean family size was 5.8. The modal class of number of years in horticultural production is between 10 and 20 years (41.1 %) with 87.8 % belonging to informal working exchange group. Fifty-six percent of the respondents have low adaptation strategy score. There is significant difference in the use adaptation strategies between (t $= -3.391$; p < 0.000) citrus and tomato farmers.

O. Adebisi-Adelani (✉)
Farming Systems and Extension Research Department, National Horticultural Research Institute (NIHORT), P.M.B 5432, Jericho,
Idi-Ishin, Ibadan, Nigeria
e-mail: adelanidotol@yahoo.com

O.B. Oyesola
Department of Agricultural Extension and Rural Development, University of Ibadan,
Ibadan, Oyo State, Nigeria
e-mail: oyetoks2002@yahoo.com

Keywords Horticultural farmers • Adaptation strategies • Agricultural development programme • Climate change

1 Introduction

Climate change brings change in rainfall patterns and consequently changes in agriculture, food security and economic growth, increased temperatures, increases the prevalence of vector-borne diseases, decreased water security, sea level rise and increased variability of floods and droughts (DFID 2004). Over the past 250 years, deforestation, combustion of fossil fuels, and production of agricultural commodities such as rice and livestock have caused atmospheric concentrations of carbon dioxide (CO_2) and other greenhouse gases to rise significantly. Greenhouse gases absorb energy radiated from earth to space and warm the atmosphere. Horticultural farmers are definitely being affected on the course of production and generally in their livelihood and the adaptation strategies that these farmers are using to cope with the present situation. Adaptation includes the actions of adjusting practices, processes and capital in the different regions to the actuality or threat of climate change as well as responses in decision environments such as changes in social and institutional structures or altered technical options that can affect the potential or capacity for these actions to be realized (Howden et al. 2007). These adaptations strategies according to Howden et al. (2007) includes:

- Altering inputs such as varieties/species to those with more appropriate thermal time and vernalization requirements and/or with increased resistance to heat shock and drought altering fertilizer rates to maintain grain or fruit quality consistent with the prevailing climate, altering amounts and timing of irrigation and other water management.
- Wider use of technologies to harvest water, conserve soil moisture (e.g. crop residue retention) and use and transport water more effectively where rainfall decreases.
- Managing water to prevent water logging, erosion and nutrient leaching where rainfall increases.
- Altering the timing or location of cropping activities.
- Diversifying income through altering integration with other farming activities such as livestock production.
- Improving the effectiveness of pest, disease, and weed management practices through wider use of integrated pest and pathogen management, development and use of varieties and species resistant to pest and diseases and monitoring or improving quarantine capabilities and monitoring programs.
- Using climate forecasting to reduce production risk.

The general objective of the study is to examine selected horticultural farmers' adaptation strategies towards the effect of climate change on their production in Nigeria. The specific objectives of the study are to:

- Identify the personal characteristics of the citrus and tomato farmers,
- Ascertain the adaptation strategies of citrus and tomato due to climate change,
- Evaluate the factors that influence the choice of adaptation strategies employed by the
- horticultural farmers

2 Research Hypothesis

$H_0 1$: There is no significant difference in the level of use of adaptation strategies to climate change by citrus and tomato farmers in the selected agricultural zones.

3 Methodology

3.1 Area of Study and Sampling Procedure

The study was carried out in Nigeria. Multi-stage sampling technique was used to arrive at four hundred and forty-one respondents.

The first stage consists of selection of Central, Northeastern, Southeastern and Southwestern zones in Nigeria using purposive sampling technique due to their prominence in horticultural crop production. While the second stage involves selection of one state Benue (Central zone), Oyo (Southwestern zone), Gombe (Northeastern) and Imo (Southeastern)) each from each of the selected agricultural zones using random sampling technique and the third stage was the purposive selection of ADP zones that are known for mass production of selected horticultural crops within the state. Thus in the fourth stage, there was selection of 25 % of ADP blocks within the selected zones using random sampling technique and the fifth stage was randomly selection of 50 % of cells within the selected blocks. Finally, the sixth stage involves the identification of Citrus and Tomatoes farmers by going through the generated list from fruits and vegetable farmers producers association in the communities selected, thus through proportional sampling technique ten percent of the respondents from the list were randomly selected. Citrus and tomatoes being frontline tree, fruit tree and vegetable crops among horticultural crops respectively were purposively selected for the study.

4 Results and Discussion

4.1 Personal Characteristics of Respondents

The result (Table 10.1) reveals that the respondents had mean age of 46.0 years. This shows that most of the respondents are still in their active years and they contribute significantly to agricultural production of the country. This finding is

Table 10.1 Distribution of selected horticultural farmers' personal characteristics

Variable description	Freq	%	Parameters
Age (years)			
35–45	249	56.5	Mean = 45.97
46–55	127	28.7	Median = 43.00
56–65	52	11.8	Mode = 40.00
66–75	7	1.6	SD = 9.26
>75	6	1.4	
Sex			
Male	370	83.9	Mode = male
Female	71	16.1	
Marital status			
Single	25	5.7	
Married	392	88.9	Mode = married
Divorced	5	1.1	
Widowed	19	4.3	
Educational qualification			
No formal	52	11.8	
Primary education	101	22.9	Mode = secondary
Secondary education	190	41.1	
Tertiary education	83	18.8	
Others	15	3.4	
Number of years in horticultural production			
>10	64	14.5	
10–20	207	46.9	Mode = 10–20 years
21–30	89	20.2	
31–40	50	11.3	
41–50	29	6.6	
>50	2	0.5	
Farm size			
<1	99	22.4	
1–3	280	63.5	
3.1–5	43	9.8	
5.1–7	10	2.2	
7.1–9	7	1.6	
>9	2	0.5	

Source: Field survey 2011

consistent with that of Yekinni (2010) and Salimonu (2007) who reported a mean age of 43.2 and 48.1 years for farmers in different studies carried out across agricultural zones of Nigeria. Distribution of respondents by sex shows that 83.9 % were male, 5.7 % were single, 88.9 % married, and 41.1 % had secondary education, This implies that majority of the respondents were literates. Discussants during the FGDs corroborated this when they stated that "they can read and write". It reveals that 46.9 % of respondents had between 10 and 20 years in horticultural production, This implies that majority of the respondents had been into horticultural

Fig. 10.1 Distribution of categorization of adaptation strategies of respondents (Source: Field survey 2011)

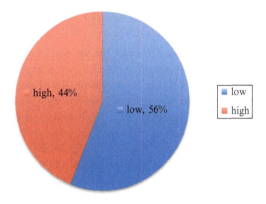

Table 10.2 Frequency distribution of zonal categorization of adaptation strategies

Adaptation categories	Southwestern Fre	%	Southeastern Fre	%	Northeastern Fre	%	Northcentral Fre	%
High adaptation (mean and above)	48	24.7	30	15.6	95	48.9	21	10.8
Low adaptation (below the mean)	62	25.1	30	12.2	6	2.4	149	60.3

Max = 71.0
Mean = 31.023

Source: Field survey 2011

production for over a decade; meaning that they are highly knowledgeable and experienced in horticultural farming this will help them to discuss better on the effect of climate change on their production over the years and be better informed about its general effects on livelihood. It further reveals that majority (54.4 %) of the horticultural farmers have between 1 and 3 ha.

4.2 Climate Change Adaptation Strategies of Respondents

The result in Fig. 10.1 shows that 56.0 % of respondents in the study area have low adaptation strategy score with 44.0 % having high adaptation strategy score. However, the study went further to see what happens to the adaptation strategies within the zones (Table 10.2). High and low adaptation scores were recorded for southwest (24.7 %; 25.1 %), southeast (15.6 %; 12.2 %), northeast (48.9 %; 2.4 %), north central (10.8 %; 60.3 %) respectively. It therefore implies that generally farmers had low adaptation strategies.

4.3 Adaptation Strategies Used by Respondents

The study intended to know the various adaptation strategies being used by farmers to cope with climate change. Thirty-one adaptation strategies were presented in six headings. The result shows (Table 10.3) that considering farmers' adaptation to climate change under Crop management (30.4 %) of farmers always used altering input such as varietal/species followed by use of different planting days (25.6 %), then use of varieties resistant to pest and diseases (23.4 %), monitoring or improving quarantine capabilities (16.8 %), planting drought resistant varieties (14.5 %), shortening length of growing period and crop relocation are both used always at the rate of (14.3 %) and (13.8 %) respectively. Adaptation under soil fertility management results shows that respondents always used barrier hedges along contour to the soil erosion (26.3 %), followed by Soil protection through tree planting (21.5 %), soil conservation (15.2 %), change amount of land (9.5 %). In water management techniques respondents always use the adaptation strategies as follows: managing water to prevent water logging erosion and run off (24.9 %), wider use of technologies to harvest water (24.3 %), conserve soil moisture (13.8 %), Increase irrigation (13.4 %) increase water conservation (13.4 %), expansion of rainwater harvesting (11.3 %), water storage and conservation technique (11.3 %), water re-use (7.7 %) and desalination (6.3 %). Adaptation under insect and pest management include wider use of integrated pest and pathogen management (16.8 %) and Planting pest and diseases resistant varieties (11.3 %). Diversification as an adaptation strategy in the study reveals that diversifying income through altering integration with other farming activities (31.1 %), Move to different site (15.6 %), farming to non-farming (12.0 %) changes from crop to livestock (10.2 %) are always used. Other categories of adaptation always used are Prayer (49.0 %), Change use of chemicals, fertilizers and pesticides (20.9 %) and use of weather insurance (6.3 %). The result implies that farmers have been using one form of adaptation strategy or the other though at a minimal rate and depending on location. During FGD discussants in North-central stated that they alter planting dates and build barriers along contour, those in the South-east stressed the fact that they plant trees and try to control erosion and flood, while discussants in the North-east maintained the use of irrigation and respondents in the Southwest talked of prayers which was also mentioned by others. Findings of Ayanwuyi (2011) on farmers' perception of impact of climate change on food crop production corroborated this finding as he indicated that increased water conservation, planting of different crops and change row orientation are common strategies adaptation by farmers in this order (68.3 %), (74.7 %) and (54.4 %) respectively. Thus, the findings indicated that any type of adaptation strategy introduced will be welcome by farmers since they are already practicing some, which should be improved upon. Nigerian agricultural research Institutions and NGOs are already collaborating with farming organizations to find adaptation strategies that will be within the farmers reach.

Table 10.3 Distribution of respondents according to adaptation strategies by selected horticultural farmers and frequency of use

			Frequency of use					
Adaptation strategies	No		Rarely		Occasionally		Always	
Crop management	F	%	F	%	F	%	F	%
Altering inputs such as varieties/species	173	39.3	47	10.7	87	19.7	134	30.4
Monitoring or improving quarantine capabilities	299	67.8	26	5.9	42	9.5	74	16.8
Use of varieties and species resistant to pest and diseases	233	52.8	20	4.5	85	19.3	103	23.4
Altering the timing or location of cropping activities	182	41.3	46	10.4	114	25.9	99	22.4
Different planting dates	169	38.3	52	11.8	107	24.3	113	25.6
Shorten length of growing period	253	57.4	45	10.2	80	18.1	63	14.3
Crop relocation	253	57.4	42	9.5	85	19.3	61	13.8
Planting drought resistant varieties	257	58.3	40	9.1	80	18.1	64	14.5
Soil fertility management								
Barriers hedges along contours to the soil erosion	166	37.6	41	9.3	118	26.8	116	26.3
Change amount of land	286	64.9	44	10.0	69	15.6	42	9.5
Soil protection through tree planting	208	47.2	49	11.1	89	20.2	95	21.5
Soil conservation	233	52.8	56	12.7	85	19.3	67	15.2
Water management								
Expansion of rainwater harvesting	297	67.3	38	8.6	47	10.7	59	13.4
Water storage and conservation technique	306	69.4	25	5.7	60	13.6	50	11.3
Water re-use	348	78.9	27	6.1	32	7.3	34	7.7
Desalination	360	81.6	23	5.2	30	6.8	28	6.3
Wider use of technologies to harvest water, conserve soil moisture	231	52.4	36	8.2	67	15.2	107	24.3
Managing water to prevent water logging, erosion and run-off	191	43.3	37	8.4	103	23.4	110	24.9
Increase irrigation	263	59.6	46	10.4	71	16.1	61	13.8
Planting flood resistant varieties	276	62.6	31	7.0	84	19.0	50	11.3
Increase water conservation	243	55.1	55	12.5	84	19.0	59	13.4
Pest and insect management								
Planting pest and diseases resistant varieties	243	55.1	25	5.7	60	13.6	50	11.3
Wider use of integrated pest and pathogen management, development	236	53.5	55	12.5	76	17.2	74	16.8
Diversification								
Move to different site	198	44.9	71	16.1	103	23.4	69	15.6
Changes from crops to livestock	293	66.4	44	10.0	59	13.4	45	10.2
Farming to non-farming	298	67.6	26	5.9	64	14.5	53	12.0
Diversifying income through altering integration with other farming activities	165	37.4	54	12.2	85	19.3	137	31.1
Others								
Prayer	122	27.7	48	10.9	55	12.5	216	49.0
Change use of chemicals, fertilizers and pesticides	226	51.2	38	8.6	85	19.3	92	20.9
Use of weather insurance	347	78.7	17	3.9	49	11.1	28	6.3

Source: Field survey 2011

Table 10.4 Frequency distributions of respondents according to factors that influence adaptation strategies

S/N	Factors	Yes Fre	%	No Fre	%	Mean ranking
1	Farming experience	420	95.2	21	4.8	1st
2	Tenure right	286	64.9	155	35.1	2nd
3	Household size	284	64.4	157	35.6	3rd
4	soil fertility	281	63.7	160	36.3	4th
5	Access to water	252	57.1	189	42.9	5th
6	Off farm activities	236	53.5	205	46.5	6th
7	Access to improve seeds/seedlings	233	52.8	208	47.2	7th
8	Access to fertilizer	217	49.2	244	50.8	8th
9	Wealth	209	47.4	232	52.6	9th
10	Access to extension	188	42.6	253	57.4	10th
11	Access to credit	161	36.5	280	63.5	11th

Source: Field survey 2011

4.4 Factors That Influence Adaptation Strategies of Respondents

Result on Table 10.4 shows that farming experience ranked 1st (95.2 %) as a major factor that influences adaptation strategies used by horticultural farmers followed by tenure right (64.9 %), household size (64.4 %), soil fertility (63.7 %) access to water (57.1 %), off farm activities (53.5 %) and access to improved seedlings (52.8 %). Less than half of the respondents refer to other factors as influencing adaptation strategies which are wealth (47.4 %), access to credit (36.5 %), access to extension (42.6 %), and access to fertilizer (49.2 %). This study is consistent with the findings of Nhemachena and Hassan (2008) who found out that farming experience increases the probability of using of all the adaptation strategies. Also in their study they opine that access to credit, markets, and free extension services also significantly increase the likelihood of farmers adopting various adaptation measures.

4.5 Test of Difference in the Use of Adaptation Strategies Between Citrus and Tomato Farmers Across the Agricultural Zones

The result in Table 10.5 shows that there is significant difference in the use of adaptation strategies between (t = −3.391; p < 0.000) citrus and tomato farmers. This is an indication that tomato and citrus farmers make use of different adaptation strategies in coping with climatic changes. This may be the reason for significant relationship that exists between tomato farmers adaptation strategies use and production level.

Table 10.5 Significant difference in the use of adaptation strategies between citrus and tomato farmers across the agricultural zones

Index of adaptation	Mean	T	P = value	df
Citrus	28.167	−3.931	0.000	393.764
Tomatoes	34.401			

Source: Computations from field survey, 2011

5 Conclusion and Recommendations

Adaptation strategies employed by horticultural farmers was low and farming experience is a major factor responsible for the adaptation strategies being used by the respondents. There is significant difference in the use of adaptation strategies between citrus and tomato farmers. Information about climate change can be in prints since majority are literates. There is still the need for more sensitization and awareness about climate change adaptation among citrus and tomato farmers since majority had low adaptation strategies score. Finally adaptation strategies should be crop specific despite the fact that both are horticultural crops. The recommendations above will bring a lot of development to horticulture industry in Nigeria.

References

Ayanwuyi E, Kuponiyi F, Ogunlade A, Oyetoro JO (2011) Farmers' perception of impact of climate change on food crop production in Ogbomosho Agricultural zone of Oyo state Nigeria. GJHSS 10(7):34–39. Added January 05, 2011

DFID Department for International Development (2004) The impact of climate change on the vulnerability of the poor; Policy Division, Global Environmental Assets, Key sheet 3, 6 pp. http://www.unpei.org/sites/default/files/PDF/resourceefficiency/KM-resource-DFID-impact-climatechange-vulnerability.pdf

Howden S, Soussana MJ, Francon J, Francesco T, Chhetri Ntra N, Michael D, Meinke (2007) Adapting Agriculture to climate change edited by William Easterling, Pennsylvania state University Park P.A. an accepted by the Editorial board 16 Aug 2007 (received for review March 2010)

Nhemachena C, Hassan RM (2008) Micro-level analysis of farmers' adaptation to climate change in Southern Africa international food policy research institute sustainable solutions for ending hunger and poverty, Center for Environmental Economics and Policy in Africa

Salimonu KK (2007) Attitude to risk in resource allocation among food crop farmers in Osun State, Nigeria. PhD thesis submitted to Agricultural Economics Department, University of Ibadan, Ibadan

Yekinni OT (2010) Determinants of utilization of information communication technologies for agricultural extension delivery in Nigeria. A PhD thesis submitted to Agricultural Extension and rural development Department, University of Ibadan, Ibadan, 226 pp

Chapter 11
Ex-ante Impact Assessment of 'Stay-Green' Drought Tolerant Sorghum Cultivar Under Future Climate Scenarios: Integrated Modeling Approach

Swamikannu Nedumaran, Cynthia Bantilan, P. Abinaya, Daniel Mason-D'Croz, and A. Ashok Kumar

Abstract An integrated modeling framework – IMPACT – which integrates partial equilibrium economic model, hydrology model, crop simulation model and climate model was used to examine the ex-ante economic impact of developing and disseminating a drought tolerant sorghum cultivar in target countries of Africa and Asia. The impact of drought tolerant sorghum technology on production, consumption, trade flow and prices of sorghum in target and non-target countries were analyzed. And also we estimated the returns to research investment for developing the promising new drought tolerant cultivars and dissemination in the target countries. The analysis indicates that the economic benefits of drought tolerant sorghum cultivar adoption in the target countries outweighs the cost of developing this new technology. The development and release of this new technology in the target countries of Asia and Africa would provide a net economic benefit of about 1,476.8 million US$ for the entire world under no climate change condition. Under climate change scenarios the net benefits derived from adoption of new drought tolerant sorghum cultivar is higher than the no climate change condition. This is due to higher production realized by sorghum under climate change scenarios. The results imply that substantial economic benefits can be achieved from the development of a drought tolerant sorghum cultivar. And also this technology will perform better than the existing cultivars in future climate change condition.

S. Nedumaran (✉) • C. Bantilan • P. Abinaya
Research Program – Markets, Institutions and Policies, International Crops Research Institute for the Semi-Arid Tropics (ICRISAT), Hyderabad, Andhra Pradesh, India
e-mail: s.nedumaran@cgiar.org; c.bantilan@cgiar.org; abhinaya.schoolofdance@yahoo.com

D. Mason-D'Croz
Environment and Production Technology Division, International Food Policy Research Institute (IFRI), Washington DC, USA

A. Ashok Kumar
RP-Dryland Cereals, International Crops Research Institute for the Semi-Arid Tropics (ICRISAT), Hyderabad, India

Keywords Africa and Asia • Climate change • Drought tolerant sorghum cultivar • Economic benefits

1 Introduction

More than 650 million of the poorest and most food-insecure people live in dryland[1] areas around the world. The farm families in the drylands cultivate the world's hardiest and least risky cereals – millets and sorghum. These cereals are consumed primarily on-farm and as much as 80 % is consumed directly by the poorest people (put reference citing the variation on consumption % by region and crop). Trapped in subsistence farming, these farm families suffer from hunger and malnutrition which is prevalent among children. By improving the productivity and thereby the production of these crops in the marginal environments where the poorest people live, they should be able to significantly gain a large share of the potential food security benefits. This report will highlight the ex-ante evaluation of producer's and consumer's benefit achieved through the development of promising technologies of sorghum cultivars with farmer preferred traits and qualities.

Sorghum [*Sorghum bicolor* (L.) Moench] is grown in the hot and dry agro ecologies of Asia, Africa, the America and Australia. It is the fifth most important cereal crop globally after rice, wheat, maize and barley and is the dietary staple of more than 500 million people in 30 countries. It is grown on 40 m ha in 105 countries of Africa, Asia, Oceania and the Americas. Africa and India account for the largest share (>70 %) of global sorghum area. India, Nigeria, Sudan, USA, Niger, Mexico and Ethiopia are the major sorghum producers. Other sorghum producing countries include Burkina Faso, Tanzania, Mali, Brazil, Chad, Australia, Cameroon, Egypt and Argentina.

Sorghum is a staple cereal in sub-Saharan Africa, its primary center of genetic diversity. It is most extensively cultivated in zones of 600–1,000 mm rainfall, although it is also important in areas with higher rainfall (up to 1,200 mm), where poor soil fertility, soil acidity and aluminum toxicity are common. Sorghum is extremely hardy and can be produced even under very poor soil fertility conditions (where other crops like maize fails). The crop is adapted to a wide range of temperatures, and is thus found even at high elevations in East Africa, alternating with barley. It has good grain mold resistance and thus has a lower risk of contamination by mycotoxins.

The grain is used mostly for food purposes (55 %), consumed in the form of flat breads and porridges (thick or thin with or without fermentation). Stover is an important source of dry season maintenance rations for livestock, especially in drylands; it is also an important feed grain (33 %), especially in the Americas. The sorghum grain has quite a high level of iron (>40 ppm) and can complement the ongoing effort on food fortification to reduce micronutrient malnutrition globally.

---

[1] Dryland is characterized by harsh agro-climatic conditions with low and erratic rainfall, high temperature, poor and saline soils with high risk of drought.

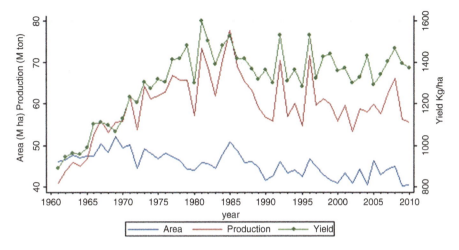

Fig. 11.1 World Sorghum area, production and yield (1961–2010) (Source: FAOSTAT 2012)

In addition to food and feed, it is used for a wide range of industrial purposes, including starch for fermentation and bio-energy. The stover of sorghum is also a significant source for construction material and fuel for cooking in rural Africa and Asia.

Sweet sorghum is emerging as a multi-purpose crop. It can provide food, feed, fodder and fuel (ethanol), without significant trade-offs among any of these uses in a production cycle. ICRISAT has pioneered the sweet sorghum ethanol production technology, and its commercialization.

Globally, sorghum production has remained more or less stable over the past 30 years, although there are notable regional differences. The area under sorghum cultivation has increased from about 40 m ha in 1960s to 51 m ha in 1980s (Fig. 11.1). Later on, there was a fluctuation by 4–10 m ha in area but reached 43.7 m ha by 2009. The productivity increased from 900 kg ha^{-1} in 1960s to 1,400 kg ha^{-1} in 2009. Adoption of improved sorghum cultivars and management practices contributed to the productivity gains though large differences exist in different parts of the world in sorghum productivity.

In 2008–10 sorghum was grown on 40 million ha in all regions of the world (Table 11.1). India has the largest sorghum area with 8.3 million ha. The second largest sorghum growing country is Nigeria, followed by Sudan, Niger and USA. Africa accounts for 61 % of global sorghum area and its cultivation is distributed all over the continent mostly in Nigeria, Sudan, Ethiopia, Burkina Faso, Mali, Egypt, Niger, Tanzania, Chad and Cameroon, where it is a key staple. In Asia, which accounts for 22 % of global area, sorghum cultivation is more concentrated in India and China accounting for 90 % of the regional area under the crop. Latin America accounts for nearly 9 % of the global sorghum acreage, with Argentina and Mexico dominating the regional area share.

Developed countries contribute 22 % of the global sorghum output, despite accounting for only 8 % of the global sorghum area (Table 11.1). United States is the most significant producer accounting for 17 % of the global output while

Table 11.1 Sorghum area, yield and production (Three years average)

Country/region	Area ('000 ha) 1980–82	1993–95	2008–10	Yield (kg/ha) 1980–82	1993–95	2008–10	Production ('000 t) 1980–82	1993–95	2008–10
World	45,170	42,905	41,943	1,469	1,334	1,413	66,376	57,225	59,332
Developed countries	6,329	4,201	3,312	3,364	3,692	4,106	21,291	15,510	13,599
Europe	278	149	195	2,691	4,469	3,975	749	666	736
North America	5,440	3,514	2,374	3,565	3,948	4,314	19,392	13,874	10,167
Oceania	610	538	743	1,885	1,802	3,543	1,151	970	2,696
Developing countries	38,841	38,704	38,628	1,161	1,078	1,184	45,085	41,716	45,716
Africa	13,840	21,451	25,711	927	801	893	12,825	17,185	22,985
LAC[a]	4,567	2,871	3,870	2,881	2,734	3,259	13,159	7,847	12,610
Asia	20,434	14,383	9,047	935	1,160	1,119	19,100	16,685	10,122
East Asia	2,715	1,342	553	2,518	4,208	3,246	6,837	5,648	1,786
China	2,698	1,333	532	2,525	4,230	3,307	6,815	5,638	1,749
South Asia	16,655	12,258	7,899	679	828	954	11,309	10,146	7,538
India	16,262	11,850	7,655	681	836	965	11,082	9,902	7,385
South East Asia	253	148	30	1,041	1,424	1,833	264	210	55
West Asia	810	626	565	853	1,063	1,309	691	665	742

Source: FAOSTAT (2012)
[a]Latin American countries

accounting for only 6 % of global area. Sorghum is grown in the central and southern plains where rainfall is low and variable. Australia has been increasing production of sorghum since the mid-nineties, while it continues to account for only 2 % of global area, it now accounts for 4 % of the global sorghum production.

In India, sorghum was grown on 7.6 million ha in 2008–10. It is grown both in the rainy and post-rainy seasons. In the rainy season, the crop is grown under rainfed conditions, while in the post rainy season it is grown on residual soil-moisture. Only 9 % of the sorghum crop area is irrigated. Sorghum cultivation is concentrated in the states of Maharashtra, Karnataka and Andhra Pradesh jointly accounting for 75 % of the national area.

2 Sorghum Crop Improvement: Research Focus on Drought Tolerance

Sorghum improvement research program at ICRISAT has been given high priority on a range of promising traits like high yield, large grain with biotic stress resistance (shoot fly, midge and grain mold) and abiotic stress tolerance (drought and salinity), grain micronutrient (Fe and Zn) density and sweet stalk traits. In this study we focused on evaluating the potential welfare benefits derived from sorghum technology with high yield and drought tolerance which is highly adapted in rainfed farming of the semi-arid tropics (SAT) region of Africa and Asia.

2.1 Drought Tolerance in Sorghum

Four growth stages in sorghum have been considered as vulnerable to drought: germination and seedling emergence, post-emergence or early seedling stage, midseason or pre-flowering, and terminal or post flowering. Terminal drought is the most limiting factor for sorghum production worldwide. In sub-Saharan Africa drought at both seedling establishment and terminal stages is very common. In India, sorghum is grown during both rainy and postrainy seasons. The variable moisture availability at both pre-flowering and post-flowering stages during the rainy season can have severe impact on grain and biomass yield (Nagy et al. 1995; Bidinger et al. 1996). Drought and/or heat stress at the seedling stage often results in poor emergence, plant death and reduced plant stands. Severe pre-flowering drought stress results in drastic reduction of grain yield. Post-flowering drought stress tolerance is indicated when plants remain green and fills grain normally. A stay-green trait has been associated with post-flowering drought tolerance in sorghum. Genotypes with the stay-green trait are also reported to be resistant to lodging and charcoal rot (Reddy et al. 2007).

2.2 Sorghum Drought Tolerant Research Process

The drought research at ICRISAT is an on-going activity for the last two decades and screening techniques for selecting drought tolerant cultivars have been developed. A large number of germplasm sources and breeding lines have been screened for different growth stages of sorghum. Some of the drought tolerant sources identified in sorghum at ICRISAT include Ajabsido, B35, BTx623, BTx642, BTx3197, El Mota, E36Xr16 8/1, Gadambalia, IS12568, IS22380, IS12543C, IS2403C, IS3462C, CSM-63, IS11549C, IS12553C, IS12555C, IS12558C, IS17459C, IS3071C, IS6705C, IS8263C, ICSV 272, Koro Kollo, KS19, P898012, P954035, QL10, QL27, QL36, QL41, SC414-12E, Segaolane, TAM422, Tx430, Tx432, Tx2536, Tx2737, Tx2908, Tx7000 and Tx7078 (www.icrisat.org).

To reduce the research lag, biotechnology[2] tools like marker-assisted selection (MAS) is being used for genetic enhancement of drought tolerance in sorghum. Six

---

[2] The conventional breeding approach will take about 12–14 years to enhancing drought tolerance in sorghum due to the quantitative inheritance of drought tolerance and yield coupled with the complexity of the timing, severity and duration of drought.

stable and major QTLs[3] (Qualitative Trait Loci) have been identified for the stay-green trait and are being introgressed through MAS into elite genetic backgrounds at ICRISAT (Ashok Kumar et al. 2011).

2.3 Research Cost for Developing Drought Tolerant Cultivars

ICRISAT has been and continues to work on the drought related research for more than two decades and over the years the researchers characterized and evaluated different traits (root system, stay green, etc.) contributing to drought from wide range of available germplasm material. Based on the past experience and availability of proof of concept, ICRISAT, in the recent years, has been focusing on exploiting stay green trait to develop drought tolerance in sorghum. In this study we assumed that ten million US$ is made available to ICRISAT and NARS to fund further research to develop drought tolerant cultivars. The annual cost will include salary component of the researchers, field and laboratory costs and other operational costs. For evaluation and validation at different location and environments, the NARS partners in target countries will be involved. Extension costs for multiplication and dissemination of seeds in the target countries would be borne by the NARS partners to the tune of US$0.25 million each. This was spread over the period until maximum adoption starting from 2019. Table 11.2 provides the breakup of the budget among ICRISAT and NARS partners over 7 years. This budget allocation is inclusive of the extension cost which is indicated against the year 2018 as 1 million for all target countries put together.

3 Technology Dissemination and Adoption Pathway

Since sorghum is mostly grown in marginal environments under rainfed conditions especially in sub-Saharan Africa and Asia, the new technology with drought tolerant traits is expected to produce higher yield than the baseline cultivars which farmers currently grow in the rainfed farming system. The new technology would also increase the resilience of the crop, so that yield would not be affected in

---

[3] Integration of the sorghum genetic map developed from QTL information with the physical map will greatly facilitate the map-based cloning and precise dissection of complex traits such as drought tolerance in sorghum. Sorghum has a compact genome size (2n = 20) and can be an excellent model for identifying genes involved in drought tolerance to facilitate their use in other crops. It was reported that with respect to withstanding drought, sorghum has four copies of a regulatory gene that activates a key gene family which is present in a wide variety of plants. Sorghum also has several genes for proteins called expansins, which may be involved in helping sorghum to recover from droughts. In addition, it has 328 cytochrome P450 genes, which may help plants respond to drought stress, whereas rice has only 228 of these genes.

Table 11.2 Proposed budget allocation for ICRISAT and NARS partners (million US$)

S. No	Year	Research activities	ICRISAT	NARS partners
1	2012	Transfer of QTLs into farmers preferred varieties/elite lines	1.8	0.2
2	2014	Developing introgression lines	1.8	0.2
3	2015	Evaluation and validation in the fields and labs	2	0.3
4	2016	Evaluation in multi-location trails	1	1.2
5	2018	Seed Multiplication ad dissemination	0.5	1

Table 11.3 The ceiling adoption level and year of maximum adoption in target countries

Region	Target countries	Ceiling adoption level	Year of release of technology	Year of maximum adoption
West and Central Africa (WCA)	Burkina Faso	20	2019	2029
	Mali	60	2019	2027
	Nigeria	60	2019	2027
Eastern and Southern Africa (ESA)	Eritrea	40	2019	2027
	Ethiopia	40	2019	2027
	Sudan	40	2019	2030
	Tanzania	40	2019	2027
South Asia	India	80	2019	2025

drought or less rainfall regimes. Hence, the drought tolerant technology would help to sustain the sorghum production even in the drought year.

The technology dissemination process and adoption pathways vary among countries and mainly depend on infrastructure, governance and policy environment, the adaptive capacity of the NARS partners, and involvement of private seed companies in technology development and dissemination (Table 11.3). For example, in India ICRISAT develops the hybrid parental line with desired traits (e.g. drought tolerant lines) while the Hybrid Parental Research Consortium (HPRC) partners including the public institutions and private companies select the parental lines and cross these selected parental lines in the locally adapted cultivars preferred by farmers; then release new hybrids in the target regions. Using the HPRC novel approach of harnessing the synergies with the public and private sectors, ICRISAT is successful in catalyzing the fast diffusion of new promising technologies across the sorghum growing states in India. Furthermore, the seed to seed multiplication ratio for sorghum is high so that private companies are extensively involved in the development of hybrid seeds and its effective distribution through retail marketing. In contrast, the other poor and low income countries in Africa and Asia who lack adaptive capacity in their national breeding programs, still constrained with poor infrastructure and having not gained the technical skills in producing hybrid seed, have remained dependent on the development of open pollinated varieties (OPVs) with the desired promising traits.

The drought tolerant cultivars developed by ICRISAT along with partners are expected to be released in the target countries in different regions globally as shown in Table 11.3 and Fig. 11.2. To estimate the *ex-ante* welfare benefits of the research

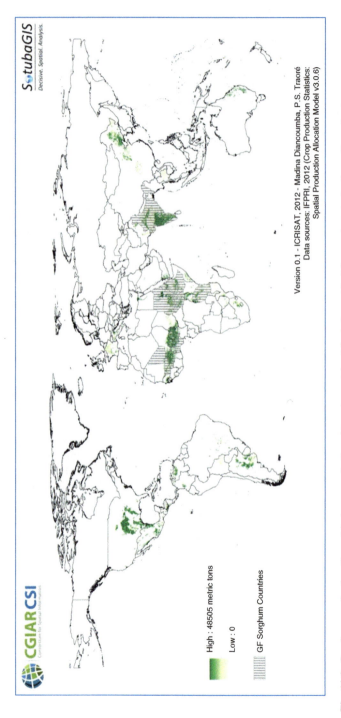

Fig. 11.2 Sorghum production (in metric tons per pixel) and target countries (*shaded* countries) for sorghum technology dissemination

investments in the target countries, we need some critical inputs like the maximum area planted with the new cultivars (i.e. ceiling adoption level) and number of years it will take to reach the maximum adoption level. These parameters would determine how fast the farmers adopt the new technologies in a target country. The adoption of the new technologies by farmers will be influenced by the profitability of the technology (depends on unit cost reduction of the new technology compared to the best available technology to the farmers), availability of the seeds to farmers at the time of sowing, government policy environments like input subsidies and infrastructures (like road networks, communication, etc.).

4 Methodology

4.1 IFPRI's IMPACT Modeling Framework

The IMPACT model combines a partial equilibrium model that has global coverage with hydrology and water supply and demand models and the DSSAT crop modeling suite (Nelson et al. 2010). The IMPACT model is a partial equilibrium agricultural model for 40 commodities of crop and livestock, including cereals, soybeans, roots and tubers, meats, milk, eggs, oilseeds, oilcakes/meals, sugar/sweeteners, and fruits and vegetables. The IMPACT model includes 281 spatial units, called Food Production Units (FPUs) based on 126 major river basins within 115 regions or country boundaries. The model links the various countries and regions through international trade using a series of linear and nonlinear equations to approximate the underlying production and demand functions. World agricultural commodity prices are determined annually at levels that clear international markets. Growth in crop production in each country is determined by crop and input prices, the rate of productivity growth, investment in irrigation, and water availability. Demand is a function of prices, income, and population growth. IMPACT contains four categories of commodity demand – food, feed, biofuels feedstock, and other uses. The IMPACT model incorporates climate effects from the DSSAT modeling results as a shifter in the supply functions (Richard et al. 2013).

4.2 Integrating Technology Adoption and Welfare Estimation in IMPACT Framework

To allow for area and yield of multiple cultivars to respond to the price of a single commodity, some minor structural changes are made in the IMPACT modeling suite. These include the addition of a nested activity structure for the cultivars. In the IMPACT model the cultivar set is named, cul, and the members of the set are called crop1, crop2, crop3, etc. To integrate the promising and existing cultivars into the activity framework, area and yield equations must be adapted.

4.2.1 Harvested Area

To achieve the unique shares of the cultivar areas while maintaining the same total activity area, the shares of area are applied for the cultivars accordingly. Currently in the IMPACT model, the equation for area is a function of the price of the activity, the own and cross price elasticities of the activity, and the exogenous area growth rate, described in the equation below.

$$Area_{j,FPU} = (1 + Areagrowth_{j,FPU}) * PPV_{j,cty}^{AreaElast_{j,jj}} * Areaint,$$

Where,

$Area_{j,FPU}$ = the total area by activity, j
$Areagrowth_{j,FPU}$ = the total rainfed area growth over time
$PPV_{j,cty}$ = the producer price
$AreaElast_{j,jj}$ = the own- and cross-price elasticities for the supply response
$Areaint$ = the area intercept

To incorporate the nested cultivar shares of the area by food production unit, the equation is adapted as follows:

$$Area_{cul,FPU} = CulShare_{cul,FPU} * (1 + Areagrowth_{j,FPU}) * PPV_{j,cty}^{AreaElast_{j,jj}} * Areaint$$

Subject to: $\quad Area_{j,FPU} = \sum_{cul} Area_{cul,FPU}$

Where,

$Area_{j,FPU}$ = the total area by activity, j
$Area_{cul,FPU}$ = the total area by cultivar, cul, for activity, j
$CulShare_{cul,FPU}$ = the share of the total area by cultivar
$Areagrowth_{j,FPU}$ = the total rainfed area growth over time
$PPV_{j,cty}$ = the producer price
$AreaElast_{j,jj}$ = the own- and cross-price elasticities for the supply response
$Areaint$ = the area intercept

4.2.2 Yield

The initial yield for each of the cultivars will be determined by using the yield of the activity for that food production unit which is calculated as the total production per hectare of area. The yield of the cultivars will respond to the prices of the activity, fertilizers, and wages based on the activity elasticities for each. The cultivar yield will also grow over time according to the exogenous yield growth rate.

4.2.3 Exogenous Yield Growth Rate

The exogenous yield growth rate for each cultivar will be determined using the intrinsic yield growth rate for the activity as a starting point for the growth over the time period. In the equation below this grow rate as, *a*. For the additional exogenous yield growth that is possible with the promising cultivars, called *b* in the equation. The adjustments will be added to the intrinsic yield growth rates along with the productivity effect from climate change, called *c*, to create the rate of growth for the promising cultivars.

$$Y_{cul,FPU,t} = Y_{t-1}\left(1 + \left(a_{j,FPU} + b_{cul,fpu} + c_{j,fpu}\right)\right) * \left[PPV_{j,cty}^{YieldPriceElast} * PFER_{j,cty}^{YieldFertElast} * PWAG_{j,cty}^{YieldWageElast}\right],$$

Where,

Y	=	the yield for the cultivar of j in each FPU
PPV	=	the producer price
PFER	=	the price of fertilizer
PWAG	=	the cost of wages
a	=	the intrinsic productivity growth of yield
b	=	the cultivar specific productivity growth of yield
c	=	the biophysical effects on productivity growth due to climate change
YieldPriceElast	=	the own-price irrigated supply elasticity
YieldFertElast	=	the elasticity of the supply response with respect to fertilizer
YieldWageElast	=	the elasticity of the supply response with respect to wages
FPU	=	the food production unit index
cty	=	the country index
cul	=	the cultivar index
j	=	the activity index

4.3 Welfare Analysis

The welfare component of the calculations follows a traditional economic welfare analysis approach to estimate the benefits to society on the consumer- and producer-side. On the consumer-side this is straightforward, as the IMPACT model has a demand curve with demand elasticities, which allows us to calculate the consumer surplus. On the producer-side, it is not as straightforward, as the quantity supplied of each commodity is an area-yield equation, and does not represent the traditional supply curve that reflects the producer's marginal cost curve. Therefore, we have had to create synthesized supply-curves by land-type (irrigate, rainfed, other) for each activity and then calculate the producer surplus for each of these supply-curves and then aggregate to the national level. The total changes in consumer and producer surplus, when combined, provide us with a benefit flow, which we could use in a benefit-cost analysis, to compare a technology's overall impact in the agriculture sector.

4.3.1 Consumer Surplus

The demand curves in the IMPACT model has income and price elasticities, and is in the following general form:

$$QF_{c,cty} = \prod \left[(PCV_{c,cty})^{FDelas_{c,cty,c}} \right] * (pcGDP_{cty})^{IncDmdElas_{c,cty}} * pop_{cty} * dmdint_{c,cty}$$

Where,

$QF_{c,cty}$	=	Quantity demanded for commodity c
$PCV_{c,cty}$	=	Consumer price for commodity c
$pcGDP_{cty}$	=	National per capita GDP
pop_{cty}	=	National Population
$dmdint_{c,cty}$	=	Food Demand Intercept
$FDelas_{c,cty,c}$	=	Own-price elasticity for commodity c
$IncDmdElas_{c,cty}$	=	Income demand elasticity for commodity c

For each year and commodity, we compute the slope, m, in the equation below, of the straight line from the equilibrium point of the reference scenario (designated as subscript ref in the equations below) to the price axis using the food demand elasticity. In this calculation of the slope, we use the total quantity of food demand (QF) and the consumer prices (PC).

$$m_{ref} = \frac{1}{\varepsilon_{ref}} * \frac{p_{ref}}{q_{ref}}$$

Using this slope we can now calculate the price intercept of this line. The price intercept is the upper bound of price on consumption.

$$PInt_{ref} = p_{ref} - m_{ref} * q_{ref}$$

With the price intercept, we can now calculate the consumer surplus of the reference scenario, which will be used for all comparisons with different simulations.

$$CS_{ref} = \frac{1}{2} * (PInt_{ref} - p_{ref}) * q_{ref}$$

We envision changes between simulations and the reference scenario to be parallel shifts of the line formed by m_{ref} and the simulations' equilibrium point.

$$P_{simulation} = m_{ref} * q_{simulation} + PInt_{simulation}$$

We solve for $PInt_{simulation}$, which then allows us to compute the consumer surplus in the technology simulation.

$$CS_{simulation} = \frac{1}{2} * (PInt_{simulation} - p_{simulation}) * q_{simulation}$$

The change in consumer surplus between the simulation and the reference scenario is the difference of these two triangles.

To decompose the price and income effects we have to calculate the demand of the new simulation demand curve, but at the reference scenario prices, which we will call Q*

$$Q^* = \frac{p_{ref} - PInt_{simulation}}{m_{ref}}$$

Now, using Q* we can compute the areas of the price and income effects. First, we calculate the hypothetical consumer surplus if the equilibrium was at reference scenario prices and Q*.

$$CS_{Q^*} = \frac{1}{2} * (PInt_{simulation} - p_{ref}) * Q^*$$

Then we subtract triangles to calculate the price and income effects.

$$Price\ Effect = CS_{Q^*} - CS_{simulation}$$
$$Income\ Effect = CS_{Q^*} - CS_{ref}$$

To test if this decomposition is correct we can check to see if the following holds:

$$\Delta CS = Income\ Effect - Price\ Effect$$

4.3.2 Producer Surplus

To calculate the producer surplus we need to be able to calculate the area above the supply curve and under the equilibrium price. In effect, we calculate the agricultural revenue at the equilibrium point and subtract the total cost of production, which is the area under the supply curve. Without a traditional supply curve, derived directly from a marginal cost curve, we have to derive a supply-curve from IMPACT's area-yield functions, which generally speaking give us the quantity supplied (QS) in the following way.

$$QS = Area \times Yield$$

To calculate the total cost, we need to make QS a function of price. First the area and yield[4] equations as functions of their own-price (PP).

$$Area = K_{area} * PP^{\varepsilon_{area}}$$
$$Yield = K_{yield} * PP^{\varepsilon_{yield}}$$

Now we can make QS a direct function of its own-price.

$$QS = K * PP^{\varepsilon}, where$$
$$K = K_{area} \times K_{yield} \ and$$
$$\varepsilon = \varepsilon_{area} + \varepsilon_{yield}$$

We then get the inverse supply function.

$$PP = P(Q) = K^{(\frac{1}{\varepsilon}-1)} \times QS^{\frac{1}{\varepsilon}}$$

Now the inverse supply function we are ready to calculate the producer surplus (PS), which is agricultural revenue (AR), less the total cost (TC) of production, which is the area under the inverse supply function, which we can calculate by taking the integral of P(Q).[5]

---

[4] K_{yield} is a constant that includes growth rates, the IMPACT yield intercept, and the effects of input costs.

[5]
$$\int_0^{Q_0} P(Q) = \int_0^{Q_0} K^{\left(\frac{1}{\varepsilon}-1\right)} \times QS^{\frac{1}{\varepsilon}} = \frac{K^{\left(\frac{1}{\varepsilon}-1\right)}}{\frac{1}{\varepsilon}+1} \times QS^{\left(\frac{1}{\varepsilon}+1\right)}$$

$$= \left(QS \frac{1}{\frac{1}{\varepsilon}+1}\right) \times \left(K^{\left(\frac{1}{\varepsilon}-1\right)} \times QS^{\frac{1}{\varepsilon}}\right) = P(Q) \times \left(QS \frac{1}{\frac{1}{\varepsilon}+1}\right)$$

$$PS = AR - TC, \text{where}$$
$$AR = P \times QS \text{ and}$$
$$TC = \int_0^{Q_0} P(Q) = \frac{1}{\left(\frac{1}{\varepsilon}+1\right)} \times (P \times QS), so$$

$$PS = (P \times QS) - \left[\frac{1}{\left(\frac{1}{\varepsilon}+1\right)} \times (P \times QS)\right] = \left[1 - \frac{1}{\left(\frac{1}{\varepsilon}+1\right)}\right] \times P \times QS$$

$$= \left[\frac{\left(\frac{1}{\varepsilon}\right)}{\left(\frac{1}{\varepsilon}+1\right)}\right] \times P \times QS = \frac{1}{1+\varepsilon} \times P \times QS = \frac{P \times QS}{1+\varepsilon}$$

Using this equation, the producer surplus for all of the scenarios is calculated and compares change using the reference case to the new technology scenario.

$$\Delta PS = PS_{simulation} - PS_{ref}$$

4.4 Cost

The cost of developing and implementing a new crop cultivar is differentiated by the source of the funding, whether it is at the global or national level. Global costs are the costs of research and development that cannot be tied directly to any specific country. The role of research and development at CG centers is a good example of global costs, as the research done in developing new crop varieties is done for the benefit of many countries.

National costs are broken up into two different types of expenditures. First there is the cost of adapting a new crop variety or technology to the country-specific conditions. The cost is born at the country-level, often by national research institutions and universities. Second, the cost of agricultural extension required for diffusion of the new technology.

This bifurcation of the costs allows for a more nuanced analysis of benefit-costs at both the national and global level. The national cost cash flow does not include global costs. This makes the assumption that from the perspective of the country all work done at the global level (in CG centers) is a public good and is received by national research institutions free of charge. Global costs include both the global costs and the national costs.

4.5 Benefit-Cost Analysis

The Benefit-cost measures can only be used in simulations, where there is a cost component and a defined discount rate associated with a new technology. These measures can be broken up into indicators that compare simulations with their respective costs and observed changes in:

- Food Security
- Welfare

4.5.1 Food Security Measures

There are three food security measures, which provide insight into the effects of different simulations on food security. These measures compare simulations to find the greatest positive returns in improving food security. The following equations describe these measures:

- Food Availability: $\frac{Kcal_{simulation} - Kcal_{ref}}{NPV(Cost_{investment})}$
- Malnourished Children: $\frac{Malnourished_{simulation} - Malnourished_{ref}}{NPV(Cost_{investment})}$
- Share at Risk of Hunger: $\frac{Share_{simulation} - Share_{ref}}{NPV(Cost_{investment})}$

4.6 Welfare Measures

4.6.1 Net Benefits and Benefit-Cost Ratio

To allow for better comparisons between the benefits of different technologies, we need to discount the benefits over time and compute the present value of change in consumer surplus and agricultural revenue between simulations. We do this by discounting future benefits at a given discount rate (r) for the years that the simulation is run.

$$NPV(CS_{simulation}) = \sum_{i=1}^{n} \frac{\Delta CS^{i}_{simulation}}{(1+r)^{i}}$$

$$NPV(AR_{simulation}) = \sum_{i=1}^{n} \frac{\Delta AR^{i}_{simulation}}{(1+r)^{i}}$$

$$NPV(Total\ Benefits_{simulation}) = NPV(CS_{simulation}) + NPV(AR_{simulation})$$

We then need to do the same with cash flow of costs for implementing the changes in technology.

$$NPV(Cost_{simulation}) = \sum_{i=1}^{n} \frac{Cost^i_{simulation}}{(1+r)^i}$$

Once we have a total benefits measure and a total cost measure we can create the Benefit-Cost ratio and calculate the Net Benefits of the technology for each crop and country.

Benefit-Cost Ratio: $\dfrac{NPV(Total\ Benefits_{simulation})}{NPV(Cost_{simulation})}$

Net Benefits: $NPV(Total\ Benefits_{simulation}) - NPV(Cost_{simulation})$

Summing over countries or commodities provides measures by crop and country, globally by crop, national totals, and global total.

4.6.2 Internal Rate of Return

In addition to the net benefits measures, we can also compute the internal rates of return (IRR) of the technology simulations. The internal rate of return of the technology is the discount rate (r),[6] which makes the NPV of total cash flows (benefits – costs) equal zero.

$$NPV = \sum_{i=1}^{n} \frac{(\Delta CS^i_{simulation} + \Delta AR^i_{simulation}) - Cost^i_{simulation}}{(1+r)^i} = 0$$

5 Scenario Results

5.1 Economic and Social Benefits of the Sorghum Drought Tolerant Technology

The welfare benefit of the adoption of new drought tolerant cultivars of sorghum in the target countries/regions and its impact on world price, production, consumption, change in malnutrition and poverty is assessed using the IMPACT model. For this analysis, it is assumed that the drought tolerant technology will have 20 % higher yield advantage over the baseline technology (earlier used by farmers). A framework illustrating the technology development, dissemination, adoption pathway and it outcomes and impacts in the target countries as well as in the world is given in the Fig. 11.3.

---

[6] Traditionally, solving for r would require using a root solving algorithm (i.e. Secant Method, or Müller's Method). However, we can let the GAMS solver do the work for us, and solve for r by creating a basic model representing the previous relationship. As we are solving for a root, there is an additional requirement for computing the IRR. In addition to a cash flow, the time discounted benefits must be non-negative, meaning no IRR can be calculated for any simulations where the benefits do not at least match the cost of investment.

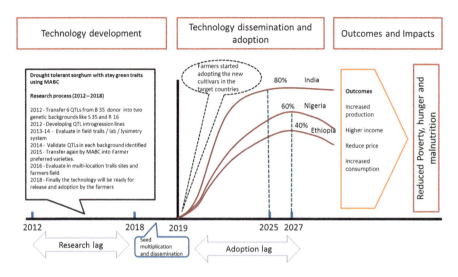

Fig. 11.3 The framework of technology development and adoption pathways linked with outcomes and impacts

In this study, we also assessed the change in the welfare benefits by adopting the new promising drought tolerant technology in different climate change scenarios. These climate scenarios used in the analysis are the MIROC (MIR A1B and B1) scenarios representing warmer and wetter climates while the CSIRO (CSI A1B and B1) scenarios represent the dry and relatively cool climates (Nelson et al. 2010).

5.2 Global Welfare Benefits Under Different Climate Change Scenarios

The likely global welfare benefits due to the adoption of drought tolerant sorghum cultivars under different climate change scenarios are given in the Table 11.4. The net welfare gains under the no climate change scenario are about $1,481 million. The global producers lose because of decrease in world market price, the negative producer surplus from the non- target countries (mainly the big exporting counties like USA, Australia, etc.) where new technology is not adopted is offsetting the positive producer surplus gained in the target countries where the new technology is adopted. The global consumers are gaining significantly due to decrease in the consumer price in the world market caused by the increased production. The net benefits under CSIRO climate scenario are higher than the without climate change (Table 11.4). Under both CSIRO climate scenarios, the adoption of new technology increased the global net welfare benefits and also higher than the no climate change condition.

Table 11.4 Global welfare benefits of drought tolerant sorghum technologies

Welfare and returns on investment	Climate change scenario				
	No climate change	MIROC 369 A1B	MIROC 369 B1	CSIRO 369 A1B	CSIRO 369 B1
Net Welfare change (NPV, m US$)	1,481.93	1,443.33	1,450.48	1,769.04	1,662.57
Cost (NPV, m US$)	5.06	5.06	5.06	5.06	5.06
Benefit-Cost ratio	292.79	285.16	286.58	349.52	328.48
Net benefits (NPV, m US$)	1,476.87	1,438.27	1,445.42	1,763.98	1,657.50
IRR (%)	59.03	58.35	58.52	60.56	60.01

Source: Authors' calculation

369 is the assumed CO_2 concentration by 2050 in ppm. A1B and B1 refer to the corresponding SRES climate change scenarios (Nelson et al. 2010)

Reported changes are over baseline, represented by the respective climate change scenario without the promising technology

The IRR for the sorghum drought research investment under no climate change condition is about 59 % and BC ratio is 292:1.

5.3 Economic Benefits in the Target Countries

The estimated net benefits of a sorghum drought tolerant cultivar developed and to be released in 2019 in the target countries under no climate change condition ranges from 692.8 US$ to 16.4 US$ depending upon the adoption rates and period until maximum adoption (Table 11.5). The net benefits are high in the larger sorghum producing countries like India, Nigeria and Sudan. The rate of return on research investments (i.e., IRR) for developing and releasing the new drought tolerant sorghum cultivar ranges from 134.2 % in India to 49.3 % in Eritrea (Table 11.5). These return on investment measures suggest that the returns to the development and release of drought tolerant sorghum variety in the target countries are worth the costs incurred. The net benefits of the sorghum drought tolerant cultivars in the target countries under climate change are higher than under no climate change condition. Even in the dryer CSIRO climatic scenario, the net benefits are higher than under no climatic change scenario. It reveals that the sorghum drought tolerant cultivar has produced higher yield and maintained the production level even in the drier climate in the target countries. In the wetter and high temperature MIROC climate scenario the economic benefit of the new technology is lower than the no climate change condition in the target region. The higher precipitation in the climate scenario contributed to the increased yield in rainfed sorghum, where more than 80 % of the sorghum area is under rainfed farming around the world.

Table 11.5 Net economic benefit drought tolerant sorghum cultivars in the target countries under climate change scenarios

Regions	Target countries	No climate change Net benefits in million US$	No climate change IRR (%)	MIROC 369 A1B Net benefits in million US$	MIROC 369 A1B IRR (%)	MIROC 369 B1 Net benefits in million US$	MIROC 369 B1 IRR (%)	CSIRO 369 A1B Net benefits in million US$	CSIRO 369 A1B IRR (%)	CSIRO 369 B1 Net benefits in million US$	CSIRO 369 B1 IRR (%)
WCA	Burkina Faso	90.7	75.8	82.4	74.1	80.0	73.9	99.8	76.4	100.8	76.6
	Mali	61.0	72.1	60.1	71.3	57.9	71.0	66.4	50.7	66.3	72.7
	Nigeria	479.6	116.0	526.7	116.7	513.6	116.4	584.6	90.0	528.1	116.6
ESA	Eritrea	16.4	49.3	15.8	49.5	16.8	49.5	19.2	134.2	18.2	50.3
	Ethiopia	138.8	89.6	121.4	87.5	117.8	87.2	150.1	72.5	140.1	89.1
	Sudan	187.6	87.4	205.0	87.2	200.8	87.1	226.0	117.7	218.1	88.6
	Tanzania	54.7	68.0	45.7	65.7	50.5	66.9	63.2	89.1	57.8	68.4
South Asia	India	692.8	134.2	547.1	133.6	578.3	133.8	776.3	69.4	722.3	134.1

Source: Authors' calculation

5.4 Change in Production, Consumption and Net Trade in the Target Countries

The IMPACT model projections of production, consumption, and net trade of sorghum in 2050 for no climate change scenario with and without drought tolerant technology intervention are presented in Table 11.6.

The model results suggest that in 2050 sorghum production and consumption in the target countries will be higher after a drought tolerant sorghum cultivar is developed and adopted as compared to the case where the variety was not developed and adopted by the farmers in the target countries. The results show that the percentage increase in production (i.e. change in production after the new technology adopted compare to baseline) ranges from 14.6 % in Eritrea to 4.9 % in Burkina Faso. It reveals that production in the target countries would have been smaller if the new drought tolerant cultivars are not developed and disseminated.

Table 11.6 Change in production, consumption and net trade in the target countries in 2050 after the adoption of drought tolerant sorghum in target countries

Regions	Target countries	Particulars	2010	Projected value in 2050 without new technology	Projected value in 2050 with new technology	% change
WCA	Burkina Faso	Production ('000 t)	1,637.3	3,985.1	4,179.0	4.9
		Consumption ('000 t)	1,480.7	3,930.0	3,986.9	1.4
		Net trade ('000 t)	159.7	58.2	195.3	
	Mali	Production ('000 t)	821.4	2,299.3	2,527.0	9.9
		Consumption ('000 t)	733.1	1,861.7	1,887.8	1.4
		Net trade ('000 t)	64.3	413.7	615.2	
	Nigeria	Production ('000 t)	9,428.5	18,103.5	19,590.2	8.2
		Consumption ('000 t)	8,801.1	21,660.1	22,049.0	1.8
		Net trade ('000 t)	488.6	−3,695.5	−2,597.7	
ESA	Eritrea	Production ('000 t)	109.3	241.2	276.9	14.8
		Consumption ('000 t)	235.4	539.6	548.5	1.6
		Net trade ('000 t)	−72.0	−244.3	−217.4	
	Ethiopia	Production ('000 t)	2,482.2	6,332.4	6,663.8	5.2
		Consumption ('000 t)	2,015.6	4,805.6	4,884.7	1.6
		Net trade ('000 t)	676.1	1,736.3	1,988.6	
	Sudan	Production ('000 t)	4,881.7	10,396.4	10,954.9	5.4
		Consumption ('000 t)	3,890.6	7,796.2	7,923.9	1.6
		Net trade ('000 t)	1,201.2	2,810.3	3,241.1	
	Tanzania	Production ('000 t)	997.6	3,492.8	3,675.4	5.2
		Consumption ('000 t)	767.5	2,443.9	2,516.5	3.0
		Net trade ('000 t)	232.8	1,051.6	1,161.6	
South Asia	India	Production ('000 t)	7,953.4	10,345.0	11,713.6	13.2
		Consumption ('000 t)	8,028.3	10,223.2	10,430.3	2.0
		Net trade ('000 t)	−371.6	−174.9	986.7	

Source: Authors' calculation

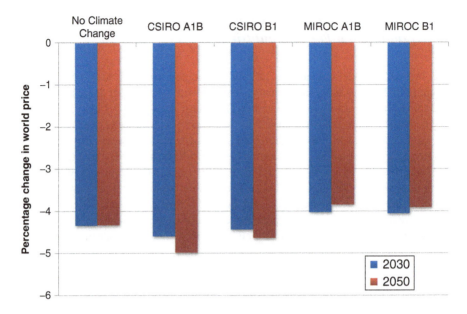

Fig. 11.4 Change in the World price of sorghum under different climate scenarios

Due to increase in the sorghum production of the target countries after the adoption of new technology, the world price of sorghum reduces by 4.3 % in 2050 under no climate change condition (Fig. 11.4). Among the climate change scenarios, CSIRO A1B scenario experiences a reduction in world price by 4.97 %. The lower sorghum world price has reduced the consumer price in the target countries as well as in other non-target countries. Because of decrease in the country level consumer price, the demand for sorghum consumption has slightly increased in the target countries ranging from 1.4 % in Burkina Faso and Mali to 3 % in Tanzania (Table 11.6).

5.5 Spillover Benefits of the Sorghum Technology Intervention in Non-target Countries

The estimated positive net economic benefits for countries[7] after the technology intervention under no climate change in presented in the Fig. 11.5. The decrease in consumer price of sorghum after the technological intervention has benefitted consumer around the world. The positive net benefits for some of the non-targeting countries has revealed that the consumers gained by the price

---

[7] The countries with net benefits greater other five million US$ is presented in the Fig. 11.5.

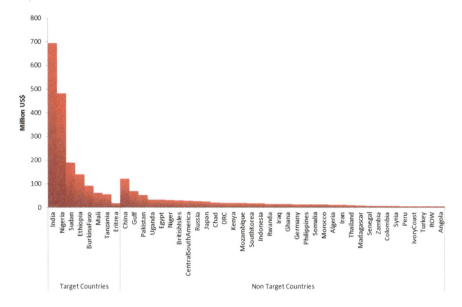

Fig. 11.5 The net welfare benefits (million US$) for the target countries and non-target countries under no climate change

spillover effect of the drought tolerant sorghum technology adoption. Countries like Niger, Chad, Somalia, etc., where sorghum is consumed as staple food has benefited due to price spillover effects of the technology intervention.

6 Summary and Conclusion

In this study we used the integrated modeling framework – IMPACT – which integrates partial equilibrium economic model, hydrology model, crop simulation model and climate model to examine the ex-ante economic impact of developing and disseminating a drought tolerant sorghum cultivar in target countries of Africa and Asia under no climate change and two different climate change scenarios (MIROC and CSIRO GCMs). Specifically, we estimated the potential yield advantage of the promising new drought tolerant sorghum cultivars over the baseline cultivar using crop simulation model and its impact on production, consumption, trade flow, prices of sorghum and welfare indicators like change in poverty, malnourished children and change in the number people under hunger risk in target countries and as well as the non-target countries. And also we estimated the returns to research investment for developing the promising new drought tolerant cultivars and dissemination in the target countries.

The analysis indicates that the economic benefits of drought tolerant sorghum cultivar adoption in the target countries outweighs the cost of developing this new technology. The development and release of this new technology in the target countries of Asia and Africa would provide a net economic benefit of about 1,476.8 million US$ for the entire world under no climate change condition. Under climate change scenarios the net benefits derived from adoption of new drought tolerant sorghum cultivar is higher than the no climate change condition. This is due to higher production realized by sorghum under climate change scenarios.

In addition, results IMPACT model projections suggest that the new technology intervention reduced the children malnourished under the age group of 5 years in the target countries ranging from 97,114 in Nigeria to about 2,198 children in Eritrea for a million UD$ investment.

These results imply that substantial economic benefits can be achieved from the development of a drought tolerant sorghum cultivar. And also this technology will perform better than the existing cultivars in future climate change condition. Thus, we strongly encourage policy makers and donors to fund the sorghum research to develop more tolerant to droughts, so that farmers can better cope and adapt to changing climate in the future.

References

Ashok Kumar A, Reddy BVS, Sharma HC, Hash CT, Srinivasa Rao P, Ramaiah B, Sanjana Reddy P (2011) Recent advances in Sorghum genetic enhancement research at ICRISAT. Am J Plant Sci 2011(2):589–600

Bidinger FR, Hammer GL, Muchow RC (1996) The physiological basis of genotype by environment interaction in crop adaptation. In: Cooper M, Hammer GL (eds) Plant adaptation and crop improvement. CAB International, Wallingford, pp 329–347

Nagy ZZ, Tuba F, Soldos Z, Erdei L (1995) CO_2-exchange and water relation responses of Sorghum and maize during water and salt stress. J Plant Physiol 145(4):539–544

Nelson GC, Rosegrant MW, Palazzo A, Gray I, Ingersoll C, Robertson R, Tokgoz S, Zhu T, Sulser TB, Ringler C, Msangi S, You L (2010) Food security, farming, and climate change to 2050: scenarios, results, policy options. Climate monographs, International Food Policy Research Institute, Washington, DC

Reddy BVS, Ramaiah B, Kumar AA (2007) Evaluation of Sorghum genotypes for stay-green trait and grain yield. E-J SAT Agric Res 3(Suppl 1). http://www.icrisat.org/journal/

Richard R, Nelson GC, Thomas T, Rosegrant M (2013) Incorporating process based crop simulation models into global economic analyses. Am J Agric Econ 95(2):228–235

Part III
Impacts of Climate Change on Water Resources: Relevant Adaptation Practices

Chapter 12
Impact of Climate Changes on Water Resources in Algeria

Nadir Benhamiche, Khodir Madani, and Benoit Laignel

Abstract This work outlines the methodology followed in the study of climate change impact on water resources in North Algeria and presents the results on the vulnerability of the water resources. It focuses on research efforts of old data and homogenization of long series of climate data since 1926 at stations in the watersheds of North Algeria. We were able to establish a database homogenized precipitation and temperatures. Recent trends show mixed but generally increasing, except in the south basin that reflects a sharpening of the drought. We then conducted a study of impact of climate change on the hydrological behavior of watersheds. Of all the simulation models used in the world, only two global models UKHI-EQ and ECHAM3TR give an acceptable results. Throughout this paper, it clear that uncertainty surrounds our understanding of future climate change and its impacts.

Keywords Climate change • Modeling • Water resources • Climate scenarios • Algeria

1 Introduction

An analysis on a worldwide scale of the annual and seasonal changes of precipitation suggest that the two most outstanding features of the second half of the twentieth century would be the increase in precipitations, up to 20 % in Northern Russia, and especially the reduction of rains in West Africa (D'Orgeval 2006) and East Africa

---

N. Benhamiche (✉) • K. Madani
Laboratoire de Biomathématique, Biophysique, Biochimie, et Scientométrie, Faculté des Sciences de la Nature et de la vie, Université de Bejaia, 06000 Bejaia, Algérie
e-mail: nadirbenhamiche@yahoo.fr; madani28dz2002@yahoo.fr

B. Laignel
UMR M2C CNRS 6143, Département de Géologie, Université de Rouen,
76 821 Mont-Saint-Aignan, Cedex France
e-mail: benoit.laignel@univ-rouen.fr

(sahel) (Hulme 1992, 1994; Hulme et al. 1992, 1998, 2001; Viner & Hulme 1994) ranging between 20 and 50 % (Bradley et al. 1987). Regarding Algeria, with a total area of approximately 2.4 million km^2, the pluviometry concerns only 10 % of this area, which is divided into three zones: The Northern one (500 mm/year), the High plateaus zone (300 mm/year) and the Southern Atlas zone (250 mm/year) (ANRH and ONM); in average, the country receives 100 billion m^3 of rain per annum, of which 85 % evaporate and the remaining 15 % runs on the surface to rivers and the sea, or infiltrates inside the underground layers (Sari 2009). The economically mobilizable quantities of water for the various uses of the population are evaluated to 5.7 billion m^3 of surface water, 1.8 billion m^3 of subterranean water in the North, 4.9 billion m^3 in the South, which makes a total of 12.4 billion m^3 (Sari 2009). Because of the limited water resources, and the need to meet the demands for the desired quantity and quality of water by the planners must develop reasonable alternatives that take into account multiple purposes and objectives (Boudjadja et al. 2003) Algeria, though very little emitting of greenhouse gases (1.5–3.5 TE CO_2/Inhabitant/year), is vulnerable to climate change because of its geographical location in a semi-arid to arid zone (Viner and Hulme 1993a, b). The aim of work is highlighting of the impact of climate change on water resources in North Algeria and presents the results on the vulnerability of the water resources.

2 Study Area

The Northern Algeria was the subject of this study. It covers the basins slopes of Oranie, Algiers and Constantine. This zone presents climatic characteristics. It extends from the west to the east over a length of more than 1,000 km, whereas it extends to 450 km between north and south (Fig. 12.1)

3 Methodology

3.1 The Climate Changes and the Simulation Models in Algeria

In Algeria, studies based on analysis of experimental data have quantified the relationship between cause and effect between climate disruption and the recent availability of water resources. In relation to future climate change, an attempt of calculating a shortfall of water supplies due to climate change in 2020 was conducted as part of developing the national strategy and national action plan for climate change (IPCC 2001). This study is based on the exploitation of two global climate models: the model UKHI – EQ, designed in 1989 by the British Meteorological Service, and ECHAM3TR model, developed in Germany in 1995 by the Max Planck Institute. In a first step, climate projections were developed at a seasonal scale (temperature and precipitation) in the form of cards. In a second step,

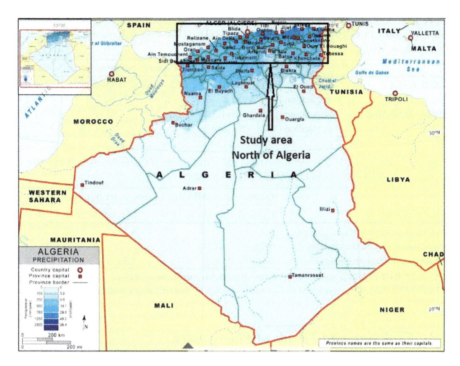

Fig. 12.1 Location of study area

and from a statistical analysis on the relationship Rainfall – runoff and climate scenarios, an estimate of the deficit in water supplies, based on total amount of surface water mobilized was made.

4 Results and Discussion

4.1 Climatic Projections

These are not forecasts but estimates of the possible evolution of the climate. For Algeria, the results give, by 2020, temperature risings of 0.5–1.5 °C and precipitation decrease of 5–20 %, and by 2050, temperature risings of 1.5–2.5 °C and a precipitation decrease of 10–50 %. An increase of 1–2 °C in temperatures can induce a reduction of 10 % in precipitations (Table 12.1). This will have important effects on the mobilization of dam waters up to 0.64 billion m^3 (Table 12.2).

4.2 Disturbances of Seasons

This disturbance will be expressed as a rising in temperatures and a concentration of precipitations over one short period increasing thus risks of flooding and an elevation in the frequency of droughts.

Table 12.1 Seasonal climate projections of temperature and rainfall in Algeria by 2020 for two GIEC models (IPCC 1990), (UKHI and ECHAM3TR) (Ministry of Regional Planning and Environment, 2001)

		Autumn	Winter	Spring	Summer
UKHI Model	Temperature (°C)	0.8–1.1	0.65–0.8	0.85–0.95	0.85–1.05
	Precipitation (%)	−6 à −8	−10	−5 à −9	−8 à −13
ECHAM3TR Model	Temperature (°C)	0.8–1.3	0.9–1	0.95–1.1	0.95–1.45
	Precipitation (%)	No change	−5	−7 à −10	−5

Table 12.2 Projection by 2020 of mobilizable surface water quantities in Algeria (ANRH 1993) according to UKHI (1989) and ECHAM3TR (1995) models**

	Quantity of mobilizable surface water (billion m^3)	
Type of projection	Optimal scenario (−10 %)	Optimal scenario (−20 %)
Projection without climate changes	6.4	6.4
Projection with climate changes	5.76	5.12
Incidences of climate changes	0.64	1.26

* Model UKHI-EQ, designed in 1989 by the British Meteorological Service
* Model ECHAM3TR, developed in Germany in 1995 by Max Planck Institute

4.3 Evolution and Trends of Pluviometry

4.3.1 Temporal Evolution of Pluviometry (1923–2006)

The tendency from 1923 through 1938 is to an overflow in rainfalls. This surplus is of 9 % in the West (Station of Oran: station 1), 17.6 % in the East (Station of Constantine: station 3) and 12.3 % in the Center (Station of Algiers: station 2). In 1939, started a dry period that stretched until 1946 to reach deficits of 14.5 % in the West and 10.2 % in the Center. In the East a surplus of 6.7 % was recorded. The period 1947 to 1973 was characterized by a wet period with an excess of 13.1 % in the Center. The driest periods were observed during 1949–1956 and 1960. Whereas in the West, the surplus is of 17.9 %. For the East of the country, the pluviometric surplus is a little lesser (4.5 %). Starting from 1974, the tendency is clearly to the dryness. The difference between the East and the other parts of the Country lies mainly in the intensity of this dryness. The rainfall deficit is of about 13 % in the East, 13.6 % in the Center and 16.1 % in the West. Since 2000, the pluviometry remained deficient for all the areas except in the East and the Center where, afterwards (2002), the tendency was to wetness with 0.4–2.7 % average deviation. In the Center and the East, this deviation is of 0–51.1 %. For the West, the deficits are of 3.2–17.7 % in comparison to the mean values. From 2002 until now, the pluviometry approaches the normal level for the Western and Central areas. On the other hand, in the East the pluviometry tends to increase. The general tendency in the North of the Country is to a decrease of pluviometry to become more marked since the middle of the 1970s (Béthoux and Gentili 1996, 1999).

The examination of the maps established by Chaumont for the period 1913–1963 (Fig. 12.1) and those of ANRH for the periods 1922–1989 (Fig. 12.2) and 1965–2004

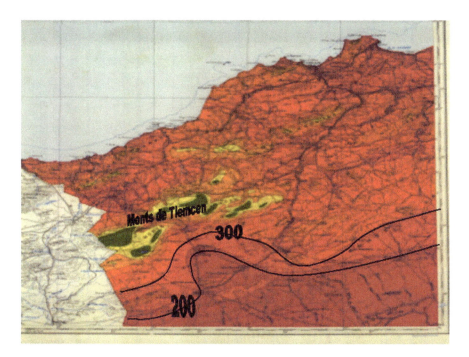

Fig. 12.2 Map according to Chaumont for the period 1913–1963

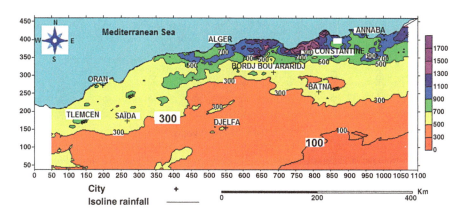

Fig. 12.3 Map according to ANRH for the period 1922–1989 (ANRH)

(Fig. 12.3), show that the isolines evolve in a significant way towards the North. This evolution is an indicator of climate change in Algeria. Indeed, the examination of the isolines 100, 200 and 300 mm shows that displacement towards the North can reach distances of more than 100 km.

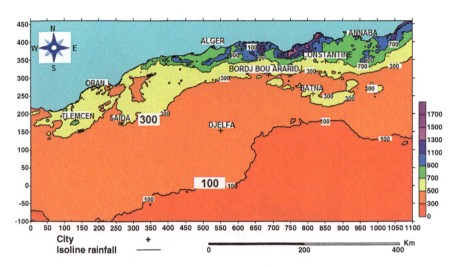

Fig. 12.4 Map according to ANRH for the period 1965–2004 (ANRH)

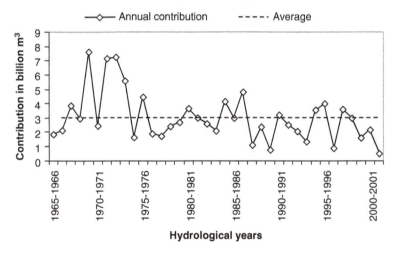

Fig. 12.5 Evolution of liquid contributions of catchment areas around Algiers (*Center*) (1965–2003) (ANRH)

4.4 Impact on Overland Flows

The climate is characterized by an important recurrence of phases of dryness and floodings. It influences directly the overland flows marked by the extreme seasonal and inter-annual irregularity of the liquid contributions characterized by the violence and the speed of rising waters and the severity of the low water levels. The impact on water flows is very significant since 1974, particularly in the west (Figs. 12.4, 12.5, 12.6, and 12.7). For this purpose, the surface

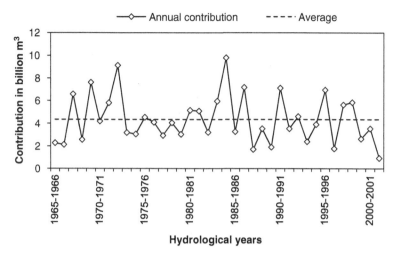

Fig. 12.6 Evolution of liquid contributions of catchment areas around Constantine (East) (1965–2003) (ANRH)

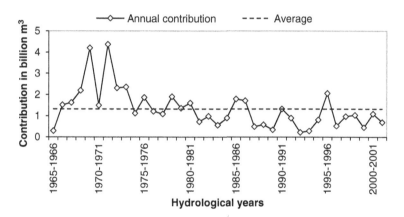

Fig. 12.7 Evolution of liquid contributions of catchment areas around Cheliff (Center-West) (1965–2003) (ANRH)

water resource, at the end of the 1970s, was evaluated to 13.5 billion m^3. From 1980 to 1990, it was estimated at 12.4 billion m^3. Currently, it is of 9.7 billion m^3 (Sari 2009).

4.5 Impact on Liquid Contributions to Dams

The impact of climatic change in Algeria during the last 25 years is expressed by a diminution of dam levels. Overflow volumes decreased in certain cases by more than 50 % (Dam of Ksob in M'sila: Fig. 12.8 – Cheffia in El Tarf Fig. 12.9 – Algeria).

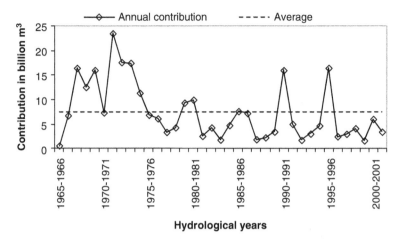

Fig. 12.8 Evolution of liquid contributions of catchment areas around Oran (West) (1965–2003) (ANRH)

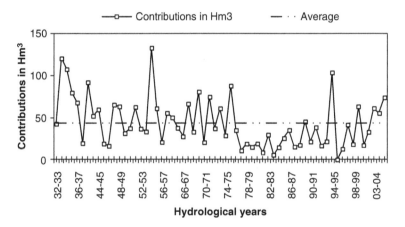

Fig. 12.9 Decreased volume in the dam of Ksob: area of M'sila (South-East of Algeria) (1932–2006) (ANRH)

Both dams, as an example, are located in a region with semiarid climate. The evapotranspiration estimated by the following formula of M. Coutagne, giving a mean value of 314 mm. The annual means rains are 341 mm. 92 % of this rain is which returned to the atmosphere (Gouaidia 2008).

4.6 Impact on Erosion and Sedimentation

The climate changes have significant consequences on erosion on height slopes and wadi beds (Achite 1999, 2002; Achite and Meddi 2004). This is expressed by a

12 Impact of Climate Changes on Water Resources in Algeria   201

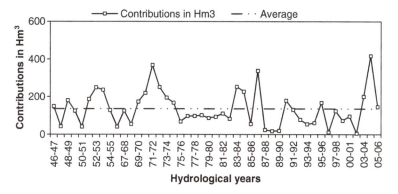

Fig. 12.10 Decreased volume in the dam of Cheffia: area of Tarf (East of Algeria) (1946–2006) (ANRH)

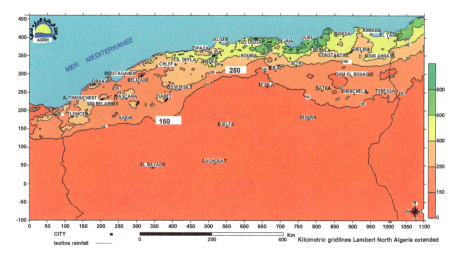

Fig. 12.11 Map of average annual rainfall in Northern Algeria – Decennial frequency **dry** (1965–**2001**) (ANRH)

tendency to an increase in the erosion phenomenon. Specific erosion reached 2,000 t/km^2/year in our basins (Tellian areas representing 45 % of the farming grounds (12 M – ha)). Also, it is estimated about 120 million tons of sediments rejected into the sea each year (Touaibia 2010).

4.7 Impact on the Extreme Phenomena

The climate changes are accompanied by extreme phenomena like brutal and devastating floodings (Figs. 12.12 and 12.13). In order to highlight the importance of these phenomena, the maps of decennial average rains over the dry and wet

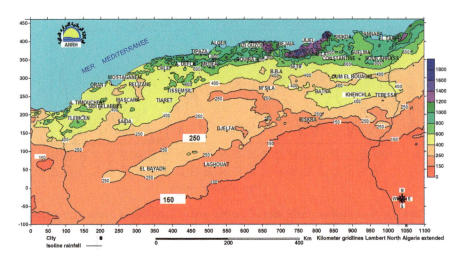

Fig. 12.12 Map of average annual rainfall in Northern Algeria – Decennial frequency **wet** (1965–2004) (ANRH)

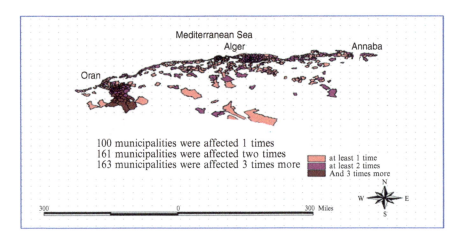

Fig. 12.13 Municipalities victim of floodings during the 3 last years (2003–2005) (ANRH)

periods were taken into account. The examination of these maps (Figs. 12.10 and 12.11) reveals a great fluctuation and a significant difference (decrease) between the dry and wet periods (shift of some isolines by more than 200 km). This instability implies that the phenomena of dryness and floodings have a severe impact on some areas of Algeria. These extreme variations translate extreme climatic conditions compared to the average deviation of the climatic system, which often provoke natural disasters (Figs. 12.12, 12.13, and 12.14).

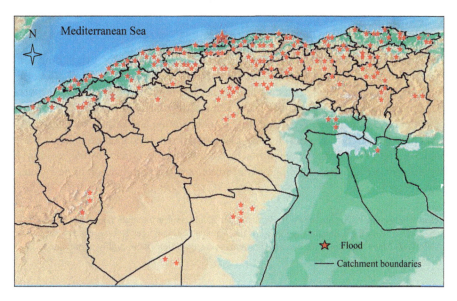

Fig. 12.14 Localization of the most important floodings during 10 last years (1973–2005) (ANRH) Behlouli (2009)

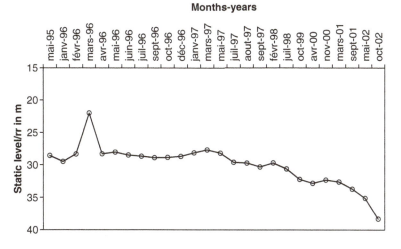

Fig. 12.15 Evolution of the groundwater depth of Mitidja – Area of Algiers (1995–2002) – Piezometer PZ N°1 Hamiz/21 (ANRH 2004)

4.8 Impacts on Underground Water Resources

The reduction in pluviometry and the increase in temperatures were directly influence the groundwater recharge and generate a significant decrease of water contributions leading to a folding back of groundwater levels (Fig. 12.15).

5 Conclusion

The water, now considered as a rare and invaluable food product, constitutes an essential element for the life and the balance of the individual. It represents a determining factor for the economic and social development of a country. Because of its precariousness, vulnerability and, even its irregularity, this resource requires in particular a special attention regarding its mobilization and with its management. The climate changes represent, for Algeria, a major risk if adaptation measures are not taken right now through the optimal use of the water resources, by the economy, the efficient use and the recycling of wastewater and the mobilization of new water resources.

Acknowledgments The authors warmly thank Mr Sahnoune for translation assistance, Mr Ould mara Arezki from the Climatology Office of the Agence Nationale des Ressources Hydrauliques (ANRH) in Alger, and reviewer whose comments and suggestions were highly appreciated.

References

Achite M (1999) Analyse multivariee de la variable Erosion Spécifique: Cas du bassin versant de l'Oued Mina. Magister thesis report, Ecole Nationale Supérieure d'Hydraulique, Blida, Algérie, p 206

Achite M (2002) Statistical approach for evaluation of solid transport in the catchment area of the Wadi Mina (Algeria). In: Proceedings of the international conference on water in the mediterranean basin, Monastir (Tunisia), 10–13 Oct 2002, pp 894–899

Achite M, Meddi M (2004) Estimation du transport solide dans le bassin versant de l'oued Haddad (Nord-Ouest algérien). Sécheresse 15(4):367–373

ANRH (1993) Carte pluviométrique de l'Algérie du Nord, Ministère de l'Equipement, PNUD/ALG/88/021, Alger, 1 carte et sa notice de 54p

ANRH-GTZ (2004) Étude de synthèse sur les ressources en eaux de surface de l'Algérie du Nord (Rapport d'étude). Agence nationale des ressources hydrauliques (ANRH), Alger

Behlouli L (2009) Crues et inondations en Algérie (ANRH). Actes de l'atelier scientifique et technique sur l'outil spatial au service du développement. Session 1: Prévention et gestion des risques majeurs. Palais de la culture Moufdi Zakaria, Alger, 28–29 March 2009, pp 17–25

Béthoux JP, Gentili B (1996) The Mediterranean Sea, coastal and deep-sea signatures of climatic and environmental changes. J Mar Syst 7:383–394

Béthoux JP, Gentili JP (1999) Functioning of the Mediterranean Sea: past and present changes related to freshwater input and climate changes. J Mar Syst 20(1–4):33–47

Boudjadja A, Messahel M, Pauc H (2003) Ressources hydriques en Algérie du Nord. J Water Sci 16(3):285–304

Bradley RS, Diaz HF, Eischeid JK, Jones PD, Kelly PM, Goodess CM (1987) Precipitation fluctuations over northern hemisphere land areas since the mid-19th century. Science 237:171–175

d'Orgeval T (2006) Impact du changement climatique sur le cycle de l'eau en Afrique de l'Ouest: Modélisation et Incertitudes. Thèse de doctorat de l'Université de Paris VI, Sciences de la terre, p 176

Gouaidia L (2008) Influence de la lithologie et des conditions climatiques sur la variation des paramètres physico-chimiques des eaux d'une nappe de Meskiana – Nord Est algérien. Thèse de doctorat és-sciences, pp 20–37

Hulme M (1992) A 1951–80 global land precipitation climatology for the evaluation of general circulation models. Clim Dyn 7:57–72

Hulme M (1994) Regional climate change scenarios based on IPCC emissions projections with some illustrations for Africa. Area 26(1):33–44

Hulme M, Marsh R, Jones PP (1992) Global changes in humidity index between 1931–1960 and 1961–1990. Clim Res 2:1–22

Hulme M, Osborn TJ, Johns TC (1998) Precipitation sensitivity to global warming: comparison of observations with HadCM2 simulations. Geophys Res Lett 25:3379–3382

Hulme M, Doherty R, Ngara T, New M, Lister D (2001) African climate change: 1900–2100. Clim Res 17:145–168

IPCC (Intergovernmental Panel on Climate Change) (1990) Climate change – the IPCC scientific assessment. Cambridge University Press, Cambridge, Uk

IPCC (2001) Climate Change 2001 impacts, adaptation, and vulnerability, ch 8. Report of WG II

Ministère de l'Aménagement du Territoire et de l'Environnement (2001) Elaboration de la stratégie et du plan d'action national des changements climatiques. Projet national ALG/98/G31. Rapport de la Direction générale de l'Environnement

Sari A (2009) Hydrologie de surface, cours d'initiation. Edition distribution Houma, pp 15–17

Touaibia B (2010) Problématique de l'érosion et du transport solide en Algérie septentrionale. Sécheresse 21(1e):1–6

Viner D, Hulme M (1993a) The UK met office high resolution GCM equilibrium experiment (UKHI). University of East Anglia, Climate Research Unit, Norwich, UK, Technical note no. 2, 19 p

Viner D, Hulme M (1993b) Construction of climate change scenarios by linking GCM and STUGE output. University of East Anglia, Climate Research Unit, Norwich, 24 p

Viner D, Hulme M (1994) The climate impacts link project, University of East Anglia, Climate Research Unit, Report to Department of the Environment, Norwich, 24 p

Chapter 13
Drought Tolerance of Different Wheat Species (*Triticum* L.)

Petr Konvalina, Karel Suchý, Zdeněk Stehno,
Ivana Capouchová, and Jan Moudrý

Abstract Water deficiency is one of the factors limiting the efficiency wheat growing all over the world. In the context of global climate changes, a question of the tolerance to drought has become more important in regions where this aspect was never taken into account before in the breeding process. Our paper presents a comparison of various wheat landraces (einkorn, emmer, spelt) and bread wheat varieties. It is based on the results of $\delta^{13}C$, chlorophyll content and quantity of stomata. The research was carried out in a greenhouse in different soil humidity (10 and 20 %). The results of our research have shown a relation between low values of $\delta^{13}C$, high proportion of chlorophyll and low quantity of pores (the predisposition to the tolerance to drought). In general, einkorn varieties, emmer and spelt wheat varieties have been the most predisposed to the tolerance to drought. On the other hand, intermediate wheat forms and modern bread wheat varieties have been supposed to be less tolerant to drought. The emmer wheat variety Rudico has become the most prospective variety of all ($\delta^{13}C = -26.93$ ‰; proportion of chlorophyll = 1.55 mg/dm; quantity of pores = 21.67 per square mm).

Keywords Droughtiness • Wheat • ^{13}C discrimination • Chlorophyll content • Stomatal characteristics

---

P. Konvalina (✉) • K. Suchý • J. Moudrý
Faculty of Agriculture, University of South Bohemia in České Budějovice, Studentská 13, 370 05 České Budějovice, Czech Republic
e-mail: konvalina@zf.jcu.cz; petr.konvalina@gmail.com; suchy@zf.jcu.cz; moudry@zf.jcu.cz

Z. Stehno
Faculty of Agriculture, Crop Research Institute, Drnovská 507, 161 06 Prague 6, Czech Republic
e-mail: stehno@vurv.cz; stehnoz@seznam.cz

I. Capouchová
Faculty of Agrobiology, Food and Natural Resources, Department of Crop Production, Czech University of Life Sciences, Kamýcká 120, 165 21 Prague 6, Czech Republic
e-mail: capouchova@af.czu.cz

1 Introduction

The agricultural sector is highly sensitive to climate change (Swearingen 1992). The most important climatic element is precipitation, particularly seasonal drought and the length of the growing season (Watson et al. 1997). Deficiency of water is one of the main factors limiting the productivity of crops (Peleg et al. 2005). The effects of climatic variations on African agriculture have been well established through decades of field experiments, statistical analyses of observed yields, and monitoring of agricultural production (Watson et al. 1997). The risk of water shortage to food security in Morocco is increasing due to the dual pressure of drought and domestic and industrial demands (Balaghi et al. 2007). Also in Europe the occurrence of serious drought has been relatively low. Global climate change has been modifying the perspectives and water requirements now need to be considered in the development of a new generation of cultivars. Droughtiness has been consequently poorly considered in breeding programs (Bucur and Savu 2006). Not only potential yield but also stability of yield under a large range of climatic conditions have to be increasingly taken into account (Konvalina et al. 2010). It is necessary to use new genetic and physiological technologies which will restore wheat productivity in developing world areas vulnerable to climate-change-induced heat and drought stress (CIMMYT 2011).

In important crops such as wheat (*Triticum* L.), a screening of cultivars and landraces leading to an identification of potential donors of drought tolerance is relevant and urgent. This requires a comprehensive exploration of potential genetic resources and an in-depth understanding of their adaptation mechanisms and responses to water stress (Peleg et al. 2005). Wheat landraces are often tolerant to local stress factors (Reynolds et al. 2007). Genetic resources of wheat are used as donors of genes in the wheat breeding process. Tolerance to drought, which is often presented in scientific literature, is one valuable characteristic that landraces possess.

Because the direct testing of genotypes of crops in dry conditions is quite expensive, other indirect methods of selection and evaluation have been proposed such as the discrimination of $^{13}CO_2$ (Farquhar and Richards 1984; O'Leary 1993; Araus et al. 2002). The values of discrimination $^{13}CO_2$ is possible to present as a carbon isotope ratio *(δ) (negative value) or carbon isotope discrimination* (Δ) (positive value). The discrimination $^{13}CO_2$ may be represented by the difference between ^{13}C and ^{12}C content in the standard and experimental sample. A negative correlation between the carbon isotope discrimination and the efficiency of the integrated transpiration has been reported (Rebetzke et al. 2002). This method has consequently been proposed as an indirect method for estimating water-use efficiency evaluation (WUE) (Araus et al. 2002).

2 Objective of Study

The objective of this study was to explore wheat genetic resources as a source for improving drought resistance in cultivated wheats. In this paper we report (1) the results of carbon isotope discrimination, represented by ratios as a tool for the selection of prospective drought resistant varieties and (2) a chlorophyll content and stomatal characteristics comparison between genetic resources of wheat and modern varieties of bread wheat.

3 Materials and Methods

3.1 Plant Material

The plant material used in the present study came from the Gene Bank of the Research Institute of Crop Production in Prague-Ruzyně (Prague). It comprised three accessions of einkorn (*Triticum monococum* L.), emmer wheat (*Triticum dicoccum* (SCHRANK) SCHUEBL), spelt wheat (*Triticum spelta* L.), intermediate form of bread wheat (*Triticum aestivum* L.) and two modern varieties of bread wheat (*Triticum aestivum* L.). More information about eco-geographical origin are given in Table 13.1. Varieties were sown in three replications in a greenhouse pot experiment with and without water deficit conditions. Concerning the abundance of water, the soil humidity of 20 % was maintained there. Concerning the deficiency of water, the soil humidity of 10 % was maintained there. A ECH2O check hygrometer was used to indicate the soil humidity percentage.

Table 13.1 List of varieties

Field number	ECN identifier	Name of variety	Origin
Einkorn (*Triticum monococcum* L.)			
J1	01C0204038	T. monococcum	Georgia
J4	01C0204542	T. monococcum No.8910	Denmark
J6	01C0204053	Schwedisches Einkorn	Sweden
Emmer [*Triticum diccocum* (Schrank)Schuebl]			
Rudico	01C0200948	Rudico	Czech Republic
D11	01C0203993	Weisser Sommer	Germany
D14	01C0204016	T. dicoccon (Dagestan. ASSR)	Russian Federation
Spelt (*Triticum spelta* L.)			
SP6	01C0204865	T. spelta (VIR St. Petersburg)	Czech Republic
SP7	01C0200982	Spalda bila jarni	Czech Republic
SP9	01C0200984	T. spelta (Kew)	Not known
Bread wheat (*Triticum aestivum* L.) – intermediate forms			
P2	01C0200051	Rosamova ceska cervena presivka	Czech Republic
P3	01C0100124	Cervena perla	Czech Republic
P4	01C0200031	Kasticka presivka 203	Czech Republic
Bread wheat (*Triticum aestivum* L.) – control			
Jara	01C0200100	Jara	Czech Republic
SW Kadrilj	01C0204877	SW Kadrilj	Sweden

3.2 ^{13}C Analysis

Samples (grains) were taken from fully-ripened harvested plants from small-plot trials. They were dried till they achieved a constant weight. These dry samples were powderized and burnt in an oxygenized atmosphere. The produced gases were separated, cleaned and reduced. Thanks to the Thermo-conductive detector (TCD), the gases were analysed in the elementary analyzer (EuroEA 3028-HT called Eurovector), and indicated the total proportion of elements. The separated gases went through the mass spectrometer (IRMS Isoprime) where each gas type containing the isotopes was diverted into the magnetic field according to its molecular weight. The analyses of isotopes, found in the plant samples, were carried out with the following instruments: Eurovector Elemental Analyser for IRMS (EuroEA 3028-HT) and Isotope Ratio Mass Spectrometer (IRMS Isoprime). The international standards, PDB for $^{13}C/^{12}C$, were applied to the indication of isotopes in the samples. $\delta^{13}C$ (‰) characteristic was used as an indicator of ^{13}C discrimination (Wright and Nageswara Rao 1994); the values were compared to the standard (Pee Dee limestone from South Carolina was compared to $^{13}C/^{12}C = R$ standard $= 0.011237$ ($= 1.1237$ %). $\delta^{13}C = (R$ sample/R standard $- 1) \times 1,000$ ($\delta^{13}C$ in ‰). $\delta^{13}C = -8.0$ ‰ in CO_2 atmosphere, $\delta^{13}C = -25$ to -30 ‰ in the dry matter of C3 plants). The plants, grown in water-friendly and favorable conditions, are usually characterised by $\delta^{13}C$ ratio close to -30‰ (Farquhar and Richards 1984).

3.3 Proportion of Chlorophyll

The proportion of chlorophyll was colorimetrically set up at Spekol. Leaves were sampled in the flowering period, mixed with silica sand, $MgCO_3$ and acetone and pulverized together. 3 ml of acetone (80 %) was added into the mixture and sieved through S3 sinter. Furthermore, the flask (10 ml) was filled with acetone (80 %). The absorbance of the chlorophyll extracts was measured at Spekol (663 and 645 nm). The total proportion of chlorophyll in leaves was counted by the particular empirical formula (Arnonova's equation): chlorophyll $a + b = (8.02 * A_{663} + 20.2 * A_{645})$.

3.4 Stomatal Characteristics

Leaves were sampled in the flowering period for the indication of the quantity of stomata. A drop of Dentracryle in the chloroform was applied on an adaxial side of the leaves. Three minutes later, a thin film of the acrylate resin was taken away from the leaves and the copied pores were counted in the calibrated microscope's field of view.

3.5 Statistical Analysis of Data

The significance of the effect of the cultivar, year and location on the evaluated parameters was evaluated by analysis of variance (ANOVA). The *LSD* test was used for division of the cultivars into statistically different groups. All statistical analyses were done using Statistica 9.0 software (StatSoft, Inc.).

4 Results and Discussion

Cereals are steppe grasses in origin and primitive landraces may serve as a valuable source for breeding and efficient water management. As alternatives to the testing of genotypes in dry conditions (e.g. with a special umbrella) (Matuz et al. 2008) which is quite an expensive, indirect, method of selection and evaluation, using the discrimination of $^{13}CO_2$ in photosynthesis, has proved to be efficient (the values are in negative correlation to the transpiration efficiency) (Farquhar and Richards 1984). Several authors recommend the ^{13}C carbon isotope discrimination method as a simple and efficient instrument of the evaluation of plant water management efficiency (Boutton 1991; Farquhar and Lloyd 1993; O'Leary 1993; Araus et al. 2002).

Drought tolerance is subject to a complexity of features. Nevertheless, the regulation of gas exchange through stomata plays a major role. Monneveux et al. (2005) or Ferrio et al. (2007), postulated that the methods based on carbon isotope discrimination can provide useful information on stomatal regulation. $\delta^{13}C$ values were influenced ($P<0.01$) by the factor of locality in particular (79 %) (Table 13.2). The water stress played no role there. Therefore, the varieties having low $\delta^{13}C$ values in dry conditions reached also low values if they are grown in an abundance of precipitation too. The following varieties which were characterised by the lowest $\delta^{13}C$ values are they were most predisposed to the tolerance to drought in dry conditions: emmer wheat called Rudico (−26.93‰), einkorn called No. 8910 (J1 T. monococcum) (−27.02‰) and spring spelt wheat variety (SP9 – T. spelta Kew) (−27.16 ‰). On the other hand, landraces of the bread wheat, intermediate forms (Cervena perla, Kasticka presivka 203) and control varieties of

Table 13.2 Examination of the contribution of factors (variety, locality, year) and its interactions to $\delta^{13}C$ via the analysis of variance (ANOVA)

Factor		DF	$\delta^{13}C$ MS	% TV	Chlorophyll content MS	% TV	Number of stomata MS	% TV
Variety	(1)	13	2.01**	79	0.90**	71	552.9**	48
Irrigation	(2)	1	0.01ⁿˢ	0	0.34**	27	485.8**	42
1 × 2		13	0.29ⁿˢ	11	0.03**	2	114.1**	10
Error		56	0.23ⁿˢ	10	0.00**	0	11.5	0

*Statistically significant $P<0.05$; **highly statistically significant $P<0.01$; ⁿˢ not significant

the bread wheat (Jara, SW Kadrilj) are supposed to be characterised by a reduced tolerance to drought. The division of the varieties into statistically different groups ($P < 0.01$) which was based on the results of the *LSD* test (Table 13.3) has also revealed a predisposition of the spring spelt wheat varieties, einkorn varieties and two emmer wheat varieties (Rudico, Weisser Sommer) to the tolerance to drought.

The proportion of chlorophyll may be one of the criteria of the selection of varieties' tolerant to drought. Mohammadi et al. (2009) describes the correlations between the total grain yield rate and a higher proportion of chlorophyll if wheat plants are grown in a hot dry place. The proportion of chlorophyll was influenced ($P < 0.01$) by the factor of variety (71 %) and the deficiency of precipitation in tested and evaluated varieties. The variety played the most important role in that case too (Table 13.2). Table 13.3 shows that the varieties have been characterised by the lowest $\delta^{13}C$ values (which are most predisposed to the drought tolerance at the same time) (e.g. Rudico emmer wheat, SP9 – T. spelta Kew spelt wheat) have had a high proportion of chlorophyll in leaves. On the other hand, the intermediate forms and control bread wheat varieties have been characterised by a reduced proportion of chlorophyll. Table 13.3 also shows that the proportion of chlorophyll is a stable characteristic despite the fact that there is an absence or abundance of precipitation. Several varieties have had the same proportion of chlorophyll in different water content in the soil; low values of the standard deviation also confirm this fact (0.01 in the majority of varieties).

Stomatal density plays a very important role in drought resistance that generally increases with increasing water stress (Nayeem 1989). Reduction in stomatal number under drought helps it to prevent water loss (Hammed et al. 2002). The quantity of stomata was influenced ($P < 0.01$) by the factor of variety (48 %) and deficiency of precipitation (42 %). The majority of varieties were characterised by an increased quantity of stomata in the absence of precipitation (Table 13.3) which could lead to losses of water (Hammed et al. 2002). The spring spelt wheat (SP6 – T. spelta VIR St. Petersburg) was the only variety characterised by a reduction of the quantity of pores in the deficiency of precipitation. It was a variety with a high proportion of chlorophyll and a low $\delta^{13}C$ value at the same time (Table 13.3). Besides the low $\delta^{13}C$ values and a high proportion of chlorophyll in the leaves, the predisposition to the tolerance to drought of Rudico emmer wheat could also be connected with a lower quantity of pores (21.67 per square mm) than the other varieties had (mean of 40.10 per square mm).

The prediction of drought tolerance using the method of $\delta^{13}C$ discrimination and other characteristics has been used and it has brought valuable observations and knowledge on the evaluated genetic resources of different wheat species and control bread wheat varieties. The results show that some varieties may be a valuable source of drought tolerance in the breeding process (e.g. emmer wheat Rudico, einkorn No. 8910, etc.). On the other hand, $\delta^{13}C$ discrimination is an indirect method of evaluation of the resistance to drought which should be accompanied by one of the direct methods, such as osmotic adaptation (Gloser and Prášil 1998). The biotechnological method, based on the analysis of a stress dehydrin proteins in

Table 13.3 Differences in droughtiness in genus *Triticum* L. (mean + SD), LSD test ($P < 0.01$)

Variety	δ^{13}C (‰) Dry	Normal	Mean	Chlorophyll content (mg·dm^{-1}) Dry	Normal	Mean	Number of stomata per mm^2 Dry	Normal	Mean
Einkorn (*Triticum monococcum* L.)									
J1	−27.19 ± 0.27	−27.50 ± 0.97	−27.35 ± 0.62^{def}	0.61 ± 0.00	0.69 ± 0.01	0.65 ± 0.01^{cd}	34.00 ± 1.73	28.67 ± 3.51	31.33 ± 2.62^{b}
J4	−27.16 ± 0.49	−26.89 ± 0.16	−27.02 ± 0.33^{f}	0.38 ± 0.01	0.49 ± 0.00	0.43 ± 0.01^{ab}	33.00 ± 1.73	27.67 ± 2.31	30.33 ± 2.02^{ab}
J6	−27.55 ± 0.64	−27.64 ± 0.46	−27.59 ± 0.55^{def}	0.50 ± 0.01	0.64 ± 0.01	0.57 ± 0.01^{bc}	36.33 ± 2.31	31.00 ± 1.73	33.67 ± 2.02^{b}
Emmer [*Triticum diccocum* (Schrank)Schuebl]									
Rudico	−27.28 ± 0.38	−26.58 ± 0.49	−26.93 ± 0.44^{f}	1.51 ± 0.01	1.58 ± 0.04	1.55 ± 0.03^{g}	28.67 ± 3.51	14.67 ± 1.15	21.67 ± 2.33^{a}
D11	−27.14 ± 0.77	−27.57 ± 0.46	−27.36 ± 0.62^{def}	0.75 ± 0.01	0.82 ± 0.01	0.79 ± 0.01^{de}	33.00 ± 3.46	30.00 ± 1.73	31.50 ± 2.60^{b}
D14	−28.30 ± 0.21	−28.57 ± 0.36	−28.43 ± 0.29^{abc}	0.84 ± 0.01	0.91 ± 0.00	0.88 ± 0.01^{e}	36.67 ± 4.04	34.67 ± 2.31	35.67 ± 3.18^{bc}
Spelt (*Triticum spelta* L.)									
SP6	−27.71 ± 0.04	−27.55 ± 0.33	−27.63 ± 0.19^{def}	1.43 ± 0.01	1.51 ± 0.01	1.47 ± 0.01^{g}	44.67 ± 5.69	59.33 ± 9.61	52.00 ± 7.65^{de}
SP7	−27.55 ± 0.15	−27.40 ± 0.86	−27.48 ± 0.51^{def}	1.40 ± 0.01	1.45 ± 0.01	1.42 ± 0.01^{g}	48.33 ± 2.31	44.33 ± 3.51	46.33 ± 2.91^{de}
SP9	−27.21 ± 0.18	−27.11 ± 0.85	−27.16 ± 0.52^{ef}	0.86 ± 0.01	1.38 ± 0.03	1.12 ± 0.02^{f}	42.00 ± 1.73	44.33 ± 3.51	43.17 ± 2.62^{cd}
Bread wheat (*Triticum aestivum* L.) – intermediate forms									
P2	−28.25 ± 0.30	−27.38 ± 0.70	−27.82 ± 0.40^{cde}	0.61 ± 0.01	0.70 ± 0.01	0.66 ± 0.01^{cd}	34.67 ± 2.31	33.33 ± 2.31	34.00 ± 2.31^{bc}
P3	−28.19 ± 0.54	−29.05 ± 0.39	−28.62 ± 0.47^{ab}	0.63 ± 0.01	0.65 ± 0.01	0.64 ± 0.01^{cd}	37.33 ± 2.31	31.00 ± 1.73	34.17 ± 2.02^{bc}
P4	−28.70 ± 0.31	−28.82 ± 0.33	−28.76 ± 0.32^{a}	0.66 ± 0.01	0.72 ± 0.01	0.69 ± 0.01^{cd}	55.00 ± 1.00	37.33 ± 4.62	46.17 ± 2.81^{de}
Bread wheat (*Triticum aestivum* L.) – control									
Jara	−27.95 ± 0.30	−27.97 ± 0.29	−27.96 ± 0.30^{bcd}	0.59 ± 0.02	0.85 ± 0.05	0.72 ± 0.04^{cd}	48.33 ± 3.06	49.00 ± 3.46	48.67 ± 3.26^{de}
SW Kadrilj	−28.16 ± 0.58	−27.93 ± 0.13	−28.05 ± 0.36^{abcd}	0.26 ± 0.01	0.44 ± 0.02	0.35 ± 0.02^{a}	64.67 ± 3.51	44.00 ± 2.65	54.33 ± 3.08^{e}

crops confronted by drought, is another possible method of the evaluation of the resistance to drought. The expression of certain genes to this group of proteins is subject to the regulation of the increased concentration of abscise acid (Gloser and Prášil 1998) which causes the closing of pores in leaves. This method, based on the evaluation of the content of the dehydrin proteins, has been consistently verified (Lopez et al. 2001, 2003; Badr et al. 2005; Brini et al. 2007). The necessity for the establishment of the seedlings in dry conditions is a disadvantage of this method (compared with the method of $\delta^{13}C$ discrimination) (Lopez et al. 2003); the differences between the content of dehydrin proteins in grains of mature crops confronted by drought are negligible.

Wheat landraces are often tolerant to local stress factors (Ehdaie et al. 1988; Skovmand and Reynolds 2000; Davood et al. 2004; Reynolds et al. 2007). Experiments carried out confirm that some landraces are adapted to the regular types of stress occurring in their growing areas. Landraces, when confronted with drought are usually characterised by a higher level of production of plant matter (Skovmand and Reynolds 2000) probably because they have a better ability to absorb water through the root system (Reynolds et al. 2007). The differences in the rate of drought tolerance between the wheat landraces are also demonstrated in the study by Ehdaie et al. (1988) who found many more or less tolerant genotypes among the landraces and bread wheat varieties. Certain genetic resources may also serve as donors of drought tolerance in the breeding process.

5 Conclusions

Tolerance to drought of newly bred varieties has not been taken into account in the regions that are not jeopardized by any serious drought. Results of our research show that the modern bread wheat varieties may be more affected by drought than various wheat landraces. Results of $\delta^{13}C$ analyses of the proportion of chlorophyll and quantity of pores have shown that the emmer wheat variety Rudico, einkorn No. 8910 and spring spelta wheat T. spelta Kew are the most predisposed to becoming tolerant to drought. Particular wheat genetic resources may be valuable for the breeding process, for the enhancement of the tolerance to drought for most worldwide spread bread wheat varieties.

Acknowledgment This work was supported by the Ministry of Agriculture of the Czech Republic – NAZV, Grants No. QH82272 and No. QJ1310072.

References

Araus JL, Slafer GA, Reynolds MP, Rovo C (2002) Plant breeding and drought in C-3 cereals: what should we breed for? Ann Bot 89:925–940

Badr S, Bahieldin A, Abdelgawad B, Badr A (2005) Construction of a dehydrin gene cassette for drought tolerance from wild origin for wheat transformation. Int J Bot 1:175–182

Balaghi R, Jlibene M, Tychon B, Mrabet R (2007) Risk management of agricultural drought in Morocco. Sci Chang Planét/Sécheresse 3:169–176

Boutton TW (1991) Stable carbon isotope rations of natural materials: 1: sample preparation and mass spectrometer analysis. In: Coleman DC, Fry B (eds) Carbon isotope techniques. Academic Press Inc., New York, pp 155–171

Brini F, Hanin M, Lumbreras V, Irar S, Pagés M, Masmoudi K (2007) Functional characterization of DHN-5, a dehydrin showing a differential phosphorylation pattern in two Tunisian durum wheat (*Triticum durum* Desf.) varieties with marked differences in salt and drought tolerance. Plant Sci 172:20–28

Bucur D, Savu P (2006) Considerations for the design of intercepting drainage for collecting water from seep areas. J Irrig Drain Eng-ASCE 132:597–599

CIMMYT (2011) Wheat – global alliance for improving food security and the livelihoods of the resource-poor in the developing world, Texcoco, Mexico, 5 p

Davood A, Ashkboos F, Sadeghi M, Bahram DN (2004) Evaluation of salinity tolerance in landrace wheat germplasms of Cereals'. Research Collection Department, Isfahan Agricultural and Natural Resources Research Center Publisher, Isfahan, 14 p

Ehdaie B, Waines JG, Hall AE (1988) Differential responses of landraces and improved spring wheat genotypes to stress environments. Crop Sci 28:838–842

Farquhar GD, Lloyd L (1993) Carbon and oxygen isotope effects in the exchange of carbon dioxide between terrestrial plants and the atmosphere. In: Ehleringer JR, Hall AE, Farquhar GD (eds) Stable isotopes and plant carbon-water relations. Academic Press Inc., New York, pp 47–70

Farquhar GD, Richards RA (1984) Isotopic composition of plant carbon correlates with water use efficiency of wheat. Aust J Plant Physiol 11:539–552

Ferrio JP, Mateo MA, Bort J, Abdalla O, Voltas J, Araus JL (2007) Relationships of grain d13C and d18O with wheat phenology and yield under water-limited conditions. Ann Appl Biol 150:207–215

Gloser J, Prášil I (1998) Fyziologie stresu (Physiology of stress). In: Procházka S, Macháčková I, Krekule J, Šebánek J (eds) Fyziologie rostlin (Plant physiology). Academia, Praha, pp 412–431

Hammed M, Mansor U, Ashraf M, Rao AUR (2002) Variation in leaf anatomy in wheat germplasm from varying drought-hit habitats number of stomata generally increased under water stress. Int J Agric Biol 4:12–16

Konvalina P, Moudrý J, Dotlačil L, Stehno Z, Moudrý J Jr (2010) Drought tolerance of landraces of emmer wheat in comparison to soft wheat. Cereal Res Commun 38(3):429–439

Lopez CG, Banowetz G, Peterson CJ, Kronstad WE (2001) Differential accumulation of a 24-kd dehydrin protein in wheat seedlings correlates with drought stress tolerance at grain filling. Hereditas 135:175–181

Lopez CG, Banowetz G, Peterson CJ, Warren E, Kronstad WE (2003) Dehydrin expression and drought tolerance in seven wheat cultivars. Crop Sci 43:577–582

Matuz J, Cseuz L, Fonad P, Pauk J (2008) Wheat breeding for drought tolerance by novel field selection methods. Cereal Res Commun 36(Suppl):123–126

Mohammadi M, Karimizadeh RA, Naghavi MR (2009) Selection of bread wheat genotypes against heat and drought tolerance on the base of chlorophyll content and stem reserves. J Agric Soc Sci 5:119–122

Monneveux P, Reynolds MP, Trethowan P, Gonzáles-Santoyo H, Pena RJ, Zapata F (2005) Relationship between grain yield and carbon isotope discrimination in bread wheat under four water regimes. Eur J Agron 22:231–242

Nayeem KA (1989) Genetic and environmental variation in stomatal frequency and distribution in wheat Triticum spp. Cereal Res Commun 17:51–57

O'Leary MH (1993) Biochemical basis of carbon isotope fractionation. In: Ehleringer JR, Hall AE, Farquhar GD (eds) Stable isotopes and plant carbon-water relationships. Academic, New York, pp 19–28

Peleg Z, Fahima T, Abbo S, Krugman T, Nevo E, Yakir D, Saranga Y (2005) Genetic diversity for drought resistance in wild emmer wheat and its ecogeographical associations. Plant Cell Environ 28:176–191

Rebetzke GJ, Condon AG, Richards RA, Farquhar GD (2002) Selection for reduced carbon isotope discrimination increases aerial biomass and grain yield of rainfed bread wheat. Crop Sci 42:739–745

Reynolds M, Dreccer F, Trethowan R (2007) Drought-adaptive traits derived from wheat wild relatives and landraces. J Exp Bot 58:177–187

Skovmand B, Reynolds MP (2000) Increasing yield potential for marginal areas by exploring genetic resources collections. In: 11th regional wheat workshop for Eastern, Central and Southern Africa, Addis Ababa, Ethiopia, 18–22 Sep 2000, 436 p

Swearingen WD (1992) Drought hazard in Morocco. Geograp Rev 82:401–412

Watson RT, Zinyowera MC, Moss RH (eds) (1997) The regional impacts of climate change: an assessment of vulnerability. IPCC/Cambridge University Press, Cambridge, 517 p

Wright GC, Nageswara Rao RC (1994) Carbon isotope discrimination, water-use efficiency, specific leaf area relationships in groundnut. In: Wright GC, Nageswara Rao RC (eds) Selection for water-use efficiency in grain legumes. Report of a workshop held at ICRISA Centre, Andhra Pradesh, India, 5–7 May 1993. ACIAR technical reports no. 27, pp 52–58

Chapter 14
Coping with Climate Change Through Water Harvesting Techniques for Sustainable Agriculture in Rwanda

Suresh Kumar Pande, Antoni Joseph Rayar, and Patrice Hakizimana

Abstract It is believed that Climate change is no longer an issue but an undisputable reality for the distant future. Climate change is already taking place, and its impacts are being felt particularly in South Asian countries, mostly affecting the poorest people. Developing countries are particularly vulnerable because their economies are closely linked to agriculture, and a large proportion of their populations depend directly on agriculture and natural ecosystems for their livelihoods. Thus, climate change has the potential to act as a 'risk multiplier' in some of the poorest parts of the world, where agricultural and other natural resource-based systems are already failing to keep pace with the demands on them. The availability of and access to freshwater is an important determinant of patterns of economic growth and social development. This is particularly the case in Africa where most people live in rural areas and are still heavily dependent on agriculture for their livelihoods. It is, therefore, essential for sustaining all forms of life, food production, economic development, and for general well being. It is impossible to substitute for most of its uses, difficult to de-pollute, expensive to transport, and it is truly a unique gift to mankind from nature. The prevailing water crisis demands an adequate and appropriate water resources management. It is time to abandon the obsolete divide between irrigated and rainfed agriculture. In the new policy

S. Kumar Pande (✉)
Department of Soil and Water Management, College of Agriculture,
Animal Sciences and Veterinary Medicine, University of Rwanda,
Busogo, Post Box-210, Ruhengeri, Musanze, Rwanda
e-mail: sureshpande21@yahoo.com

A.J. Rayar
Higher Institute of Agriculture and Animal Husbandry, University Crescent Burnaby,
201, 9188, Vancouver, B.C., Canada, V5A 0A5
e-mail: ajrayar@yahoo.co.in

P. Hakizimana
USAID, Kigali, Rwanda
e-mail: phakiza@yahoo.co.uk

approach rainfall will be acknowledged as the key freshwater resource, and all water resources, green and blue, will be explored for livelihood options at the appropriate scale for local communities. Therefore, water conservation/harvesting would be the most appropriate option to utilize the rainwater for optimum productivity. The paper emphasizes present scenario of water resources, water crisis and the immediate need for the adoption of water harvesting through appropriate soil and water conservation techniques to face the challenges posed by climate change on agricultural production systems.

Keywords Climate change • Water harvesting • Water resource management

1 Introduction

The specter of climate change has been with us for a long time. As early as 1896, the Swedish chemist and Nobel Prize winner Svante Arrhenius published a paper discussing the role of carbon dioxide in the regulation of the global temperature, and calculated that a doubling of CO_2 in the atmosphere would trigger a rise of about 5–6 °C. In more recent years we have moved to a better understanding of what this means for our planet and its people, and we have developed some plausible approaches to tackling this bedeviling problem. However, we have yet to implement most of them (Moorhead 2009).

We know that climate change will mean higher average temperatures, changing rainfall patterns and rising sea levels among many other hidden manifestations. Obviously, there will be more and more intense, extreme events such as, unprecedented snow fall, droughts, floods and hurricanes. Although there is a lot of uncertainty about the location and magnitude of these changes, there is no doubt that they pose a major threat to agricultural systems, particularly in the fragile ecosystems of the world. The ironic problem of excessive rain, provoking uncontrollable flooding and extreme drought will be manifesting scenario, particularly in Sub-Saharan Africa. The single worst African drought disaster killed 300,000 people in Ethiopia in 1984. In 2002, 14.3 million people were affected by drought in the same country. In economic terms, the cost of droughts in Africa is enormous. For example, the economic impacts of the 1991/1992 drought in Southern Africa resulted in a GDP reduction of $3 billion, reduced agricultural production, increased unemployment, heavy government expenditure burden and reduced industrial production due to curtailed power supply (NEPAD 2006).

Warmer air temperatures are expected to have several impacts on water resources including diminishing snow pack and increasing evaporation, which affects the seasonal availability of water (Field et al. 2007, p. 619). The significant surface and water temperatures will impact evaporation, vegetation and aquatic life.

Of all the planet's renewable resources, water has a unique place. It is essential for sustaining all forms of life, food production, economic development, and for general well being. It is impossible to substitute for most of its uses, difficult to de-pollute, expensive to transport, and it is truly a unique gift to mankind from nature (Rao et al. 2010).

2 Global Water Cycle

The Fig. 14.1 shows how rainwater is used globally and the services each use provides. The main source of water is rain, falling on the earth's land surfaces (110,000 km^3). The arrows express the magnitude of water use, as a percentage of total rainfall, and the services provided. So, for example, 56 % of green water is evapotranspired by various landscape uses that support bioenergy, forest products, livestock grazing lands, and biodiversity, and 4.5 % is evapotranspired by rainfed agriculture supporting crops and livestock. Globally, about 39 % of rain (43,500 km^3) contributes to blue water sources, important for supporting biodiversity, fisheries, and aquatic ecosystems. Blue water withdrawals are about 9 % of total blue water sources (3,800 km^3), with 70 % of withdrawals going to irrigation (2,700 km^3). Total evapotranspiration by irrigated agriculture is about 2,200 km^3 (2 % of rain), of which 650 km^3 are directly from rain (green water) and the remainder from irrigations water (blue water). Cities and industries withdraw 1,200 km^3 but return more than 90 % to blue water, often with degraded quality. The remainder flows to the sea, where it supports coastal ecosystems. The variation across basins is huge. In some cases people withdraw and deplete so much water that little remains to flow to the sea (Molden 2007; Rockström et al. 1999; Scanlon et al. 2007).

3 Facing Up to the Water Crisis

Water is the defining and intricate link between the climate and agriculture. But even without climate change, we are in serious trouble. Among *competing demands combined with mismanagement of this critical resource, water availability* has

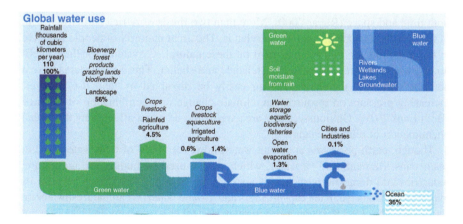

Fig. 14.1 Global water use pattern (Source: Calculations for the Comprehensive Assessment of Water Management in Agriculture based on data from Oki and Kanae (2006), UNESCO–UN World Water Assessment Programme (2006))

become an urgent issue facing the farmers (and other users) the world over. And typically, the most extreme shortages are experienced by those least able to cope with them – the most impoverished inhabitants of developing countries. Climate change will exacerbate an already critical and perplexing situation. Warmer air temperatures are expected to have several impacts on water resources including diminishing snow pack and increasing evaporation, which affects the seasonal availability of water (Field et al. 2007).

Agriculture is responsible for 70 % of the world's water withdrawals and is therefore its primary consumer of water. Municipal needs account for 10 % and industry 20%. The respective figures for Africa are 85, 9 and 6%. Sub-Saharan Africa only uses 2.9 % of its water resources (FAO 2004).

4 The Increasing Demand of Water

4.1 As the World Gets Wealthier, It Gets Thirstier

In Asia, wheat consumption has tripled between 1961 and 2002; meat consumption grew by a factor of seven over the same period. Globally, food demand is expected to grow by 70–90 % by 2050. With no changes in the dietary pattern, water for food requirements from rain-fed and irrigated lands could almost double. Therefore, water productivity gains in agriculture are critical if other growing demands for water are also to be met.

About 40 % of the earth's land surface is covered by semi-arid and arid ecosystems, otherwise known as drylands. Undoubtedly, one of the major crushing problems that militate against environmentally sustainable agricultural production in Sub-Saharan Africa is the extreme variations in the distribution of rainfall over both time and space (Rayar 2000). More than two thirds of Africa and virtually all of the Middle East are classified as drylands. The great majority of people who live in these regions are highly dependent on these natural ecosystems for their livelihoods, which provide them with sustenance and shelter. These ecosystems have a delicate state of balance, which can easily be disrupted by land use changes, increasing pressure on resource use, climate change or a combination of all three (Both Ends 2010). Obviously, these area are/would be worst affected by the global phenomenon of climate change.

Though, earth possesses abundant water, however, 97 % of total water exist in sea/oceans. Remaining 3 % water, which can be used by human for agriculture, industries and domestic purposes, is also not easily available. This is because its major proportions lye in ice caps/glaciers and ground water. Of all the freshwater on Earth, only about 0.3 % is contained in rivers and lakes – yet rivers and lakes are not only the water we are most familiar with, it is also where most of the water we use in our everyday life.

4.2 Sea Level Rise

"Global mean sea level has been rising", according to the IPCC. "From 1961 to 2003, the average rate of sea level rise was 1.8 ± 0.5 mm [per year]. For the 20th century, the average rate was 1.7 ± 0.5 mm [per year]. There is high confidence that the rate of sea level rise has increased between the mid-19th and the mid-20th centuries" This will also affect the land and water quality adjacent to sea areas. Also gradual inundation of natural systems and human infrastructure in coastal and estuarine areas will be a stark reality (EPA 2008). Sea-level rise poses a particular threat to deltaic environments, especially with the synergistic effects of other climate and human pressures. This is particularly important since many of the largest deltas are home to large urban and rural farming populations and providers of important environmental services. The impacts of sea level rise and increased storm surges on these areas must be seen in the context of the existing and ongoing impacts of human induced land changes. Whereas present rates of sea-level rise are contributing to the gradual diminution of many of the world's deltas, most recent losses of deltaic wetlands are attributed to human development. An analysis of satellite images of 14 of the world's major deltas indicated a total loss of 15,845 km^2 of deltaic wetlands over the past 14 years. Every delta showed land loss, but at varying rates, and human development activities accounted for over half of the losses. In Asia, for example, where human activities have led to increased sediment loads of major rivers in the past, the construction of upstream dams is now seriously depleting the supply of sediments to many deltas with decreased land formation and increased coastal erosion a widespread consequence. Large reservoirs constructed on the Huanghe River in China have reduced the annual sediment delivered to its delta from 1.1 billion metric tons to 0.4 billion metric tons (Li et al. 2004, cited in IPCC Working Group II 2007, Ch 6).

5 Climate Change and Agriculture

We can isolate some critical challenges that agriculture will face as the climate changes. Many of these are amplifications of the substantial challenges that the current climate already imposes; water availability is at the top of the list. Already scarce in many regions, increasing demand and competition for water will combine with changing and less predictable rainfall and river flows. In Asia, changes in the monsoon rains (quantum and pattern) and, in glacier and snow melt are probably the greatest threats. In Africa, where a huge population rely directly on the rain for their food and livelihoods, any change in rainfall pattern will present a major risk. Indeed, the IPCC's Fourth Assessment Report suggests that some African countries may see yields from rainfed agriculture fall by as much as 50 % by 2020, if production practices remain unchanged. Further, water quality is also at stake – saline water will increasingly affect agricultural systems due to seawater intrusion, over-exploited aquifers and unsustainable irrigation practices.

Unequivocally, higher temperatures will challenge many agricultural systems for the worst. Plants are sensitive to high temperatures during critical stages, such as flowering and seed development. Often combined with drought, high temperatures can mean disaster to farmers' fields. Raised carbon dioxide levels also have implications for crop plants, although impacts are complex and need further research (Moorhead 2009).

Many pests and diseases of crops, animals and humans are sensitive to climate, and we can expect these to change in unpredictable ways. For example, some will become prevalent in areas where they were previously unknown, when the climate becomes favorable in those areas. The danger is that there is usually low immunity to a disease, and poor knowledge of pest or disease management, in areas where they have not occurred before.

The impacts of climate change will not be felt evenly across the world, and may not all be negative. Some agricultural systems, mainly at higher latitudes and higher altitudes, may benefit at least in the short term from higher temperatures. Some dry areas may get more rainfall. But the most vulnerable – the many millions of people who survive by rainfed agriculture in the drylands of Africa, the millions more who make up the world's small-scale fishing communities, and those who make their livelihoods in low-lying regions like the Indo-Gangetic Plains, for example – look likely to face some of the most severe impacts, which will probably overwhelm their current coping capacities (Moorhead 2009). The vast potential of the rainfed agriculture need to be unlocked through knowledge-based management of natural resources for increasing the productivity and incomes to achieve food secured developing world. Long-term studies at ICRISAT have demonstrated over last 30 years that productivity of vertisols can be increased to 4.7 tonnes per hectare with a carrying capacity of 18 persons per hectare per year as against the current productivity of 0.95 tonnes per hectare with carrying capacity of four persons per hectare per year. Yield gap analysis carried out in number of countries in Asia and Africa have shown that yields of cereals, legumes and oil seeds can be increased by 2.3 times to 3 times. Successful case studies at specific locations revealed considerable opportunities to increase crop productivity by two to three folds of rain-fed systems through productivity-enhancing agricultural technologies.

5.1 How Much Water Is Used for Agriculture?

To produce enough food to satisfy a person's daily dietary needs takes about 3,000 L of water converted from liquid to vapor – about 1 L per calorie. Only about 2–5 L water are required for drinking. In the future, more people will require more water for food, fiber, industrial crops, livestock, and fish. But the amount of water per person can be reduced by changing what people consume and how they use water to produce food (Molden 2007).

Global freshwater assessments have not addressed the linkages among water vapor flows, agricultural food production, and terrestrial ecosystem services. We perform the first bottom-up estimate of continental water vapor flows, subdivided into the major terrestrial biomes, and arrive at a total continental water vapor flow of 70,000 km^3/year (ranging from 56,000 to 84,000 km^3/year). Of this flow, 90 % is attributed to forests, including woodlands (40,000 km^3/year), wetlands (1,400 km^3/year), grasslands (15,100 km^3/year), and croplands (6,800 km^3/year). These terrestrial biomes sustain society with essential welfare-supporting ecosystem services, including food production. By analyzing the freshwater requirements of an increasing demand for food in the year 2025, we discover a critical trade-off between flows of water vapor for food production and for other welfare-supporting ecosystem services. To reduce the risk of unintentional welfare losses, this trade-off must become embedded in intentional ecohydrological landscape management.

According to Rost et al. 2009, the potential for increasing global crop production through on-farm water management strategies: (a) reducing soil evaporation ('vapor shift') and (b) collecting runoff on cropland and using it during dry spells ('runoff harvesting'). A moderate scenario, implying both a 25 % reduction in evaporation and a 25 % collection of runoff, suggests that global crop production can be increased by 19 %, which is comparable with the effect of current irrigation (17 %). Climate change alone (three climate models, SRES A2r emissions and population, constant land use) will reduce global crop production by 9 % by 2050, which could be buffered by a vapor shift level of 50 % or a water harvesting level of 25 %.

About 80 % of agricultural evapotranspiration – when crops turn water into vapor, comes directly from rain, and about 20 % from irrigation. Arid areas like the Middle East, Central Asia, and the western United States tend to rely on irrigation. There has also been large-scale irrigation development in South and East Asia, less in Latin America, and very little in Sub-Saharan Africa (Molden 2007).

6 The Costs of Climate Change

A recent study by the International Food Policy Research Institute (IFPRI), titled 'Climate change: Impact on agriculture and costs of adaptation', highlighted some of the anticipated costs of climate change:

- 25 million more children will be malnourished in 2050 due to climate change without serious mitigation efforts or adaptation expenditures
- Irrigated wheat yields in 2050 will be reduced by around 30 % and irrigated rice yields by 15 % in developing countries
- Climate change will increase prices in 2050 by 90 % for wheat, 12 % for rice and 35 % for maize, on top of already higher prices.
- At least US$7 billion a year are necessary to improve agricultural productivity to prevent adverse effects on children (Moorhead 2009)

- The above estimated figure may change but they are more relevant to developing countries where more people depend on agriculture and allied pursuits. More investment on land and water infrastructure for sustainable development remains the stark reality in present situation.

6.1 How Much More Water?

Without further improvements in water productivity or major shifts in production patterns, the amount of water consumed by evapotranspiration in agriculture will increase by 70–90 % by 2050. The total amount of water evaporated in crop production would amount to 12,000–13,500 km^3, almost doubling the 7,130 km^3 of today. This corresponds to an average annual increase of 100–130 km^3, which is almost three times, the volume of water supplied to Egypt through the High Aswan Dam every year (Molden 2007).

6.2 Can Upgrading Rainfed Agriculture Meet Future Food Demands?

At present, 55 % of the gross value of our food is produced under rainfed conditions on nearly 72 % of the world's harvested cropland. In the past, many countries focused their "water attention" and resources on irrigation infrastructure development. The future food production that should come from rainfed or irrigated agriculture is the subject of intense debate, and the policy options have implications that go beyond national boundaries. Obviously, the t option to us is to upgrade rainfed agriculture through best water management practices (BWMPs). Better soil and land management practices will go a long way in increasing water productivity, adding a component of irrigation water through smaller scale interventions such as rainwater harvesting. Integrating livestock in a balanced way to increase the productivity of livestock water is important in rainfed areas. At the global level the potential of rainfed agriculture is large enough to meet present and future food demand through increased productivity (Molden 2007).

In several regions of the world rainfed agriculture generates the world's highest yields. These are predominantly temperate regions, with relatively reliable rainfall and inherently productive soils. Even in tropical regions, particularly in the subhumid and humid zones, agricultural yields in commercial rainfed agriculture exceed 5–6 t/ha (Rockström and Falkenmark 2000; Wani et al. 2003a, b). Evidence from a long-term experiment at the International Crops Research Institute for the Semi-Arid Tropics (ICRISAT), Patancheru, India, since 1976 demonstrated the virtuous cycle of persistent yield increase through improved land, water and nutrient management in rainfed agriculture.

6.3 Where Is the Hope? Increasing the Productivity of Land and Water

The hope lies in closing the gap in agricultural productivity in many parts of the world – often today no greater than that on the fields of the Roman Empire – and in realizing the unexplored potential that lies in better water management along with no miraculous changes in policy and production techniques. The world has enough freshwater to produce food for its entire people over the next half century. But world leaders must take action now – before the opportunities to do so are lost. Nonetheless, there is a silver lining in the clouds – about 75 % of the additional food we need over the next decades could be met by bringing the production levels of the world's low-yield farmers up to 80 % of what high-yield farmers get from comparable land. Better water management plays a key role in bridging that gap.

More good news: the greatest potential increase in yields are in rainfed areas, where many of the world's poorest rural people live and adopting best water management practices is the key to such increases. Only if leaders decide to adopt better water and land management practices in these areas could reduce poverty and increase productivity.

Even more good news: while there will be some need to expand the area of land we irrigate to feed eight to nine billion people, we will have to deal with the associated adverse environmental consequences. Such efforts are attempted with focused attention, there is real scope to improve production on many existing irrigated lands. This approach would lessen the need for more water in these areas and more areas could be brought under irrigation.

In South Asia, where more than half the cropped area is irrigated but the productivity is low. With determined policy change and adoption of appropriate water management practices, almost all additional food demand could be met by improved water productivity in already irrigated areas. In general, the water table of arid and semi-arid regions of Sub-Saharan Africa is sinking at an alarming rate, due largely to over-exploitation of groundwater for crop and animal consumption, in the wake of a steadily decaline in the amount of rainfall received over these regions (Rayar 2000). Therefore, comprehensive water management policies and sound institutions would spur economic growth for the benefit of all in these regions. Despite the bad news of groundwater depletion, there is still potential in many areas for highly productive groundwater use, for example, in the lower Gangetic plains and parts of Sub-Saharan Africa (Molden 2007).

7 Action Needed

There is no time to waste in debating on such issues as, climate change, agriculture and food security. We must move to a deeper understanding of the problem as quickly as possible. Where we have already identified the ways and means to move forward, and as we identify new strategies, we must become more efficient in implementing them. There should be no room for duplication of efforts, absent or weak policy, or poor communications. – All these potential limitations must be dealt with decisively, and that too soon. Therefore, there should be no dilution of our strategic approach to tackle these multifarious constraints to sustainable agricultural production. In aiming at attaining a credible system of combating climatic change, we have to look at the vulnerability of systems, in the light of possible changes in the climatic pattern Vulnerability depends both on sensitivity of the system to the climate, and the adaptive capacity of the population. All measures have to be taken to deal with impact of climate change on agriculture specifically through water harvesting/conservation, effective management of rainfed agriculture to achieve its production potential along with irrigation water management to improve water productivity.

7.1 Assessment of the Problem

The Comprehensive Assessment of Water Management in Agriculture is a critical evaluation of the benefits, costs, and impacts for the past 50 years of water development, water management challenges communities face today, and the solutions people have developed around the world. It is a multi-institute process aimed at assessing the current state of knowledge and stimulating ideas on how to manage water resources with a view to meeting the growing needs for agricultural products, to help reduce poverty and food insecurity in addition to contributing to environmental sustainability.

The target audience of this assessment are the people who make the investment and management decisions in water management for agriculture – agricultural producers, water managers, investors, policymakers, and civil society. In addition, the assessment should inform the general public about these important issues, so that we can all help to make better decisions through our political processes.

7.2 Solutions

Water harvesting, precision irrigation and multi-use reservoirs are some of many options for eking out limited water supplies, while low quality water can be managed and used in certain systems.

Crucially, these and other sustainable water management options need to be embedded in improved stewardship of this shared resource. Improved management at the watershed level should focus on equity of access, recognizing multiple users and their interdependence in the face of limited water supplies. Only by using water much more efficiently, and considering other users in the chain, will conflict be avoided in the future.

The following actions are vital to be taken immediately:

- Change the way we think about water and agriculture – Thinking differently about water is essential for achieving our triple goal of ensuring food security, reducing poverty, and conserving ecosystems. Instead of a narrow focus on rivers and groundwater, view rain as the ultimate source of water that can be managed. Instead of blueprint designs, craft institutions while recognizing the politically contentious nature of the reform process. And instead of isolating agriculture as a production system, view it as an integrated multiple-use system and as an agroecosystem, providing services and interacting with other ecosytsems. For all sectors, water demand can be reduced by introducing participatory management approaches and applying key principles, such as the polluter pays principle. Measures are also taken for the protection of water resources and these include the establishment of monitoring networks with water quality indicators and the enforcement of laws
- Fight poverty by improving access to agricultural water and its use – Target livelihood gains of smallholder farmers by securing water access through water rights and investments in water storage and delivery infrastructure where needed, improving value obtained by water use through pro-poor technologies, and investing in roads and markets. A recent study by the International Food Policy Research Institute (IFPRI), titled 'Climate change: Impact on agriculture and costs of adaptation', highlighted some of the anticipated costs of climate change stated that 25 million more children will be malnourished in 2050 due to climate change without serious mitigation efforts or adaptation expenditures. Also Irrigated wheat yields in 2050 will be reduced by around 30 % and irrigated rice yields by 15 % in developing countries Climate change will increase prices in 2050 by 90 % for wheat, 12 % for rice and 35 % for maize, on top of already higher prices. At least US$7 billion a year are necessary to improve agricultural productivity to prevent adverse effects on children (Moorhead 2009).
- Increase the productivity of water – Gaining more yield and value from less water can reduce future demand for water, limiting environmental degradation and easing competition for water. A 35 % increase in water productivity could reduce additional crop water consumption from 80 to 20 %. Without better water management in agriculture the Millennium Development Goals for poverty, hunger, and a sustainable environment cannot be met. Access to water is difficult for millions of poor women and men for reasons that go beyond the physical resource base.

7.3 Making Agriculture and Land Use Climate-Friendly and Climate-Resilient

An agricultural landscape should simultaneously provide food and fiber, meet the needs of nature and biodiversity, and support viable livelihoods for people who live there. In terms of climate change, landscape and farming systems should actively absorb and store carbon in vegetation and soils, reduce emissions of methane from rice production, livestock, and burning, and reduce nitrous oxide emissions from inorganic fertilizers. At the same time, it is important to increase the resilience of production systems and ecosystem services to climate change (Sara and Sthapit 2009).

Now it is time to abandon the obsolete divide between irrigated and rainfed agriculture. In the new policy approach, rainfall will be acknowledged as the key freshwater resource, and all water resources, green and blue, will be explored for livelihood options at the appropriate scale for local communities. Therefore, water conservation/harvesting would be the most appropriate option to utilize the rainwater for optimum productivity.

8 Meaning and Scope of Water Harvesting/Conservation

Water conservation is a broad term involving all the measures which aim at arresting maximum amount of runoff water, storing or collecting it in safe, artificially raised structures and recycling the stored water again to the field in times of need. The ultimate aim of the water conservation is to produce maximum from the limited availability of water by using it most efficiently. An efficient system of collecting, storing and reusing runoff water is sometimes also referred as water harvesting.

The first water harvesting techniques are believed to originate from Iraq over 5,000 years ago, in the Fertile Crescent of Mesopotamia, where agriculture originated some 8000 BC. Since then, many civilizations have used such techniques. In some, the development of water harvesting probably formed the very basis for their existence (e.g., the agricultural societies of the Negev desert, the spate irrigation societies in the Horn of Africa). In India water harvesting systems have been practiced for millennia. Archaeological findings show that rainwater collection in stone rubble dams for irrigation purposes was practiced as early as 3000 BC. As shown by Agarwal and Narain (1997), these old traditional water harvesting systems developed into a multitude of different systems for domestic use and crop production. These systems formed an important backbone of the Indian farming systems in semi-arid savannah environments until the early twentieth century.

Thus, water harvesting is nothing new, but received little development attention during the decades of modernisation of agriculture from the 1940s onwards. The recent realization of the potential of small-scale water solutions for improvement of rural livelihood has resulted in renewed interest in water harvesting. For a couple of decades, farmers and development agencies in both South Asia and sub-Saharan Africa are paying more attention these technologies and methodologies (Rockström 2002).

In sub-Saharan Africa in-situ water harvesting systems, generally defined as soil and water conservation, are the most dominant indigenous water harvesting practice. Traditional so-called Zai pits, small planting pits for concentration of within-field sheet runoff, originating in Burkina Faso, have been practiced for centuries among smallholder farmers in farmed savannahs (Reij et al 1996). However, in general terms, water harvesting systems for supplemental irrigation are much less common in sub-Saharan Africa as compared, for example, to South Asia (Sivannapan 1997).

From the point of dry land agriculture, water harvesting indicates not only storing rainwater in irrigation tanks/ponds but also employing mechanical and cultural manipulations of soil to improve its rainfall absorption and moisture retention capacity. However, the measures adopted are location specific. To decide on the suitable measures to be taken a preliminary inventory of soil water and other resources, a study of land use pattern, a dialogue with the target group and a study of socio-political, economic and cultural setting are a must. This implies a scientific approach to water conservation. The need for educating, and organizing people to make them accept the necessary reforms for effective implementation of integrated water conservation programs need not be over-emphasized.

In most countries except those with large desert areas, the chief agent of soil erosion is water. The conservation of water virtually reduces the soil erosion and its bad effects. Thus, the scope of conservation is very wide, encompassing all the measures of soil conservation, moisture retention and nutrient preservation in the soil. Agricultural water management needs the sound knowledge of soils, topographical conditions, crops grown, prevailing climatic conditions and socio economic status of the population engaged in agricultural activities. The low cost, location specific. replicable technologies are recommended for effective rainwater management, whether *insitu* or *exsitu* to meet the crop water requirements.

8.1 Water Requirement of Crops

The water requirement (WR) of a crop depends upon soil, plant and climatic factors. While soil factors affect retention and transmission pattern of water, plant factors affect absorption, transmission within the plant, transpiration, and effective rainfall. Climatic factors affect the vapor pressure and energy for evapotranspiration (ET). It is the quantity of water regardless of its source required by a crop in a given period of time for its maturity. Water requirement includes losses due to ET or CU (Consumptive Use) plus losses during the application of irrigation water, which might be unavoidable and water required for special operation such as land preparation, leaching etc. (Rayar and Pande 2008). The approximate values of seasonal water needs for maize, sorghum/barley/oat/wheat, ground nut, cotton and soybean are 500–800 mm, 450–650 mm, 500–700 mm, 700–1,300 mm and 450–700 mm respectively (www.fao.org/docrep/U3160E/U3160E00.htm 1991). Based on rainfall, needed irrigation may be provided through water harvesting along with insitu moisture conservation.

8.2 Water Use Efficiency (WUE)

Water use efficiency is the ratio of crop yield to the amount of water depleted through evapotranspiration (ET). Increasing the crop yield or decreasing ET could increase WUE. Increasing crop production is achieved through an integrated use of productive inputs. Decreasing ET requires adaptation of plant varieties to soil, genetic and climatic improvements.

Studies conducted in the semi arid tropics of Sub Saharan Africa revealed that soil evaporation accounts for 30–50 % of rainfall (Cooper et al. 1987; a value that can exceed 50 % in sparsely cropped farming systems in semi-arid regions (Allen 1990). Surface runoff is often reported to account for 10–25 % of rainfall (Casenave and Valentin 1992). The characteristics in dry lands of frequent, large and intensive rainfall events result in significant drainage, amounting to some 10–30 % of rainfall (Klaij and Vachaud 1992). Studies conducted in the semi-arid region of Rwanda indicate that substantial volume of rainwater could be collected through the construction of trapezoidal ponds of top 10 m × 10 m, bottom 7 m × 7 m with a depth of 1.5 m (Ngabonziza et al. 2007).

This clearly shows that a great portion of rainfall is lost in evaporation, surface runoff and deep percolation losses as compared to water used for transpiration. If losses are reduced through appropriate techniques to 50 %, crop yields can easily be 2 to 2.5 times (considering 100 mm transpired water can produce about 700–1,000 kg/ha).

8.3 Insitu Rainwater Conservation

8.3.1 Biological Measures

Contour Cultivation – This practice is very simple and can easily be practiced by performing all operations along the contour or at least across the general slope. The main objective of contour cultivation is to increase yields through conservation of precious soil water. Contour ridges increase soil water retention, vastly improves the amount of levels of plant available water and increases crop yields by as much as 29 % (Troeh et al. 1991). Investigations carried out in India reveal that this simple practice could reduce substantially the soil and water losses, irrespective of terrain, soil, climate and other conservation practices. The extent of reduction in water and soil losses from 23 to 44 % and 32 to 63 % in Maize – Wheat and Potato was found in contour cultivation, with respect to up and down cultivation experiments conducted in Deharadun and Ootacamund in India (Bonde 1985). Hence, farmers may be advised to adopt contour cultivation with a view to using the limited rainfall and at the same time conserving soil.

8.3.2 Inter Terrace Land Management

Reduction of soil loss and runoff could be achieved by inter terrace management such as, contour furrowing, ridging and ridge and furrow methods. These methods were adopted in various places of India within the contour and graded bounded area, and a considerable reduction of runoff and soil losses was observed (Bonde 1985). Significant higher yield of seed cotton was obtained at ridges & furrows system over flat bed system. Both the Bt hybrids gave an additional yield of about 600 kg by utilizing run-off water through land configuration as ridges & furrows system over flat bed system (Jagvir et al. 2006).

8.3.3 Mulching – Effects on Runoff and Soil Loss

Mulching is not only effective in moisture conservation during the drought spell but also very effective in soil and water conservation. Studies in the volcanic soils of Rwanda, indicate that the practice of mulching greatly conserve soil moisture and increase the yield of crops, such as cabbage, onion, potato etc. (Rayar and Pande 2008). Mulching also plays a major role in increasing the organic matter build up in soils, and as such this practice is to be adopted by the farmers cultivating in the highland regions of Rwanda.

8.3.4 Haloding

It is one of the modified/refined indigenous practice for maize crop. In this practice, an inverted plow is run in inner row spacing in a month old maize crop. The practice destroys weeds, does earthing up and creates shallow ditches between the rows. The ditches intercept and detain running water. In addition, earthing up supports the plants, aerates the rooting zone and decrease resistance to growing roots. In an experiment conducted on farmers field it was observed that practice of haloding decreased weeds, conserved water and increased maize yield by 24 % than traditional farmers practice (Arora et al. 2010).

8.3.5 Vegetative Barriers

Vegetative barriers of vetiver, agave and closely grown *pennisatum* have been found effective for soil and water conservation. Vetiver has been found a climax plant to use it at appropriate vertical and horizontal distance based on rainfall and slope of land.

The vertical interval and horizontal spacing between two consecutive barriers can be determined as follows:

The vertical interval between two consecutive barriers is:

$$VI = 15S + 60 \tag{14.1}$$

Where VI is in cm and slope is in % in low rainfall area
and
VI = 10S + 60 in medium and high rainfall area

$$HS = VI \times 100/S \tag{14.2}$$

Horizontal Interval or spacing between two consecutive trenches

$$HI = VI \times 100/S$$

8.3.6 Mechanical Measures – Field Bunding

It has been experienced that rainwater is not only lost through runoff but also erodes the soil, washes nutrients and disfigure farmers land. Formation of bunds is an easy and adaptative method, which is supported by modern scientific knowledge to facilitate infiltration and percolation of rainwater to help soil and ground water recharge.

8.3.7 Progressive Terracing

In Rwanda, progressive terracing is becoming popular and recommended up to 12 % slope. It is commonly known as "Fanya juu" terraces in some countries of Africa. They are made by digging a trench, normally along the contour, and throwing the soil upslope to form an embankment. This technique has proved to be an effective way of controlling soil erosion. Planting agro forestry species like *calliandra* sp., *leuceana callothyrms* along with *pennisetum* gives fodder and fuel wood to the farmers.

Tiffen et al. (1994) present evidence from Machakos District in Kenya suggesting that the adoption of "fanya juu" terraces played an important role in reducing land degradation over the period from 1930s to 1990s when population increased more than fivefold. For instance, in a study of soil conservation methods in Kiambu District in Kenya, Mati (1984) found that "fanya juu" terraces accounted for over 50 % of all soil conservation interventions. Results from studies have shown substantial increases in yield on land with "fanya juu" terraces compared to non-terraced land (Ngigi 2003). Studies in Kangundo, Machakos District, Kenya (Lindgren 1988) measured maize yield increments on terraced land versus the yield from the non-terraced land, which was obtained as 47 % in 1984/1985, and 62 % in 1987/1988. Thus, "fanya juu" terraces increase crop productivity.

8.3.8 Bench Terracing

In the steep hill slopes, mere reduction of slope length is not sufficient to deal with intense storm and resulting runoff. Bench terracing converts the original sloping ground to level step like fields, which reduce the length as well as degree of slope. The bench terraces are usually recommended on 16 to 33 % slope. However in mountainous countries with dense population, up to 55 % slope bench terracing is done. This conserve soil and water, makes field operations easy and enhance crop production. The site requires sufficiently deep soils and appropriate soil management practices to maintain soil fertility after construction of terraces. *The Loess plateau of North China is characterized by severe soil erosion resulting deep gullies and gorges. To control land degradation about 1,080 km^2 area was covered under bench terracing in Zhuanglang county. The crop yield was substantially increased from pre-project (750–900 kg/ha) to post project (3,000–3,750 kg/Ha) within 3–4 years* (WOCAT 2007).

8.3.9 Contour Trenches

In cognizance of the rainfall pattern and topography, contour trenches could be constructed in the sloppy non-arable areas. The trenches may be staggered or continuous. The cross section of trapezoidal shaped trenches is normally kept as 0.18 m^2 (0.30 m × 0.30 m) with 1:1 side slope. The excavated soil should be placed in the down streamside of the trenches and grass seed can be seeded on the bund for protection of soil and production of fodder. The area between the trenches can be used for forestry plantations. If the soil depth is very shallow and stones are available, then a 30 cm high and 30 cm wide stonewall is recommended at appropriate vertical intervals. The length of the trenches may be 3, 5, 7 m or as long as 300 m.

A field experiment was conducted on lateritic soils in hilly region of Goa State to study the effect of different *in-situ* soil and water conservation measures on quality and production of cashew (*Anacardium occidentale*). The study reveals that an increase of 0.2–1.3 g in individual nut weight was due to the effect of *in-situ* soil and water conservation measures. Continuous contour trenches with live vegetative barriers increased the apple weight by 13.2 g and staggered contour trenches with vegetative barriers increased by 11.3 g as compared to control (Manivannan et al. 2010).

8.3.10 Negarim Microcatchments

Negarim microcatchments are diamond-shaped basins surrounded by small earth bunds with an infiltration pit in the lowest corner of each. Runoff is collected from within the basin and stored in the infiltration pit. Microcatchments are mainly used for growing trees or bushes. This technique is appropriate for small-scale tree planting in arid and semi arid areas which has a moisture deficit. The rainfall can be as low as 150 mm with soils 1.5 to 2 m deep. Besides harvesting water for the trees, it simultaneously conserves soil. Negarim microcatchments are neat and precise, and relatively easy to construct (Hai 1998; Critchley and Siegert 1991).

8.3.11 Demi Lunes or Semi Circular Bunds

In Kenya semi-circular earth bunds (demi-lunes) are found in arid and semi-arid areas (annual rainfall, 200–750 mm), for both rangeland rehabilitation and for annual crops on gently sloping lands (normally <2 %). For the establishment of fruit trees in arid and semi-arid regions (seasonal rainfall as low as 150 mm), Negarim micro-catchments are often used. These are regular square earth bunds turned 45° from the contour to concentrate surface runoff at the lowest corner of the square (Hai 1998). Similarly, large trapezoidal bunds (120 m between upstream wings and 40 m at the base) have been tried in arid areas in Kenya (e.g. Turkana) for sorghum, tree and grass growth (Hai 1998; Critchley and Siegert 1991; Duveskog 2001). They can be used in low rainfall eastern region of Rwanda.

8.3.12 Farm Ponds

Dug out type of farm ponds may be constructed at suitable locations. Based on the annual/seasonal rainfall, a farm pond may be designed to collect 60 % dependable rainfall. This rainfall can be converted in to runoff yield from Binni's table. For example, if dependable rainfall of any area is 850 mm, so the runoff percent from Binni's table is 31 %. Therefore, runoff yield from any given catchments would be $850 \times 31/100 = 263.50$ mm. Therefore, one hectare catchment area of the farm pond can yield 2,635 m^3 of runoff for storage. The water harvested during rainy season can be used during dry spells for irrigation.

Small farm ponds for supplemental irrigation can increase yields and water productivity, especially if combined with soil fertility management. In Kenya (Machakos District) and in Burkina Faso (Ouaigouya) farmers have been testing small systems where surface runoff is collected from 1 to 2 ha catchments and stored in manually dug farm ponds (100–250 m^3 storage capacity) (Barron et al. 1999; Fox and Rockström 2000).

In these systems simple gravity-fed furrow irrigation was used. Passion fruit, a cash crop in the region, was grown on the edges of the farm pond and onto a wire grid covering the pond in order to provide shading to reduce evaporation losses from the pond. Small volumes of supplemental irrigation were applied during dry spells (on average 70 mm). Typically for semi-arid agro-ecosystems, dry spells occurred every rainy season, and seasonal rainfall ranged from approximately 200–600 mm in the Kenyan location (with two rainy seasons per year) and 400–

700 mm in the location in Burkina Faso. In Kenya, one out of five rainy seasons was classified as a meteorological drought (short rains of 1998/1999) resulting in complete crop failure. One season at each site (long rains 2000 in Kenya and the rainy season 2000 in Burkina Faso) resulted in complete crop failure for most neighboring farmers, while the water harvesting system enabled the harvest of an above average yield (>1 t/ha). The seasonal long-term average yield levels for traditional farming in both areas are approximately 0.5 t/ha. In the Kenyan case, yield levels of maize using water harvesting for dry spell mitigation increased on average (for five rainy seasons 1998–2000) with 30 % compared to the experimental average (from 0.9 to 1.3 t/ha). Yield using water harvesting for dry spell mitigation increased yields compared to neighboring farmers with 240 % from an average of 0.5–1.3 t/ha. The difference between the experimental control yield (which should be representative of the farmers' normal practices) and the actual yield observed on farmers' fields, is explained by general improvements in farm management used by the farmer in conjunction with adoption of the water harvesting system. These include improved timing of operations, through early land preparation and dry planting, and improved weeding. Soil fertility management alone, by the addition of two levels of N-fertilizer (30 and 80 kg N/ha) resulted in a similar yield increase as dry spell mitigation alone (a yield increase of 40 % compared to experimental control).

This is interesting, as it demonstrates that (i) water management alone will not give full yield benefits to the farmer (at least not on soils subject to decades of soil mining and which are inherently low in soil nutrients, such as most savannah soils), and (ii) that soil nutrients can be as limiting as water even in semi-arid tropical environments. Combining water harvesting for dry spell mitigation and soil fertility management.

8.4 Ex-situ Rainwater Conservation

8.4.1 Dams/Tanks/Barrages

In many parts of the world, total rainfall seems to be sufficient for raising two crops. However, its distribution in space and time is not uniform. This results heavy runoff and soil erosion from catchments and moisture stress during dry spells. In such places, it is recommended to construct big water storage through dams, tanks and barrages which can supply irrigation water to its command during soil moisture deficit for meeting crop water requirement. They also have many indirect advantages like flood control, recharging of ground water, reclaiming land below the embankment and improvement of ecology etc. (Jindal et al. 1990).To enhance the land and water productivity, Rwanda has initiated land husbandry, water harvesting and hillside irrigation project supported by world bank. The project aims at avoiding soil erosion and reduced land productivity by revolutionizing effective land-husbandry and irrigated-agriculture as a land management culture at hillsides.

This project will demonstrate improved land-husbandry and productivity on 27,800 ha lands in 34 pilot watersheds and irrigated agriculture on 3,100 ha distributed in 32 locations. Constructing 101 valley-dam reservoirs (~370,000 m^3 each) and establishing water conveyance infrastructure to 12,000 ha irrigable fields (MINAGIRI 2009). The hillside-irrigation program is intentionally decided to be for highly economical horticultural crops such as coffee, tea, mangoes, avocado, cooking banana and pineapple.

8.4.2 Percolation Pond

The percolation pond is a multipurpose conservation structure depending on its location and size. It stores water for livestock and recharges the groundwater. It is constructed by excavating a depression, forming a small reservoir or by constructing an embankment in a natural ravine or gully to form an impounded type of reservoir. The cost of this type of structure is estimated at around US$ 5,000–10,000. The capacity of these ponds or tanks varies from 0.3 to 0.5 mcft (10,000–15,000 m^3). Normally 2 or 3 fillings are expected in a year (season) and hence the amount of water available in one year in such a tank is about 1–1.5 mcft (30,000–45,000 m^3). Normally percolation tanks are made on permeable soils to facilitate ground water recharge where well irrigation is practiced in the adjacent area of this structure. The cost for 1 mcft (30,000 m^3) varies from US$5,000 to 15,000. This quantity of water, if it is used for irrigation, is sufficient to irrigate 4–6 ha of irrigated dry crops (maize, cotton, pulse, etc.) and 2–3 ha of paddy crop. In Amravati District, three percolation tanks and ten cement plugs benefiting an area of 280 ha and 100 ha respectively have been constructed- rise in water level up to 10 m recorded.

8.4.3 Irrigation Water Management

Looking to the fact that water is precious, its wise to use those water application methods, which offers high irrigation efficiencies. Well designed surface irrigation methods in gravity flow irrigation through large storage structures and highly efficient pressure irrigation methods like drip and sprinklers have to be adopted matching crop, soil, topography and climatological requirements. The experiments conducted by Mona Reclamation Experimental Project (MREP) – International Water logging and Salinity Research Institute (IWASRI), Pakistan showed water savings of 85 % for citrus crop using drip system as compared to farmer's traditional method. Also, sprinkler irrigation system in Bhalwal area has shown water savings of 56 % for wheat crop and 59 % for maize crop compared with farmer's traditional method (Rafiq and Alam 2004).

9 Conclusion

While considering the impact of climate change on available water resources and agricultural systems, this paper has attempted to emphasize the urgent actions to integrate both land and water for enhanced productivity. Looking to low production of rainfed areas and high investment in irrigation infrastructure, water harvesting has been elaborated as an important option for efficient use of rainwater for sustainable crop production. Various soil and water conservation measures matching to location specific soil, crop, topography and socio economic conditions were discussed. Further, the urgent need to adopt cost effective and appropriate *Insitu and ex-situ* water conservation measures, in order to exploit the rainfall in different parts of globe, has been emphasized.

References

Agarwal A, Narain S (1997) Dying wisdom. Centre for Science and Environment/Thomson Press Ltd, Faridabad

Allen SJ (1990) Measurement and estimation of evaporation from soil under sparse barley crops in northern Syria. Agric For Meteorol 49:291–309

Arora S, Gupta RD, Jalali VK (2010) Paper entitled "rain water management through watershed management approach for soil conservation and productivity in sub mountainous tract of Jammu region". In: Proceedings of 3rd international conference on hydrology and watershed management at Hyderabad, India, 3–6 Feb 2010, pp 598–606

Barron J, Rockström J, Gichuki F (1999) Rain water management for dry spell mitigation in semi-arid Kenya. East Afr Agric For J 65(1):57–69

Bonde WC (1985) Marry conservation agronomy harvesting with production agronomy for sustained production. In: National seminar on soil conservation and watershed programme, New Delhi, 17–18 Sept 1985

Both Ends (2010) Briefing paper on Agriculture and Food security on Africa's dry lands, p 3

Casenave A, Valentin C (1992) A runoff classification system based on surface features criteria in semi-arid areas of West Africa. J Hydrol 130:231–249

Cooper PJM, Gregory PJ, Tully D, Harris HC (1987) Improving water use efficiency of annual crops in the rainfed farming systems of West Asia and North Africa. Exp Agric 23(1987):113–158

Critchley W, Siegert K (1991) Water harvesting. FAO, Rome

Duveskog D (2001) Water harvesting and soil moisture retention. A study guide for Farmer Field Schools. Ministry of Agriculture and Farmesa, Sida, Nairobi

EPA Climate Change Impacts on Water Resources pp 15–16. http://water.epa.gov/scitech/climatechange/upload/impacts_on_water_resources.pdf, 2008-12-29

FAO (2004) Corporate Document Repository- twenty-third regional conference for africa, Johannesburg, Integrated Water Resource Management for Food Security in Africa, South Africa, 1–5 Mar 2004

Field CB, Mortsch LD, Brklacich M, Forbes DL, Kovacs P, Patz JA, Running SW, Scott MJ (2007) North America. Climate change 2007: impacts, adaptation and vulnerability. Contribution of working group II to the fourth assessment report of the Intergovernmental Panel on Climate Change, p 619

Fox P, Rockström J (2000) Water harvesting for supplemental irrigation of cereal crops to overcome intra-seasonal dry-spells in the Sahel. Phys Chem Earth Part B Hydrol Oceans Atmos 25(3):289–296

Hai MT (1998) Water harvesting. An illustrative manual for development of microcatchment techniques for crop production in dry areas. RSCU, Nairobi

Jagvir S, Blaise D, Rao MRK, Khadi BM, Tandulkar NR (2006) Appropriate technique of rainwater management to enhance soil moisture and higher productivity of rainfed Bt cotton

Jindal PK, Singh RP, Singh I (1990) Technical feasibility and economic viability of small irrigation dams in the Kandi area of Punjab state – a case study. In: Proceedings of international symposium on water erosion, sedimentation and resource conservation, CSWCRTI, Dehradun, pp 507–515

Klaij MC, Vachaud G (1992) Seasonal water balance of a sandy soil in Niger cropped with pearl millet, based on profile moisture measurements. Agric Water Manage 21:313–330

Li CX, Fan DD, Deng B, Korotaev V (2004) The coasts of China and issues of sea-level rise. J Coast Res 43:36–47

Lindgren BM (1988) Comparison between terraced and non-terraced land in Machakos District, Kenya, 1987. Working paper no. 95. International Rural Development Centre. Swedish University of Agricultural Sciences, Uppsala

Manivannan S, Rajendran V, Ranghaswami MV (2010) Effect of *in-situ* moisture conservation measures on fruit and nut quality of cashew (*Anacardium occidentale*) in lateritic soils of Goa. Ind J Soil Conserv 38(2):116–120

Mati BM (1984) A technical evaluation of soil conservation activities in Kiambu District. Soil and Water Conservation Branch, Ministry of Agriculture, Nairobi

Ministry of Agriculture and Animal Resources (MINAGIRI) (2009) Land husbandry, water harvesting and hillside irrigation project environmental impact assessment report, Rwanda

Molden D (2007) Water for food water for life A Comprehensive Assessment of Water Management in Agriculture. IWMI, pp 5

Moorhead A (2009) Climate, agriculture and food security: a strategy for change. A CGIAR Publication

NEPAD (2006) Water in Africa – management options to enhance survival and growth

Ngabonziza JD, Nzabonimpa C, Habitegeko F, Musana B, Kurothe RS (2007) Assessment of water availability in smallholder rainwater harvesting system: experiences from Murama watershed, Eastern Rwanda. In: Proceedings of national conference on agricultural research outputs, Kigali, Rwanda, 2007, pp 417–425

Ngigi SN (2003) Rainwater harvesting for improved food security. Promising technologies in the Greater Horn of Africa. Kenya Rainwater Association, Nairobi

Oki T, Kanae S (2006) Global hydrological cycles and world water resources. Science 313(5790):1068–72

Rafiq M, Alam MM (2004) Annual report 2003–2004, International Waterlogging and Salinity Research Institute, Lahore, December 2004

Rao IS, Pande S, Kumar S (2010) Paper entitled "impact of climate change on water resources of India". In: Proceedings of international conference on land-water resources, biodiversity and climate change, BSSS, Bhopal, India, from 16–18 Feb 2010

Rayar AJ (2000) Sustainable agriculture in Sub-Saharan Africa: the role of soil productivity. AJR Publications, Chennai

Rayar AJ, Pande SK (2008) Best water management practices for sustainable agricultural production in Rwanda published in the proceedings (604) of Water Resource Management, Africa WRM 2008, Gaborone, Botswana, 8–10 September 2008

Reij C, Scoones I, Toulmin C (eds) (1996) Sustaining the soil. Indigenous soil and water conservation in Africa. Earthscan, London, p 248

Rockström J (2002) Potential of rainwater harvesting to reduce pressure on freshwater resources international water conference, Hanoi, Vietnam

Rockström J, Falkenmark M (2000) Semiarid crop production from a hydrological perspective: gap between potential and actual yields. Crit Rev Plant Sci 19(4):319–346

Rockström J, Gordo L, Folke C, Falkenma M, Engwall M (1999) Linkages among water vapor flows, food production, and terrestrial ecosystem services. Conserv Ecol 3(2):5. [online] URL: http://www.consecol.org/vol3/iss2/art5/

Rost S, Gerten D, Hoff H, Lucht W, Falkenmark M, Rockström J (2009) Global potential to increase crop production through water management in rainfed agriculture. Environ Res Lett 4:044002. Online at stacks.iop.org/ERL/4/044002

Sara JS, Sthapit S (2009) State of the world into a warming world. The World Watches Institute, Washington, DC. www.worldwatch.org

Scanlon BR, Jolly I, Sophocleous M, Zhang L (2007) Global impacts of conversions from natural to agricultural ecosystems on water resources: quantity versus quality. Water Resour Res 43, W03437. doi:10.1029/2006WR005486

Sivannapan RK (1997) State of the art in the area of water harvesting in semi-arid parts of the world. In: Paper presented at the international workshop on water harvesting for supplemental irrigation for staple food crops in rainfed agriculture. Stockholm University, Department of Systems Ecology, Stockholm, Sweden, 23–24 June 1997

Tiffen M, Mortimore M, Gichuki F (1994) More people less erosion- environmental recovery in Kenya. Acts Press, Nairobi

Troeh FR, Hobbs JA, Donahue RL (1991) Soil and water conservation. Prentice Hall, Englewood Cliffs, pp 127–129, 235–239, 247–255, 349–351

UNESCO–UN World Water Assessment Programme (2006) Water: a shared responsibility. The United Nations World Water Development report 2. UNESCO/Berghahn Books, New York

Wani SP, Pathak P, Jangawad LS, Eswaran H, Singh P (2003a) Improved management of vertisols in the semi-arid tropics for increased productivity and soil carbon sequestration. Soil Use Manage 19:217–222

Wani SP, Pathak P, Sreedevi TK, Singh HP, Singh P (2003b) Efficient management of rainwater for increased crop productivity and groundwater recharge in Asia. In: Kijne JW, Barker R, Molden D (eds) Water productivity in agriculture: limits and opportunities for improvement. CAB International/International Water Management Institute (IWMI), Wallingford/Colombo, pp 199–215

WOCAT (World Overview of Conservation Approaches and Technologies (2007) Where the world is greener- SWC technology Zhuanglang terraces China, p 40

Part IV
Impacts of Climate Change on Fisheries and Fishery-Based Livelihoods

Chapter 15
Climate Change and Biotechnology: Toolkit for Food Fish Security

Wasiu Adekunle Olaniyi

Abstract Climate change (CC) is a reality! It is a great threat to fisheries production especially in Africa that is characterized with normal climate variability phenomenon. Capture fisheries both in marine and freshwater are greatly impacted, resulting in real negative consequences on fish and fish products availability, and the livelihood of fisheries dependent individuals and communities. Some basic problems faced are genetic erosion especially on the endemic species that have adaptive features under negative effect of CC; decline in capture fisheries due to CC variability with other anthropogenic activities; and nutrition insecurity due to malnutrition or under-nutrition. The primary concern is how to mitigate the effects of CC on the decline of capture fisheries through sustainable fisheries and aquaculture production; genetic erosion through breed conservation; provision of sustainable food fish security to solve food fish demand that outstrips fish supply of ever-increasing human population. However, development of technology to improve fish health, to help restore and protect environments, to extend the range of aquatic species and to improve management and conservation of wild stock will be of great benefit to achieve fish and fish products security. Technology strategies such as animal genetic conservation, selective breeding, hybridization, tissue culture and genetic manipulation have been recently employed in developed economies as panacea to myriads of agri-food insecurity. This paper presents biotechnology approaches as toolkit to alleviate the menace of CC and increase fish and fish products for sustainable production, malnutrition solution and enhance the livelihood of fisheries dependent individuals or communities.

Keywords Climate change • Endemic species • Biotechnology

W.A. Olaniyi (✉)
Biotechnology and Wet Laboratories, Department of Animal Sciences,
Obafemi Awolowo University, Ile-Ife 220005, Nigeria
e-mail: tawakkull@yahoo.com; waolaniyi@gmail.com

1 Introduction

Climate change (CC) can be described as a change in the normal natural climate resulting from both importunate direct or indirect anthropogenic activities and natural processes in the composition of atmosphere (adapted from UNFCCC (1992) and IPCC (2007) definitions). It has great importance, positive (e.g. extended suitable planting season in areas characterized with short growing season) or negative (e.g. failed season due to drought, catastrophes) in some facet of existence. However, as its awareness is increasing so the challenges are enormous but not insurmountable. People and developing countries are both being speculated to be the first and mostly to be affected. The consequences are expected to be great as reported by IPCC (2007) due to high rising in sea levels, frequent and unpredictable floods and severe droughts, natural and anthropogenic continuous contributions to global warming from greenhouse gases among others. These are predicted to have impacts largely on agricultural systems in most developing nations especially of Africa and Asia. The African continent is expected to be greatly impacted due to its vulnerability as results of widespread poverty, low level of technical development and limited adaptation capabilities (Jones and Thornton 2009; JotoAfrika 2009). Crop yield is predicted to reduce by 10–20 %, and world human population to increase from 6.5 billion to 9.2 billion in the year 2050 (UNPP 2008; Jones and Thornton 2009), thereby making global demand for agricultural products to double and/or continue to increase significantly (Delgado et al. 1999; FAO 2009).

The overall effect will be food insecurity, health hazards, malnutrition, and increase in hunger and poverty if there is no appropriate mitigation or adaptation strategy. Analysis of Sub Saharan African (SSA) countries predicted that there will be transitional effect, perhaps as adaptation strategy, in flora and fauna agricultural productivity. Some countries in SSA will have to shift from cropping to increased dependence on livestock as a result of increased probabilities of failed seasons, severe marginal cropping land and high poverty (Jones and Thornton 2009). Such is the case of traditionally Cattle-keeping Samburu people of Northern Kenya that adopted Camel as part of their livelihood strategy in the past two or three decades. This resulted from decline in their cattle economy from 1960 onwards caused by drought, cattle raiding and epizootics (Sperling 1987; Jones and Thornton 2009). Another transitional effect is the change in herd species – the FulBe herders in Nigeria switching from grass dependent Bunaji breeds to Sokoto Gudali known to digest browse much more easily (Blench and Marriage 1999; Jones and Thornton 2009).

However, this form of mitigation perhaps have serious impact on the issue of biodiversity and conservation that is crucial for genetic improvement, poverty alleviation, food security and rural development, and most importantly as good component of CC mitigation and adaptation strategy. The animal genetic diversity is important in genetic improvement and its resources are critical tools and basic building blocks in real biology and technology (biotechnology) applications.

Genetic resources are basic keys and ultimate non-renewable resources that must be protected, conserved and developed. Its conservation, understanding and development in research for development enhance good selection from local stocks that possess some vital traits. These may be ability to utilize lower quality and quantity feed, resistance to diseases and thermal stresses, with better performances in other economic traits and/or new breeding in response to new or resurgent diseases, nutritional requirements, value addition and changing conditions including CC (Hoffman 2010). But reports (FAO 2009) showed that 20 % of reported breeds are now classified as risk, and that almost one breed per month is becoming extinct (CGRFA 2007). Hence, application of effective systems of technology with clear evaluation and recommendation is paramount in the conservation of these genetic resources.

Biotechnology applications can significantly provide effective toolkit for genetic resources conservation, mitigation of undesirable effects as well as adaptive mechanism to undesirable events like diseases, productivity and essentially on CC effects. Biotechnology tools have been used in research for development in enhancing food productivity in developed economies but its application is not evenly spread across African continent. The Food and Agricultural Organization (FAO 1996) reported that Africa is endowed with vast and diverse local gene pool that are the basic building block of biotechnology as expressed in their domesticated breeds. Africa also has potential indigenous knowledge that reveals biotechnology at work, local field ecosystem for product development and improvement, capacities and infrastructures required for possible collaborations with advanced laboratories or foreign multinationals that have necessary expertise, equipment and funding (Wambugu 1999; Sonaiya and Omitogun 2000; Omitogun 2010). With all these, Africa can successfully implement and benefit from biotechnology as toolkit not only on food fish security but also on food production generally, and effectively increase her food productive capacity through its application.

Conventional breeding techniques like selective breeding, crossbreeding or hybridization have been applied to improve genetic merit of fish till present date (Hulata 2001; Olaniyi 2008). These traditional technicalities have been proved to be potent and effective tools in improving superior traits of interest, but may take several years. Selective form of breeding enables the accumulation of superior genes of interest in the successive generations for economic traits like increase in growth and development, fillet production, disease and thermal resistance etc. Though much success has been achieved as matter of genetic improvement through these conventional technologies, greater achievement will be recorded provided there is sustainable combination of these conventional tools with the recent biotechnology techniques. Experts of food security had submitted that biotechnology can play tremendous and sustainable roles in addressing the issues of CC, and that it, biotechnology, is much more relevant for developing countries than for developed countries. This is because of the emerging consequences of CC, and because of the existing problems on food scarcity and food quality.

Biotechnology is the application of indigenous and/or scientific knowledge to the management of whole (or part of) microorganisms, or of cells and tissues of higher organisms, to make or modify a product; so that these supply of goods and services are of use to human beings (Sonaiya and Omitogun 2000; Rege 2005). It can also be explained as a continuum of technologies ranging from the simple, long established practices such as brewing, baking and breeding of plants and animals widely used by the rural people to sophisticated molecular techniques that allow direct manipulation of genetic materials, particularly Deoxyribonucleic acid (DNA). Therefore, in as much as the most of the world's genetic diversity is found in SSA then genetic conservation, employing biotechnological applications, should be accorded the highest priority in research and development.

Conservation and development of endemic species are very important because many of them possess the quantitative genetic traits – ability to utilize lower quality and quantity feed, high quality carcass/fillet production and are more resilient to climatic variability, stress, parasites and diseases; and represent a unique source for stock improvement of genes for improving health and performance traits of industrial breeds in research for development. Hence, it is important to develop and utilize local breeds that are already adapted to their environments with very limited natural and managerial input (Omitogun 2010).

Fisheries and aquaculture contribute immensely to food fish security, and serve as sources of employment for a large number of people in the world. They play important roles in enhancing foreign earnings, poverty alleviation and solving health and malnutrition problems. However the state of global food fish insecurity due to continuous decline in capture fisheries, biodiversity loss, (rise in sea level, decline in lake and ocean levels) as a result of CC or climate variability, environmental degradation with increasing demand for fish and its products among other factors needs serious attention. CC or climate variability effect on water availability, that is indispensable for fish as well other aquatic lives, cannot be overemphasized. Africa is the driest of the continents, after Australia, and yet it supports over 800 million people unlike Australia that has about 22 million, and of which many live in dry lands. Africa is characterized with climate variability as normal phenomenon which its fluctuation or deviation from the long-term meteorological average over a certain period of time e.g. over a specific month, season or year; with its period of drought and flood, warmth and cold could have occurred interchangeably in past (Ropelewski and Halpert 1987; Nicholson and Grist 2001; Tyson et al. 2002). These effects such as floods, droughts, and other extreme weather conditions have serious impacts on both terrestrial and aquatic lives. Desertification effect is becoming significant with dry lands covering 43 % of the continent and harbor a third of the population (UNDP/UNSO 1997; WWF 2000) while drought and floods are seemingly increasing in frequency and severity over the past 30 years (Andreae 1995; Duce 1995; Scholes et al. 2008). Rainfall pattern is the most crucial climate issue in the African continent for its variability at all timescales. It is established that rainfall pattern in many parts of Africa is correlated with patterns of sea-surface temperature, and this is a good tool to make predictions. But the relationship between ocean conditions and CC consequences are not well established, hence prediction can only be made for few months (Scholes et al. 2008). Moreover, evidence also revealed that rainfall patterns at a regional scale are also partly linked to land surface conditions through mechanisms such as the changes in surface albedo that result from reduced vegetation cover (Scholes et al. 2008). All these issues really have significant effects on aquatic lives and that needs serious attention.

Aquatic biotechnology (biotechnology application in fisheries and aquaculture) connotes a range of technologies that present opportunities to: increase growth rate in farmed species, improve nutrition of feeds for aquaculture, improve fish health, help restore and protect environments, extend the range of aquatic species and improve management and conservation of wild stocks (FAO 2000). It may involve automation technology, traceability and quality assurance, offshore farming technology and water treatment technology that may be indigenous and/or scientific. Improvement on breeding strategies through application of innovative biomanipulation technologies at parent, gamete, zygote, chromosome and gene levels e.g. chromosome set manipulation, cryopreservation, intergeneric and interspecific hybridization, nuclear transfer and transgenesis, besides better performance will have a great contribution to food and nutrition security, health and environmental protection. A good example is developing drought tolerance crops for instance rice, by transferring genes into it. This is one of biotech ways of solving CC effects (CropBiotech Net 2008). Another is on increasing the freeze resistance of Atlantic salmon (Fletcher et al. 1992, 1988) and cold tolerance of gold fish (Wang et al. 1995). This paper presents on how biotechnology applications with respect to CC can serve as toolkit for food fish security.

2 Biotechnology Applications

2.1 *Hybridization*

Hybridization is the breeding of related species. It is a natural phenomenon but can also be propagated artificially. It occurs in closely related species of isochronous reproductive seasons, and especially those relying on communal spawning rather than complex courtship patterns (Colombo et al. 1996). Hybridization has been employed in aquaculture and has been attractive in many ways (Chevassus 1983). This may range from production of sterile species avoiding growth loss or fragility related to sexual maturation, and this sterility reduces precocious mating and potential interactions between domestic and wild fishes. Hybridization may also lead to monosex population production which might be an advantage in case of a differential growth between males and females (that is in some fish species that males perform better than females or reciprocal in other species) or in species whose proliferation has to be avoided (e.g. in tilapias). However, if hybrids growth and survival are rarely higher than those of their parent species, they might present the growth rate of the faster growing parent species and additional characteristic (s) sought-after in the other parent e.g. robustness, salinity or thermal tolerance as the case for CC, morphology, flesh quality among others (Lenormand et al. 1988).

It may be intergeneric or interspecific and have great contribution to genetic and reproductive improvements in fish breeding programs. It is a form of mechanism whereby species exchange genetic resources with the goal and possible outcome of strengthening fitness. Hence, hybridization can offer solution to challenges in fish breeding practices by producing improved species that are more suited to specific needs and better than the parents in certain traits of interest. Moreover, if any hybridization would have had to occur for several generations with the genotype frequencies being stabilized and at equilibrium, then this will represent hybridization as a mode of evolution and speciation.

Dunham (2004) explained hybridization as a natural phenomenon that actually occurs in the aquatic world. It is spontaneous in freshwater species compared to marine species that may be as a result of environmental instability, precipitating confinement of different species within a restricted water body, reduction of spawning grounds by anthropogenic activities, massive restocking with kindred species or introductions of alien species are often cited as causes of hybridization in fishes (Colombo et al. 1996). The relative stability of marine environment and lesser importance of all these reasons for hybridization can be explanation for rareness of spontaneous hybridization in the marine species (Campton 1987; Purdom 1993).

Many studies have been conducted on artificial hybridization technological application and its awareness by fish farmers to culture fish for particular traits of importance is in practice (Lenormand et al. 1988, 1992; Madu et al. 1992; Salami et al. 1993; Adeyemo et al. 1994; Amhed and Sarder 1994; Aluko 1995; Nwadukwe 1995; Rahman et al. 1995).

2.2 Gene Transfer/Transgenesis

Gene transfer or transgenesis is the stable incorporation of a gene from another species in such a way that it functions in the receiving species and is passed on from one generation to the next (Cunningham 1999). Gene transfer was developed to complement traditional breeding programs for the improvement of both quantitative and qualitative traits. The application allows the transfer of genes for the manipulation of individual genes rather than the entire genome. This transfer is mostly done by direct injection of the foreign DNA, from a source other than parent germ plasm, into genome of other species or the nucleus at the early embryonic stage by viral vectors, microinjection, electroporation, sperm-mediated transport or gene-gun bombardment. The figures below, Fig. 15.1, depicts microinjection of DNA into salmon eggs; and, Fig. 15.2 shows electroporation of fish eggs.

Organisms containing foreign genes, homologous genes or DNA sequences inserted artificially are referred to as 'transgenics'. Many transgenic fish has been successfully developed through this technology. Six important conditions must be met for any successful transgenic organism's production, viz: firstly, the particular and appropriate gene has to be isolated, and cloned. The foreign gene must be

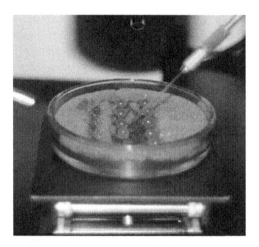

Fig. 15.1 Microinjection of DNA into salmon eggs (Source: Dunham 2004)

Fig. 15.2 Electroporation of fish eggs. Eggs (15–50) are placed in a DNA solution inside the cuvette and then placed between two electrodes to receive rapid pulses of high voltage (Source: Dunham 2004)

transferred and integrated into the host's genome. The transgene must be expressed since transfer of a particular gene does not guarantee that it will express and function. Then a positive biological effect must result from expression of the foreign DNA, and no adverse biological or commercial effects should occur. Finally, the foreign gene must be inherited by subsequent generations (Dunham 2004). This has been achieved in aquatic and major livestock species since its success in 1985. Dunham (2004) reported that prior to the early 1990s, very few fish genes had been isolated, but with the advent of functional genomics and new molecular genetic techniques, such as Expressed Sequence Tag (EST) analysis and the generation of huge gene databases, thousands of fish genes have been isolated, while a long list of fish genes is growing for the study of gene expression and potential transfer and manipulation in fish and aquatic invertebrates. But in livestock, more than 50 different transgenes have been inserted into farm animals (Cunningham 1999).

Transgenesis is one of the recent advances in research for development and offers significant opportunities by conferring the ability to insert new genes of important traits such as resistance or tolerance to environmental stresses, fecundity, disease resistance, fillet or carcass yield and other quantitative traits. Other importance of transgenesis is the ability for the production of medically important proteins such insulin and clotting factors in the milk of domestic livestock. The genes coding for these proteins have been identified and the human factor IX construct has been successfully introduced into sheep and expression achieved in sheep milk (Clark et al. 1990). Moreover, the founder animal has been shown to be able to transmit the trait to its offspring (Niemann et al. 1994). Further advancement in gene transfer is tissue culture and culturing of embryos whereby mRNA and DNA are being transferred into eggs. The first and landmark research on this was the transfer of mRNA and DNA into mouse eggs (Palmiter and Brinster 1986) leading to fish tissue culture or genetic engineering (Dunham 2004).

2.2.1 Trangenesis to Improve Food Fish Production

Fish is a good organism for transgenesis unlike mammalian systems of manipulation that involves *in vivo* culturing of embryos and further transfer into fostering mothers (Powers et al. 1992a, b). Transgenesis application especially in warm water species of fish and shellfish is not complex due to their large quantity of eggs production that are easy to obtain, fertilize, manipulate, incubate and hatch rapidly as a result of their *in vitro* fertilization. Transgenesis is harnessed for fish culture enhancement and has been successfully practiced in freshwater-spawning teleosts, like cyprinids, salmonids, cichlids, esocids and ictalurids (Colombo et al. 1996), for research and development, and significantly in food fish production. These applications followed the pioneering work of Zhu et al. (1985) on the transfer of a growth hormone (GH)-encoding genetic construct into the fertilized eggs of gold fish to enhance body growth. Most researches have been conducted on growth enhancement by microinjection of GH-encoding genetic constructs into fertilized eggs and to confer disease resistance (Jiang 1993; Anderson et al. 1996), and freeze/cold resistance (Fletcher et al. 1992; Wang et al. 1995). These will be of great importance in adaptation measures to CC effects by conferring genes of tolerance to extreme thermal conditions that perhaps arise from CC or climate variability. Several reviews have been reported that elucidates progress of gene transfer in fish in both basic and applied researches (Powers et al. 1992a, b; McEvoy et al. 1992; Jiang 1993; Maclean and Rahman 1994; Chen et al. 1995; Gong and Hew 1995; Knibb et al. 1996; Yiengar et al. 1996; Knibb 1997).

2.3 Genome Maps/Genetic Markers and Marker Assisted Selection

A genetic map, also called linkage map, is a diagram of the order of genes on a chromosome in which the distance between adjacent genes is proportional to the rate of recombination between them (Dunham 2004). Gene map offers the opportunity to identify genetically superior organism in a very direct way. A gene, which is a stretch of DNA that codes a particular protein, has simplified and direct effects like coding for coat color, meaning such animal carrying them can easily be identified and selected for such trait. But for many traits such as fillet production, there is interaction of many genes to produce the eventual result and not just a single gene effect. Therefore, in order to identify such animal with particular gene of superior trait, it becomes necessary to measure the performance of large numbers of their relations including their progenies and compare the results with similar data for other individuals to be able to accurately identify individuals with these superior genes, hence genetic markers.

A genetic marker for a trait is a DNA segment which is associated with a particular trait of importance, and hence segregates in a predictable pattern (Rege 2005). Genetic markers facilitate the identification and/or detection of individual genes or chromosome segments that influence particular traits of interest in a much more direct way. Since an individual animal has up to a hundred thousand of genes, the functional ones in each individual constitute a small fraction, less than 5 %, of that individual's total DNA sequences while the rest has no known functions (Cunningham 1999). Nevertheless, there are scattered thousands useful small pieces of DNA, called "microsatellites", and each contain a number (typically 5–15) repeats of very short DNA sequences. These very short DNA sequences can then be amplified using Polymerase Chain Reaction (PCR). Particularly if a microsatellite is located very close to a targeted useful functional gene, for example a gene of higher fillet production or cold resistance trait, then it is very possible they were inherited together; in view of this, such a microsatellite can serve as a marker for the useful functional gene.

The development of marker technology recently in aquaculture has revolutionalized the field of fisheries and aquaculture. Though the technology is still at developmental stage in aquaculture with very few examples (Agnèse and Teugels 2001; Liu and Cordes 2004), it will actually facilitate research and improve studies in aquaculture genomics. Markers are vital tools in aquaculture and in breeding and genetics to study genomes, progeny testing, sex determination, conducting gene-linkage mapping, locating genes on chromosomes, genes isolation, and to determine gene expression. Other applications are: to study biochemical and molecular mechanisms of performance, conducting population genetic analysis and in marker-assisted selection. With the technological advancement, several biochemical and molecular/genetic marker types are available and well-known for study of fisheries and aquaculture genetics like allozymes (Liu et al. 1992), mitochondrial DNA (mtDNA) (Curtis et al. 1987), Restriction Fragment Length Polymorphism (RFLP) (Miller and Tanksley 1990; Liu et al. 2003), Random Amplified Polymorphic

DNA (RAPD) (Williams et al. 1990; Welsh and McClelland 1990; Liu et al. 1998a), Amplified Fragment Length Polymorphism (AFLP) (Vos et al. 1995; Liu et al. 1998b), microsatellite (Hughes and Queller 1993; Queller et al. 1993; Liu et al. 1999b, c; Tan et al. 1999), Single Nucleotide Polymorphism (SNP), and Expressed Sequence Tag (EST) markers (Liu et al. 1999a; Ju et al. 2000); (Dunham 2004; Liu and Cordes 2004). These markers and their application have tremendous effects on the detection of major genes of interests influencing quantitative traits in aquaculture, and in studying genetic variability and inbreeding, parentage assignments, species, and strain identification and in construction of high resolution genetic linkage maps for aquaculture (Dunham 2004; Liu and Cordes 2004). Therefore, the easy usage of these markers efficiently and effectively involves the construction of a marker/linkage map covering the entire genome (Cunningham 1999). The procedure is also termed "Linkage analysis" that is the screening of the whole genome for genes with a large effect on traits of economic importance (Paterson et al. 1988; Rege 2005). The aim is to produce sufficient genetic maps that largely cover many genes of particular traits of interest. The linkage map/genetic marker will then be used, in the selection process, to develop strategies aimed at achieving more rapid genetic improvement of traits of importance called "Marker Assisted Selection (MAS)".

MAS is the selection process for particular trait of interest by employing genetic markers. It enables improvement of important economic traits and provides alternative for many other significant traits that are difficult to breed for, such as disease and thermal tolerance, fillet production, feed efficiency etc. Thus, increasing the frequency of favorable quantitative trait loci (QTL), or targeting their introgression into other lines. Several works have demonstrated that MAS enhance the rate of genetic progress by increasing accuracy of selection process and reducing the generation interval (Smith and Simpson 1986; Hetzel and Moore 1996; Davis and DeNise 1998). Moreover, the benefit of MAS is its application when there is: low – heritability and those complicated by dominant traits, and when the marker shows a larger proportion of the genetic variance than the economic traits, then the value of information on individual QTL tend to be higher. Also, when the traits of importance cannot be measured on one sex, then the marker gives the information on the basis to rank such animals of that sex; if the trait is not measurable before sexual maturity, marker information can then be used to select such at a juvenile stage, thereby reducing generation interval; and if a particular trait is difficult to measure or requires sacrificing the animal (as the case may be for many carcass traits), then marker information can be used (Hetzel and Moore 1996). However, many studies have shown that MAS has successfully been utilized for genetic enhancement in fish, many livestock and plant species, ranging from growth improvement, drought and thermal resistance, insect and diseases resistance, increased antibody resistance etc (Kashi et al. 1990; Lande and Thompson 1990; Meuwissen and Arendonk 1992; Kerr et al. 1994; Soller 1994; Stromberg et al. 1994; Meuwissen and Goddard 1996; Miklas et al. 1996; Spelman and Garrick 1998; Danzmann et al. 1999). Some MAS studies have been reported on genetic improvement potentials in aquacultural species, for example, Danzmann et al. (1999) reported 25 % of rainbow trout progenies exhibited upper-temperature tolerance after MAS for heat tolerance.

For feed efficiency, MAS has been reported to improve feed conversion efficiency by 11 % in aquacultural species compared to 4.3 % of traditional selection feed conversion efficiency (Davis and Hetzel 2000). Kincaid (1983) also reported 30 % increased body weight in six generations of traditional selection in rainbow trout. Predictions had revealed that MAS would increase the rate of genetic gain by 25–50 % over traditional animal-breeding programs (Weller 1994). Moreover, traditional selective breeding in livestock such as cattle, pigs and sheep has resulted in genetic progress of approximately 1 % per year for several traits (Korver et al. 1988). MAS typical genetic gain for fish growth rates are 6–14 % per generation (Dunham et al. 2001), which is equivalent to 2–14 % per year, with an average of about 3–4 % per year (Dunham 2004).

3 Conclusion

Measures of adaptation and mitigation on the effect of CC with respect to fisheries and aquaculture, as well as other livestock in alleviating food and nutrition insecurity, poverty, malnutrition or under nutrition can be achieved through animal genetic resources conservation, characterization and breeding technologies which biotechnology renders. This is the relationship point between CC and biotechnology. Biotechnology application renders solutions to the effect of CC or climate variability since many of these effects are certainly based on genotype and environment interactions and biotechnology actions are direct on the germ plasm which is its basic building block. Hence biotechnology can effectively serve as toolkit for food fish and nutrition security by dealing with the particular trait of interest e.g. thermo-tolerance, feed efficiency, growth and development, fillet production, disease resistance etc through its varying technologies of hybridization, transgenesis, MAS, genetic conservation strategies etc. It gives direct and faster solutions to CC effects compared to conventional strategies that may take years to achieve, due to accumulation of superior genes of interest in the successive generations. But tremendous success will be achieved if there is combination of these conventional technologies with the advanced biotechnology techniques discussed. As this paper focused on biotechnology as toolkit for food fish security; biotechnology applications and issues are immense and few are just mentioned, therefore readers are referred to consult more on references cited.

References

Adeyemo AA, Oladosu GA, Ayinla AO (1994) Growth and survival of fry of African catfish species, *Clarias gariepinus* Burchell, *Heterobranchus bidorsalis* Geoffery and Heteroclarias reared on *Moina dubia* in comparison with other first feed sources. Aquaculture 119:41–45

Agnèse JF, Teugels GG (2001) Genetic evidence for monophyly of the genus *Heterobranchus* and paraphyly of the genus *Clarias* (Siluriformes, Clariidae). Copea 2:548–552

Aluko PO (1995) Growth characteristics of first, second and backcross generations of the hybrids between *Heterobranchus longifilis* and *Clarias anguillaris*. National Institute for Freshwater Fisheries Research Annual Report, New Bussa, Nigeria, pp 74–78

Amhed GU, Sarder MRI (1994) Growth of hybrid catfishes under different supplemental diets. In: Chou LM, Munro AD, Lam TJ, Chen TW, Cheong LKK, Ding JK, Hooi KK, Khoo HW, Phang VPE, Shim KF, Tan CH (eds) The third Asian fisheries forum. Asian Fisheries Society, Manila

Anderson ED, Mourich DV, Fahrenkrug SC, LaPatra S, Leong J (1996) Genetic immunization of rainbow trout (*Oncorhynchus mykiss*) against infectious hematopoietic necrosis virus. Mol Mar Biol Biotechnol 5:114–122

Andreae MO (1995) Climate effects of changing atmospheric aerosol levels. In: Henderson-Sellers A (ed) World survey of climatology, vol 16. Future climates of the world: a modelling perspective. Elsevier, Amsterdam

Blench R, Marriage Z (1999) Drought and livestock in semi-arid Africa and southwest Asia. Working paper 117. Overseas Development Institute, London, 138 pp

Campton DE (1987) Natural hybridization and introgression in fishes. In: Ryman N, Utter F (eds) Population genetics & fishery management. University of Washington Press, Seattle, pp 161–192

CGRFA (Commission on Genetic Resources for Food and Agriculture) (2007) The state of the world's animal genetic resources for food and agriculture. FAO, Rome, 523 pp

Chen TT, Lu JK, Shamblott MJ, Cheng CM, Lin CM, Burns JC, Reimschuessel R, Chatakondi N, Dunham RA (1995) Transgenic fish: ideal models for basic research and biotechnological applications. Zool Stud 34(21):5–234

Chevassus B (1983) Hybridisation in fish. Aquaculture 33:245–262

Clark AJ, Archibald AL, McClenghan M, Simons JP, Whitelow CBA, Wilmut I (1990) The germ line manipulation of livestock: progress during the last five years. Proc N Z Soc Anim Prod 50:167–180

Colombo L, Barbaro A, Francescon A, Libertini A, Benedetti P, Balla Valle L, Pazzaglia M, Pugi L, Argenton F, Bortolussi M, Belvedere P (1996) Potential gains through genetic improvements: chromosome set manipulation and hybridization. In: Chatain B, Saroglia M, Sweetman J, Lavens P (eds) Proceedings of the international workshop on seabass and seabream culture: problems and prospects. International workshop, Verona, Italy, 16/18-10-1996. Published by European Aquaculture Society, Oostende, Belgium, pp 343–362

CropBiotech Net (2008) Climate change and biotechnology. http://www.monsanto.co.uk/news/ukshowlib.phtml?uid=13029

Cunningham EP (1999) Recent developments in biotechnology as they relate to animal genetic resources for food and agriculture. Commission on Genetic Resources for Food and Agriculture. FAO, Rome

Curtis TA, Sessions FW, Bury D, Rezk M, Dunham RA (1987) Induction of polyploidy in striped bass, white bass and their hybrids with hydrostatic pressure. Proc Ann Conf Southeast Assoc Fish Wildl Agencies 41:63–69

Danzmann RG, Jackson TR, Ferguson MM (1999) Epistasis in allelic expression at upper temperature tolerance QTL in rainbow trout. Aquaculture 173:45–58

Davis GP, DeNise SK (1998) The impact of genetic markers on selection. J Anim Sci 76:2331–2339

Davis GP, Hetzel DJS (2000) Integrating molecular genetic technology with traditional approaches for genetic improvement in aquaculture species. Aquacult Res 31:3–10

Delgado C, Rosegrant M, Steinfield H, Ehui S, Courbois C (1999) Livestock to 2020; the next food revolution. Food, agriculture and environmental discussion paper 28. IFPRI/FAO/ILRI, Washington, DC, USA

Duce RA (1995) Sources, distributions and fluxes of mineral aerosols and their relationship to climate. In: Charlson RJ, Heintzenberg J (eds) Aerosol forcing of climate. Wiley, New York, pp 43–72

Dunham RA (2004) Aquaculture and fisheries biotechnology: genetic approaches. CABI Publishing, Wallingford, pp 385

Dunham RA, Majumdar K, Hallerman E, Bartley D, Mair G, Hulata G, Liu Z, Pongthana N, Bakos J, Penman D, Gupta M, Rothlisberg P, Hoerstgen-Schwark G (2001) Review of the status of aquaculture genetics. In: Subasinghe RP, Bueno P, Phillips MJ, Hough C, McGladdery SE, Arthur JR (eds) Technical proceedings of the conference on aquaculture in the third millenium, Bangkok, Thailand, 20–25 Feb 2000. NACA/FAO, Bangkok/Rome, pp 129–157

FAO (1996) Report on world's plant genetic resource. Food and Agricultural Organization of the United Nations, Rome, May 2006

FAO (2000) Electronic forum on biotechnology in food and agriculture: how appropriate are currently available biotechnologies for the fishery sector in developing countries? Food and Agriculture Organization of the United Nations' conference from 1 Aug–8 Oct 2000

FAO (2009) The state of world fisheries and aquaculture, 2008. FAO Fisheries and Aquaculture Department, Food and Agricultural Organization of the United Nations, Rome

Fletcher GL, Shears MA, King MJ, Davies PL, Hew CL (1988) Evidence for antifreeze protein gene transfer in Atlantic salmon (*Salmo salar*). Can J Fish Aquat Sci 45:352–357

Fletcher GL, Davies PL, Hew CL (1992) Genetic engineering of freeze resistant Atlantic salmon. In: Hew CL, Fletcher GL (eds) Transgenic fish. World Scientific, Singapore, pp 190–208

Gong Z, Hew CL (1995) Transgenic fish in aquaculture and development biology. Cur Top Dev Biol 30:177–214

Hetzel DJS, Moore SS (1996) Applications of molecular genetic technologies in livestock improvement. In: Exploring approaches to research in the animal sciences in Vietnam. ACIAR proceedings no. 68. ACIAR, Canberra, pp 115–119

Hoffman I (2010) Climate change and the characterization, breeding and conservation of animal genetic resources. Anim Genet 41 (Suppl. 1):32–46. http://www.fao.org/docrep/012/al188e/al188e00.pdf

Hughes CR, Queller DC (1993) Detection of highly polymorphic microsatellite loci in a species with little allozyme polymorphism. Mol Ecol 2:131–137

Hulata G (2001) Genetic manipulations in aquaculture: a review of stock improvement by classical and modern technologies. Genetica 111:155–173

IPCC (Intergovernmental Panel on Climate Change) (2007) Climate change 2007: impacts, adaptation vulnerability. Summary for policy makers. Online at: http://klima.hr/razno/news/IPCCWG2_0407.pdf

Jiang Y (1993) Transgenic fish – Gene transfer to increase disease and cold resistance. Aquaculture 111:31–40

Jones PG, Thornton PK (2009) Croppers to livestock keepers: livelihood transitions to 2050 in Africa due to climate change. Environ Sci Policy 12:427–437

Jotoafrika (2009) Climate change and the threat to African food security, Issue 1, July 2009

Ju Z, Karsi A, Kocabas A, Patterson A, Li P, Cao D, Dunham R, Liu Z (2000) Transcriptome analysis of channel catfish (*Ictalurus punctatus*): genes and expression profile from the brain. Gene 261:373–382

Kashi Y, Hallerman E, Soller M (1990) Marker-assisted selection of candidate bulls for progeny testing programmes. Anim Prod 51:63–74

Kerr RJ, Frisch JE, Kinghorn BP (1994) Evidence for a major gene for tick resistance in cattle. In: Proceedings of the 5th world congress on genetics applied to livestock production, Guelph, Canada, 7–12 Aug 1994, vol 20. International committee for world congress on genetics applied to livestock production, Guelph, ON, Canada, pp 265–268

Kincaid HL (1983) Results from six generations of selection for accelerated growth rate in a rainbow trout population [abstract]. In: The future of aquaculture in North America. Fish Culture Section of the American Fisheries Society, Bethesda, Maryland, pp 26–27

Knibb W (1997) Risk from genetically engineered and modified marine fish. Transgenic Res 6:59–67

Knibb W, Gorshkova G, Gorshkov S (1996) Potentials gains through genetic improvement: selection and transgenesis. In: Chatain B, Saroglia M, Sweetman J, Lavens P (eds) Seabass and seabream culture: problems and prospects, International workshop, Verona, Italy, 16/18-10-1996. Published by European Aquaculture Society, Oostende, Belgium, pp 175–188

Korver S, van der Steen HAM, van Arendonk JAM, Bakker H, Brascamp EW, Dommerholt J (1988) Advances in animal breeding. Pudoc, Wageningen, 33 pp

Lande R, Thompson R (1990) Efficiency of marker-assisted selection in the improvement of quantitative traits. Genetics 124:743–756

Legendre M, Teugels GG, Canty C, Jalabert B (1992) A comparative study on morphology, growth rate and reproduction of *Clarias gariepinus* (Burchell 1822), *Heterobranchus longifilis* (Valenciermes, 1840), and their reciprocal hybrids (Pisces: Clariidae). J Fish Biol 40:59–79

Lenormand S, Slembrouck J, Pouyaud L, Subadgja J, Legendre M (1988) Evaluation of hybridisation in five Clarias species (Siluriformes, Clariidae) of African (*C. gariepinus*) and Asian origin (*C. batrachus, C. meladerma; C. nieuhofii* and *C. teijsmanni*). In: Legendre M, Pariselle A (eds) Proceedings of the mid-term workshop of the "Catfish Asia Project": the biological diversity and aquaculture of Clariid and Pangasiid catfishes in South-East Asia, Cantho, Vietnam, 11–15 May, pp 195–209

Liu ZJ, Cordes JF (2004) DNA marker technologies and their applications in aquaculture genetics. Aquaculture 238:1–37

Liu Q, Goudie CA, Simco BA, Davis KB, Morizot DC (1992) Gene-centromere mapping of six enzyme loci in gynogenetic channel catfish. J Hered 83:245–248

Liu Z, Li P, Argue BJ, Dunham RA (1998a) Inheritance of RAPD markers in channel catfish (*Ictalurus punctatus*), blue catfish (*I. furcatus*) and their F1, F2 and backcross hybrids. Anim Genet 29:58–62

Liu ZJ, Nichols A, Li P, Dunham R (1998b) Inheritance and usefulness of AFLP markers in channel catfish (*Ictalurus punctatus*), blue catfish (*I. furcatus*) and their Fl, F2 and backcross hybrids. Mol Gen Genet 258:260–268

Liu Z, Karsi A, Dunham RA (1999a) Development of polymorphic EST markers suitable for genetic linkage mapping of catfish. Mar Biotechnol 1:437–447

Liu ZJ, Tan G, Kucuktas H, Li P, Karsi A, Yant DR, Dunham RA (1999b) High levels of conservation at microsatellite loci among ictalurid catfishes. J Hered 90:307–312

Liu ZJ, Tan G, Li P, Dunham RA (1999c) Transcribed dinucleotide dicrosatellites and their associated genes from channel catfish, *Ictalurus punctatus*. Biochem Biophys Res Commun 259:190–194

Liu ZJ, Karsi A, Li P, Cao D, Dunham RA (2003) An AFLP-based genetic linkage map of channel catfish (*Ictalurus punctatus*) constructed by using an interspecific hybrid resource family. Genetics 165:687–694

Maclean N, Rahman A (1994) Transgenic fish. In: animals with Novel Genes, transgenes in fish. Transgenic Res 5:147–166

Madu CT, Mohammed S, Mezie A, Issa J, Ita EO (1992) Comparative growth, survival and morphometric characteristic of *Clarias gariepinus, Heterobranchus bidorsalis* and their hybrid fingerlings. NIFFR Annual Report, pp 56–61

McEvoy TG, Gannon F, Sreenan JM (1992) Gene transfer in fish: potential and practice. Anim Biotechnol 3(21):1–243

Meuwissen T, Arendonk J (1992) Potential improvements in rate of genetic gain from marker assisted selection in dairy cattle breeding schemes. J Dairy Sci 75:1651–1659

Meuwissen THE, Goddard ME (1996) The use of marker haplotypes in animal breeding schemes. Genet Select Evol 28:161–176

Miklas PN, Afanador L, Kelly JD (1996) Recombination-facilitated RAPD marker-assisted selection for disease resistance in common bean. Crop Sci 36:86–90

Miller JC, Tanksley SD (1990) RFLP analysis of phylogenetic relationships and genetic variation in the genus *Lycopersicon*. Theor Appl Genet 80:437–448

Nicholson SE, Grist JP (2001) A conceptual model for understanding rainfall variability in the West African Sahel on interannual and interdecadal timescales. Int J Climatol 21:1733–1757

Niemann H, Halter R, Paul D (1994) Gene transfer in cattle and sheep: a summary perspective. In: Proceedings of the 5th world congress on genetics applied to livestock production, Guelph, Canada, 7–12 Aug 1994, vol 21. International Committee for World Congress on genetics applied to livestock production, Guelph, ON, Canada, pp 339–346

Nwadukwe FO (1995) Hatchery propagations of five hybrid groups by artificial hybridization of *Clarias gariepinus* (B) and *Heterobranchus longifilis* (Val) (Clariidae) using dry, powdered carp pituitary hormone. J Aquac Trop 10:1–11

Olaniyi WA (2008) Induction of triploidy, gynogenesis and androgenesis in African catfish (*Clarias gariepinus* Burchell). M.Sc. thesis, Department of Animal Sciences, Obafemi Awolowo University, Ile-Ife, Nigeria

Omitogun OG (2010) Biotechnology: the silver bullet for agricultural productivity, Inaugural lecture series 228 (Obafemi Awolowo University). Obafemi Awolowo University Press Ltd., Ile-Ife

Palmiter RD, Brinster RL (1986) Gene – Line transformation of mice. Annu Rev Genet 20:465–499

Paterson AH, Lander ES, Hewitt JD, Peterson S, Lincoln SE, Tanksley SD (1988) Resolution of quantitative traits into Mendelian factors, using a complete linkage map of restriction fragment length polymorphisms. Nature 335:721–726

Powers DA, Chen TT, Dunham RA (1992a) Transgenic fish. In: Murray JAH (ed) Transgenesis. Applications of gene transfer. Wiley, Chichester, pp 233–249

Powers DA, Cole T, Creech K, Chen TT, Lin CM, Kight K, Dunham R (1992b) Electroporation: a method for transferring genes into the gametes of zebrafish, *Brachydanio rerio*, channel catfish, *Ictalurus punctatus*, and common carp, *Cyprinus carpio*. Mol Mar Biol Biotechnol 1:301–309

Purdom CE (1993) Genetics and fish breeding. Chapman & Hall, London/Glasgow/New York/Tokyo/Melbourne/Madras, 277pp

Queller DC, Strassmann JE, Hughes CR (1993) Microsatellites and kinship. Trends Ecol Evol 8:285–288

Rahman MA, Bharda A, Begum N, Islam MS, Hussain MG (1995) Production of hybrid vigor through cross breeding between *Clarias batrachus* Lin. and *Clarias gariepinus* Bur. Aquaculture 138:125–130

Rege JEO (2005) Biotechnology options for improving livestock production in developing countries, with special reference to sub-Saharan Africa. International Livestock Centre for Africa (ILCA) Addis Ababa, Ethiopia. http://www.ilri.org

Ropelewski C, Halpert MS (1987) Global and regional scale precipitation patterns associated with the El Niño/Southern oscillation. Mon Weather Rev 115:1606–1626

Salami AA, Fagbenro OA, Sydenham DHJ (1993) The production and growth of Clariid Catfish hybrids in concrete tanks. Isreali J Aqua-Bamidgeh 45(1):18–25

Scholes B, Ajavon A, Nyong T, Tabo R, Vogel C, Ansorge I (2008) Global environmental change (including climate change and adaptation) in sub-Saharan Africa. ICSU ROA Science Plan, Seychelles. 27pp. Available at: http://www.icsu.org/africa/publications/copy_of_ICSUROASciencePlanonGlobalEnvironmentalChange.pdf

Smith C, Simpson SP (1986) The use of genetic polymorphisms in livestock improvement. J Anim Breed Genet 103:205–217

Soller M (1994) Marker-assisted selection – An overview. Anim Biotechnol 5:193–207

Sonaiya EB, Omitogun OG (2000) Regional and international cooperation for sustainable agrobiotechnology in Africa. In: Ogbadu GH, Onyenekwe DC (eds) Proceedings of the international conference on Biotechnology: commercialisation and food security, Abuja, Nigeria, 21–23 Oct 2000. pp 105–116

Spelman RJ, Garrick DJ (1998) Genetic and economic responses for within-family marker-assisted selection in dairy cattle breeding schemes. J Dairy Sci 81:2942–2950

Sperling L (1987) The adoption of Camels by Samburu cattle herders. Nomadic Peoples 23:1–18

Stromberg LD, Dudley JW, Rufener GK (1994) Comparing conventional early generation selection with molecular marker-assisted selection in maize. Crop Sci 34:1221–1225

Tan G, Karsi A, Li P, Kim S, Zheng X, Kucuktas H, Argue BJ, Dunham RA, Liu ZJ (1999) Polymorphic microsatellite markers in *Ictalurus punctatus* and related catfish species. Mol Ecol 59:190–194

Tyson P, Fuchus R, Fu C, Lebel L, Mitra AP, Odada E, Perry J, Steffen W, Virji H (eds) (2002) Global-regional linkages in the earth system. Springer, Berlin/Heidelberg

UNDP/UNSO (1997) Aridity zones and dryland populations: an assessment of population levels in the World's drylands. United Nations Sudano-Sahelian Office/United Nations Development Programme, New York

UNFCCC (United Nations Framework Convention on Climate Change) (1992) United Nations. Online at: http://unfccc.int/resource/docs/convkp/conveng.pdf

UNPP (2008) Population Division of the Department of Economic and Social Affairs of the United Nations Secretariat, World Population Prospects: the 2006 Revision and World Urbanization Prospects: the 2005 revision. http://www.un.org/esa/population/publications/WPP2006RevVol_III/WPP2006RevVol_III_final.pdf. Accessed 26 Mar 2008

Vos P, Hogers R, Bleeker M, Reijans M, van de Lee T, Hornes M, Frijters A, Pot J, Peleman J, Kuiper M, Zabeay M (1995) AFLP: a new technique for DNA fingerprinting. Nucleic Acids Res 23:4407–4414

Wambugu F (1999) Why Africa needs agricultural biotech. Nature 400(6739):15–16. www.nature.com

Wang R, Zhang P, Gong Z, Hew CL (1995) Expression of the antifreeze protein gene in transgenic goldfish (Carassius auratus) and its implication in cold adaptation. Mol Mar Biol Biotechnol 4:20–26

Weller JI (ed) (1994) Economic aspects of animal breeding. Chapman & Hall, London, 172 pp

Welsh J, McClelland M (1990) Fingerprinting genomes using PCR with arbitrary primers. Nucleic Acids Res 18:7213–7218

Williams JGK, Kubelik AR, Livak KJ, Rafalski JA, Tingey SV (1990) DNA polymorphisms amplified by arbitrary primers are useful as genetic markers. Nucleic Acids Res 18:6531–6535

WWF (2000) The Africa water vision for 2025: equitable and sustainable use of water for socioeconomic development. World Water Forum, The Hague. http://www.afdb.org/fileadmin/uploads/afdb/Documents/Generic-Documents/african%20water%20vision%202025%20to%20be%20sent%20to%20wwf5.pdf

Yiengar A, Müller F, Maclean N (1996) Regulation and expression of transgenes in fish – A review. Transgenic Res 5:147–166

Zhu Z, Li G, Chen S (1985) Novel gene transfer into the fertilized eggs of goldfish (*Carassius auratus* L. 1758). Z Angew Ichthyol 1:31–34

Chapter 16
Climate Change and Fisheries in Chile

Eleuterio Yáñez, María Angela Barbieri, Francisco Plaza, and Claudio Silva

Abstract The possible effects of climate change on Chilean marine ecosystems are considered. Relationships between the abundance of exploited species, fishing effort and environmental variables are elucidated, and conceptual models for an ecosystemic management of fisheries are proposed. A projection for anchovy fisheries in northern Chile is carried out, considering four different temporal climatic change scenarios until 2100. Finally, indications on the necessity to evaluate the spatio-temporal fisheries performance, given the climate change scenario, are suggested.

Keywords Climate change scenarios • Conceptual models • Fisheries • Temperature • Wind

1 Introduction

The biological diversity, a source of significant environmental, economic and cultural value, will be threatened by the fast climate change, which will change the composition and geographical distribution of ecosystems as species respond to these new conditions. Habitats, at the same time, may become degraded and fragmented as a result of human pressures. Species unable to adapt rapidly could become extinct, a fact representing an irreversible loss. Species and ecosystems are already responding to global warming (Calvo et al. 2000).

FONDEF D11I1137 (2012) and FONDECYT 13170 (2013) Projects.

E. Yáñez (✉) • C. Silva
Pontificia Universidad Católica de Valparaíso, Casilla, 1020 Valparaíso, Chile
e-mail: eyanez@ucv.cl; claudio.silva@ucv.cl

M.A. Barbieri • F. Plaza
Instituto de Fomento Pesquero, Blanco 839, Valparaíso, Chile
e-mail: angela.barbieri@ifop.cl; francisco.plaza.vega@ifop.cl

To address this warning scientifically the problem should be analyzed from different points of view. Climate change may impact national fisheries from differently. Understanding the implications of climate change is particularly important for countries like Chile, because of the potential social and economic impact which would result if the current productivity of the region (the average 5 million tons landed each year) was jeopardized.

To address this issue we conceptualize the climate change phenomenon, taking into account the plausible effects of climate change on the Chilean marine ecosystem resulting from changes in sea surface temperature, sea level and winds. Relationships among fishing resources exploited in Chile, environmental variations, and the conceptual models addressing the ecosystem-based approach to fisheries management (FAO 2008) are suggested. Future trends in anchovy fishery production were projected under different climate change scenarios using a predictive artificial neural network model. Finally, indications regarding the need for applications to evaluate the performance of fisheries in different spatio-temporal climate change scenarios are recommended.

2 Global and Local Effects on Climate Change

Global climate is determined by radiation balance, which may be affected by three main factors: changes in the incoming solar radiation, variations in the reflected solar fraction, and alteration of the energy longwave radiation sent to space (Jansen et al. 2007). Thus, climate changes have been observed approximately every 120 thousand years, developing with a larger or smaller extension at least five times in the last 500 thousand years. The maximum values reached by the measurement of different geological variables are similar, except for some of them, which have been extensively exceeded during the current climate change (CO_2, N_2O and CH_4). The latter is ascribed to the anthropogenic effect.

Global average temperature variation at geological scales are quite large, with glacial and inter-glacial periods varying from 21 ºC ten million years ago to 7 ºC 10,000 years ago; in the last 1,000 years between 14 and 16 ºC with varied forecasts (Sharp 2004; Duarte et al. 2006). The variations of the global average temperature in the terrestrial and oceanic surface during the twentieth century showed a significant cold period between 1880 and 1935, followed by a less intense cold period between 1945 and 1975. On the other hand, two warm periods were observed: the first between 1935 and 1945, and the second significantly stronger since 1975 (Trenberth et al. 2007).

Fuenzalida et al. (2007) and Gregory et al. (2001) forecast that through 2100 the average sea level will rise 20 cm between the 30° and 60°S Latitude and 25–30 cm between the 20° and 30°S Latitude offshore of Chile. The surface winds of the same

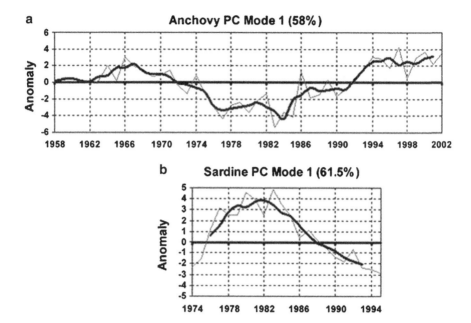

Fig. 16.1 First axis of the principal component analysis for the time series: capture, fishing effort, Ekman transport and turbulence for anchovy (**a**); recruitment, biomass, sea surface temperature and Ekman transportation for sardine (**b**) (Yáñez et al. 2008a)

area would strengthen, exceeding an average of 6.5 m/s during the period 2000–2005, to 7.5 m/s during the period 2071–2100; such effect together with the upwelling may cool down the surface waters (Garreaud and Falvey 2008). In the next 10–15 years, interdecadal variability in local ocean conditions will probably continue to dominate the system. For example, the average satellite sea surface temperature (SST) information for the period 1979–2006, shows a clear cooling at coastal stations in central and northern Chile (17°–37°S), in comparison with the widespread ocean warming (Falvey and Garreaud 2009). However, longer term predictions based on two global warming scenarios of the IPCC (Intergovernmental Panel on Climate Change) done by Fuenzalida et al. (2007) shows a warming on the Chilean coast.

Wind direction and strength significantly will probably influence the distribution and abundance of marine species. Small and coastal pelagic species, for example, show different behaviors: while anchovy maximizes recruitment at current speeds of 5.46 m/s, showing an important decrease with lower and higher values, sardine maximizes recruitment at 5.63 m/s or more (Yáñez et al. 2001). Furthermore, anchovy dominates during cold inter-decadal periods, while sardine prevail during warm inter-decadal periods (Fig. 16.1). Such inter-decadal variations also influence recruitment, a situation that has been documented in anchovies off the Peruvian coast (Cahuin et al. 2009).

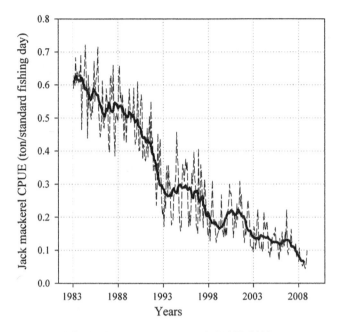

Fig. 16.2 Jack mackerel CPUE decrease during the period 1983–2008

The abundance index CPUE (catch per unit effort) for jack mackerel fisheries (*Trachurus murphyi*) at south-central Chile decreases after 1983 (Fig. 16.2), while the ocean SST at the archipelago Juan Fernandez (33°38′S 78°49′W) also tends to decrease (Falvey and Garreaud 2009), possibly due to a displacement from the fishing areas, while recognizing the likely effects of overfishing.

Fisheries and environmental relations are not only observed in pelagic fish but also in demersal resources, such as common hake (Merluccius gayi) from central Chile, where captures suffer significant decreases during long-term warm periods (Fig. 16.3). SST is also a good proxy for swordfish distribution (Yáñez et al. 1996a), as thermal fronts are for mackerel, where encounters of water masses and food accumulation are produced (Yáñez et al. 1996b).

3 Fisheries Conceptual and Forecast Models

Considering different biological and environmental variables affecting fisheries from local to global and daily to interdecadal in the spatial and temporal scales respectively a conceptual model can be deduced (Fig. 16.4). Regime shifts are observed in environmental (local and global) and biological variables. The same situation is observed in the inter-annual scale, affected by El Niño phenomena, and both the intra-seasonal scale, influenced by Equator remote events, and the daily-seasonal

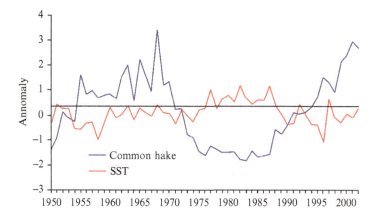

Fig. 16.3 Anomalies of common hake landings and sea surface temperature in central Chile (Yáñez et al. 2003).

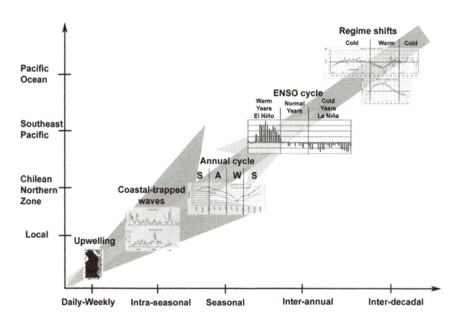

Fig. 16.4 Integrating conceptual model of local and large-scale phenomena affecting the northern area of Chile and the main fishing resources. The direction and magnitude influences of the phenomena are shown by *arrows* (Yáñez et al. 2008a)

scale, associated to the annual cycle (Yáñez et al. 2008a). A comprehensive conceptual model, involving environmental, biological and anthropogenic aspects, was established for swordfish captures offshore Chile (Fig. 16.5). Both conceptual models have a spatio-temporal setting that could be coupled to a general circulation model (GCM).

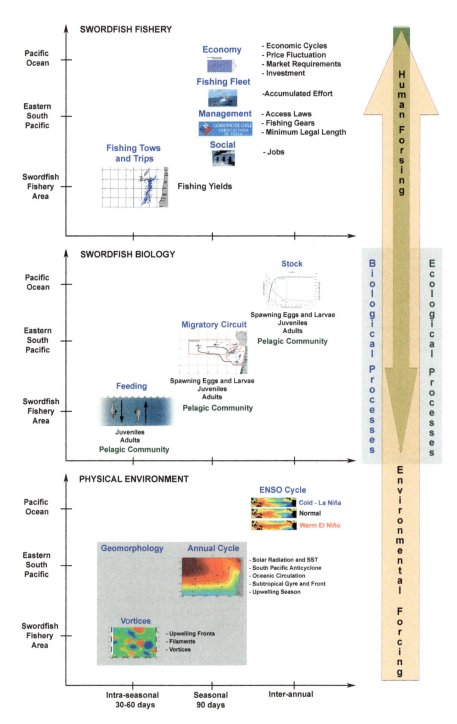

Fig. 16.5 Spatio-temporal conceptual model (3 × 3) with planes in the marine, physical, biological and human (fishing) environments, along with processes and forcing from the swordfish ecosystem (Yáñez et al. 2008b)

These conceptual models may be useful for predictive and ecosystemic modeling. Barbieri et al. (1989, 2002) and Silva et al. (2002, 2004) developed models to predict daily catch in the fishing areas for tuna (*Thunnus alalunga*), swordfish (*Xiphias gladius*) and anchovy (*Engraulis ringens*) off the coast of Chile. Plaza et al. (2008a, b) developed a model to predict the monthly abundance of anchovy using an artificial neural network model (ANN). Similar forecasts on small pelagic fisheries in northern Chile have been also carried out (Gutiérrez-Estrada et al. 2007, 2009; Rodríguez and Yáñez 2008; Yáñez et al. 2010). These ANN models may also consider the temporal and/or spatial dimension for coupling into GCMs, and discussing the possible effects of climate change.

4 Climate Change Scenarios and Anchovy Fishery Projections

Predictive models of artificial neural networks performed by Plaza et al. (2008a, b) were implemented, these models consider the anchovy catches in terms of sea surface temperature in Antofagasta lagged in 7 months affecting reproduction, the temperature in the Niño 3 + 4 region with 4 months lag affecting accessibility and anchovy catches with 6 months lag related to fishing effort, also considered as an indicator of resource abundance (Fig. 16.6).

To project the model, the average structure of catches and temperature (of Antofagasta and the region Niño 3 + 4) for the years 2005, 2006 and 2007 were used as starting point. We consider a linear increase in temperature, taking into account four climate change scenarios based on the scenarios presented by IPCC, designed for the northern part of Chile until 2100. The first scenario considers an increase in temperature of 0.034 ° C per year (Fuenzalida et al. 2007), similar to that estimated by Trenberth et al. (2007). A second scenario, more moderate, of 0,025 °C/year is also proposed by Fuenzalida et al. (2007). The third scenario is not considered a significant effect on the area, following the work of Trenberth et al. (2007). The fourth scenario is contradictory, indicating a cooling de0.02 °C/year (Falvey and Garreaud 2009). It should be noted that according to the work of Fuenzalida et al. 2007 and Falvey and Garreaud (2009) the same SST increase (or decrease) were considered for both temperatures (in Antofagasta and in the Niño3 + 4 region).

As expected, anchovy fisheries show declines associated with increases in temperature (Fig. 16.7), whereas cooling, probably produced by the intensification of the Pacific anticyclone (Falvey and Garreaud 2009), causes an increase in catches to a certain level; thereafter, winds associated with the environmental optimal window for anchovy would be outweighed (Yáñez et al. 2001), affecting recruitment at 5–6 months and causing declines in catches mostly at 6 months to 2 years (Braun et al. 1995; Castillo et al. 2002, 2008). In terms of SST, which is a good environmental proxy (Yáñez et al. 2008a), the maximum landings are obtained at 16.9 °C, while further cooling conditions would necessarily imply a

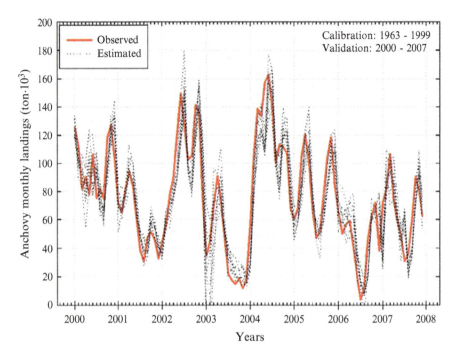

Fig. 16.6 ANN models (10) for anchovy fisheries in northern Chile calibrated (Plaza et al. 2008a, b), and validated with information from 2000 to 2007 (average R2 of 77 %)

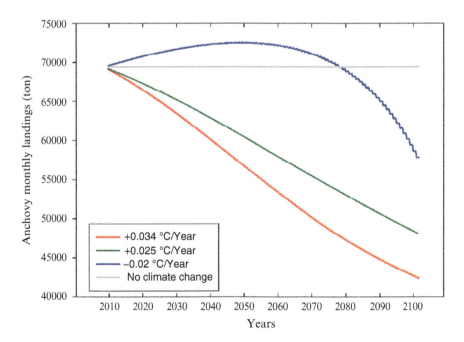

Fig. 16.7 Monthly anchovy landing projections in northern Chile up to 2100, considering four climate change scenarios

decrease in abundance mainly associated to recruitment. This agrees with acoustic surveys information, where the abundance of anchovy regarding temperature, presents a similar curve maximizing abundance at 16–16.5 °C (Castillo et al. 2008).

5 Discussion

This approach allows observation of the trend in anchovy catches considering the SST linear projection as an indicator of the environment fluctuations associated with climate change. Future applications should incorporate other environmental effects in a spatio-temporal dimension to the phenomena that occur at different scales (local, mesoscale, global), as presented in the conceptual models such as the El Niño effects and the regime shifts.

The results of this work emphasized the necessity to analyze the influence of environmental factors on fisheries from a given area and assess the capture and biodiversity predictability under a climate change framework in order to develop more effective management policies. As suggested by Brander (2010), the knowledge related to the functional relations between the fishing resources and the environment is crucial for planning future scenarios, thus facilitating the control of the plausible effects of such changes on fisheries management.

The integration of calibrated models into specific applications is needed, allowing the estimation of fisheries performance in different scenarios of climate change and obtaining a first approximation to the probable economic and social implications, taking into account medium and long term scenarios considered in fisheries management.

It is particularly important to continue the work on this type of model, in order to improve long term predictions and to plan and design fishing management strategies. Calibrated and validated models must be developed using an easy codification and application language for those environmental scenarios generated through GCM. These correspond to numerical models representing physical phenomena in the atmosphere, ocean and earth surface, and constitute advanced tools for the simulation of the general responses of the systems regarding the climate change.

GCMs are generally applied on a three dimensional grid over the globe, having a horizontal resolution of between 250 and 600 km, 10 to 20 vertical layers in the atmosphere and sometimes as many as 30 layers in the oceans. This low resolution does not allow the simulation of the effect of certain phenomena at a regional scale, thus posing a great uncertainty during simulation. Therefore, GCMs are able to simulate different responses to a single forcing, i.e. different environmental scenarios can be generated from a unique condition in the atmosphere, the ocean and the earth.

Different experiments on climate change have been carried out in the last years using many GCMs. Each experiment involves the spatial-temporal evolution of climate conditions for a 100-year period, considering annual increases of greenhouse gases equivalent to 0.5 and 1 %. Thus, the calibrated models must be integrated, i.e. they must be applied considering the climate conditions simulated by different GCMs in the cell corresponding to the studied area. Data on this subject is available for many GCM in www.ipcc-data.org.

Once all environmental scenarios are calculated and integrated, the plausible economic and social implications of each simulation may be analyzed. The latter implies the availability of socio-economic databases associated to each resource (number of fishermen, vessel type and fishing gear, economic value of fisheries, among others). These approaches will serve as basis for the principles used to develop the strategic plans for the fishing sector regarding all simulated environmental scenarios.

Acknowledgments The authors would like to express their gratitude to Dr. Humberto Fuenzalida, from the Geophisics Department at Universidad de Chile, for his suggestions related to the revised literature.

References

Barbieri MA, Yáñez E, Farías M, Aguilera R (1989) Determination of probable fishing areas for the albacore (*Thunnus alalunga*) in Chile's central zone. In: Quantitative remote sensing: an economic tool for the nineties. IGARSS '89, Vancouver, vol 4, pp 2447-2450

Barbieri MA, Nieto K, Silva C, Yáñez E (2002) Evaluation of potential swordfish (*Xiphias gladius*) fishing grounds along central Chile, by the use of remote sensed sea surface temperature satellite data. In: Remote sensing applications for fisheries science: from science to operation. UNESCO, París, Publication no. 83, 213 pp

Brander K (2010) Climate change and fisheries management. In: Grafton Q, Hilborn R, Squires D, Tait M, Williams M (eds) Handbook of marine fisheries conservation and management. Oxford University Press, New York, pp 123–136, 770 pp

Braun M, Castillo J, Blanco J, Lillo S, Reyes H (1995) Monitoreo hidroacústico y oceanográfico de recursos pelágicos en I y II regiones. Informe Final FIP/93-15:1–172

Cahuin S, Cubillos L, Ñiquen M, Escribano R (2009) Climatic regimes and the recruitment rate of anchoveta, *Engraulis ringens*, off Peru. Estuar Coast Shelf Sci 84:591–597

Calvo E, Campos M, Carcavallo R, Cerri C, Gay-García C, Malta L, Saizar A (2000) América Latina. In: Canziani O, Díaz S (eds) Impactos regionales del cambio climático. IPCC, Evaluación de la Vulnerabilidad, 45 pp

Castillo J, Córdova J, Saavedra A, Espejo M, Gálvez P, Barbieri MA (2002) Evaluación acústica de la biomasa, abundancia, distribución espacial y caracterización de cardúmenes de anchoveta en el periodo de reclutamiento. Primavera 2001. Evaluación del reclutamiento de anchoveta en I y II regiones, temporada 2001–2002. Informe Final FIP-IT/2001-11, pp 1–207

Castillo J, Saavedra A, Hernández C, Catasti B, Leiva F (2008) Evaluación hidroacústica del reclutamiento de la anchoveta entre la I y II región. Instituto de Fomento Pesquero, Valparaíso, 438 pp

Duarte M, Alonso S, Benito G, Dachs J, Monte C, Pardo M, Ríos A, Simó R, Valladares F (2006) Cambio global: impacto de la actividad humana sobre el sistema tierra. Colección Divulgación, CSIC, 187 pp

Falvey M, Garreaud R (2009) Regional cooling in a warming world: recent temperature trends in the southeast Pacific and along the west coast of subtropical South America (1979–2006). J Geophys Res 114:1–16

FAO (2008) Fisheries management 2. The ecosystem approach to fisheries. 2.1 Best practices in ecosystem modelling for informing an ecosystem approach to fisheries. FAO fisheries technical guidelines for responsible fisheries, N°˜ 4, Suppl. 2, Add, 1. FAO, Rome, 78 pp

Fuenzalida H, Aceituno P, Falvey M, Garreaud R, Rojas M, Sánchez R (2007) Study on climate variability for Chile during the 21st century. In: Technical report of the National Environmental Committee, Santiago, Chile. http://www.dgf.uchile.cl/PRECIS

Garreaud R, Falvey M (2008) The coastal winds off western subtropical South America in future climate scenarios. Int J Climatol 29:543–554

Gregory JM, Church JA, Boer GJ, Dixon KW, Flato GM, Jackett DR, Lowe JA, O'Farrel SP, Roeckner E, Russel GL, Souffer RJ, Winton M (2001) Comparison of results from several AOGCMs for global and regional sea-level change 1900–2100. Clim Dyn 18:225–240

Gutiérrez JC, Silva C, Yáñez E, Rodríguez N, Pulido I (2007) Monthly catch forecasting of anchovy *Engraulis ringens* in the north area of Chile: non-linear univariate approach. Fish Res 86:188–200

Gutiérrez-Estrada JC, Yáñez E, Pulido-Calvo I, Silva C, Plaza F, Bórquez C (2009) Pacific sardine (*Sardinops sagax*, Jenyns 1842) landings prediction. A neural network ecosystemic approach. Fish Res 100(2):116–125

Jansen E, Overpeck J, Briffa KR, Duplessy JC, Joos F, Masson-Delmotte V, Olago D, Otto-Bliesner-B, Peltier WR, Rahmstorf S, Ramesh R, Raynaud D, Rind D, Solomina O, Villalba R, Zhang D (2007) Palaeoclimate. In: Solomon S, Qin D, Manning M, Chen Z, Marquis M, Averyt KB, Tignor M, Miller HL (eds) Climate change 2007: the physical science basis. Contribution of working group I to the fourth assessment report of the Intergovernmental Panel on Climate Change. Cambridge University Press, Cambridge, UK/New York, pp 433–497, 996 pp

Plaza F, Yáñez E, Gutiérrez JC, Pulido I, Rodríguez N (2008a) Non linear forecast of anchovy (*Engraulis ringens*) catches in northern Chile: a multivariate approach. GLOBEC Int Newslett 14(1):75

Plaza F, Yáñez E, Gutiérrez-Estrada JC, Pulido-Calvo I, Rodríguez N (2008b) Predicción de capturas de anchoveta (*Engraulis ringens*) en el norte de Chile mediante redes neuronales artificiales: un enfoque multivariado. In: Gutiérrez-Estrada JC, Yáñez E (eds) Nuevas aproximaciones metodológicas para el análisis de pesquerías. Universidad de Huelva publicaciones, Huelva, España, pp 39–58, 274 pp

Rodríguez N, Yáñez E (2008) Swarn intelligence based multivariate polynomial for anchovy catches forecasting. In: Ruan D, Montero J, Lu J, Martínez L, D'hondt P, Kerre E (eds) Computational intelligence in decision and control: ISI proceedings, 8th international FLINS conference. World Scientific Co. Pte. Ltd., pp 79–84

Sharp G (2004) Future climate change and regional fisheries: a collaborative analysis. FAO, Documento Técnico de Pesca 452, 84 pp

Silva C, Nieto K, Barbieri MA, Yáñez E (2002) Expert systems for fishing ground prediction models: a management tool in the Humboldt ecosystem affected by ENSO. Investig Mar 30 (1):201–204

Silva C, Yánez E, Nieto K, Barbieri MA, Martínez G (2004) EFISAT project: use of remote sensing and GIS technologies to predict pelagic fishing grounds in Chile. In: Nishida T, Kailola P, Hollingworth C (eds) GIS/spatial analyses in fishery and aquatic sciences. Fishery GIS Research Group, vol 2, pp 311–322

Trenberth KE, Jones PD, Ambenje P, Bojariu R, Easterling D, Klein, Tank A, Parker D, Rahimzadeh F, Renwick JA, Rusticucci M, Soden B, Zhai P (2007) Observations: surface and atmospheric climate change. In: Solomon S, Qin D, Manning M, Chen Z, Marquis M, Averyt KB, Tignor M, Miller HL (eds) Climate change 2007: the physical science basis. Contribution of working group I to the fourth assessment report of the Intergovernmental Panel on Climate Change. Cambridge University Press, Cambridge, UK/New York, pp 235–336, 996 pp

Yáñez E, Catasti V, Barbieri MA, Böhm G (1996a) Relaciones entre la distribución de recursos pelágicos pequeños y la temperatura superficial del mar registrada con satélites NOAA en la zona central de Chile. Investig Mar 24:107–122

Yáñez E, Silva C, Barbieri MA, Nieto K (1996b) Pesquería artesanal de pez espada y temperatura superficial del mar registrada con satélites NOAA en Chile central. Investig Mar 24:131–144

Yáñez E, Barbieri MA, Silva C, Nieto K, Espíndola F (2001) Climate variability and pelagic fisheries in Northern Chile. Prog Oceanogr 49:581–596

Yáñez E, Barbieri MA, Silva C (2003) Fluctuaciones ambientales de baja frecuencia y principales pesquerías pelágicas chilenas. In: Yáñez E (ed) Actividad Pesquera y de Acuicultura en Chile. Escuela de Ciencias del Mar, PUCV, pp 109–122

Yáñez E, Hormazábal S, Silva C, Montecinos A, Barbieri MA, Valdenegro A, Ordenes A, Gómez F (2008a) Coupling between the environment and the pelagic resources exploited off North Chile: ecosystem indicators and a conceptual model. Lat Am J Aquat Res 36(2):159–181

Yáñez E, Vega R, Silva C, Letelier J, Barbieri MA, Espíndola F (2008b) An integrated conceptual approach to study the swordfish (*Xiphias gladius*) fishery in the eastern South Pacific. Rev Biol Mar Oceanogr 43(3):641–652

Yáñez E, Plaza F, Gutiérrez-Estrada JC, Rodríguez N, Barbieri MA, Pulido-Calvo I, Bórquez C (2010) Anchovy (*Engraulis ringens*) and sardine (*Sardinops sagax*) abundant forecast off Northern Chile: a multivariate ecosystemic neural networks approach. Progress Oceanogr. In review

Chapter 17
Livelihoods of Coastal Communities in Mnazi Bay-Ruvuma Estuary Marine Park, Tanzania

Mwita M. Mangora, Mwanahija S. Shalli, and Daudi J. Msangameno

Abstract Marine protected areas (MPAs) are created to manage people's behavior in their use of coastal and marine resources. Although MPAs have strived to deliver the objects of resource protection, they often face challenges in translating the accrued benefits into enhanced livelihoods of local communities in and around their areas of jurisdiction. We used Mnazi Bay-Ruvuma Estuary Marine Park (MBREMP) in Tanzania to appraise the scenario of pro-poor conservation. The purpose of comparison between park and non-park villages was done to verify the hypothesis that establishment and operations of MPAs impairs local socio-economic practices without robust provision of alternative livelihood safety nets. Agriculture remains a persistent livelihood occupation both in park and non-park villages. Artisanal fishing is a substantial livelihood occupation in seafront villages but a secondary activity in overall. Income and expenditure patterns indicated that non-park villages are better-off with significantly high income to expenditure ratios. Fishing make the most contribution to income in sea front villages as agriculture is doing in non-fishing villages. Impacts on livelihoods emanate from disrupted resource use patterns which significantly influence the communities' perception on need, role and overall acceptance of the marine park. Traditional access and user rights are marred by MPA operations putting at stake livelihood security of the communities therein. Alternative strategies have not yet been given due thrust and local communities remain insecure in accessing political assets such as cooperatives, community credit schemes and financial assets such as government and/or commercial banking sponsored schemes and loans. Local communities are already carrying the costs of denied access to livelihood sources, but the marine park is not quick enough to translate the accrued value and benefit of the improved resource base in enhancing local communities' livelihood and welfare. Reducing pressure on marine resources through sound management interventions will have to be accompanied by mitigating measures to safeguard household

M.M. Mangora (✉) • M.S. Shalli • D.J. Msangameno
Institute of Marine Sciences, University of Dar es Salaam, Mizingani Road,
P.O. Box 668, Zanzibar, Tanzania
e-mail: mmangora@yahoo.com; shalli@ims.udsm.ac.tz; msanga@ims.udsm.ac.tz

food security, such as compensation, and developing alternative sources of income. There is still considerable polarization between conservation and socio-economic welfare of the people. MPAs should focus on combining resource management with livelihood opportunities that provide economic benefits in the short-run to address economic disruptions emanating from disrupted access to the once common resources.

Keywords Conservation • Livelihoods • Local communities • Marine Park • Poverty

1 Introduction

Coastal and marine resources in developing countries are under increasing threats due to ever increasing numbers of resource users with competing interests (Crossland et al. 2005). The damage of these natural assets diminishes livelihood opportunities and therefore aggravates poverty. The poor who are, living in remote and marginal lands in rural areas, where the basic social services are persistently inadequate, remain at stake. In such situations, they viciously remain prone to natural resource dependency for their primary livelihood options. So, degradation and losses continue unabated albeit at the expense of the poor. To this, arrays of both institutional and operational strategies have been evolving over the past couple of decades in attempts to curb the deteriorating resource bases and the livelihood assets thereof. One of such ecosystem-based management approaches are Marine Protected Areas (MPAs) in coastal locations of Tanzania (Halpern 2003; Tobey and Torrel 2006; Pollnac et al. 2010).

The conservation concern of MPAs is that the health of these coastal and marine resources is affected by human activities, though livelihoods and prosperity of these people depend upon the condition of the same resources (IUCN 1988). Thus, MPAs are directly linked to the socio-economic environment in which the beneficiaries operate. From the institutional point of view, MPAs have evolved to manage the behavior of people in wise-use of coastal and marine resources (Mascia 2004; Pomeroy et al. 2004) and they are being advocated to win the support and participation of local stakeholders (Ruitenbeek et al. 2005; Sesabo et al. 2006). In Tanzania, there are currently three operating marine parks and 15 marine reserves (Fig. 17.1). The marine parks include Mafia Island Marine Park (MIMP), the first to be established in 1996, Mnazi Bay-Ruvuma Estuary Marine Park (MBREMP, situated along the border with Mozambique), which is used as a case study area in the present work was connoted in 2000, and the Tanga Coelacanth Marine Park (TACMP) established in 2009. The fifteen marine reserves include Dar es Salaam Marine Reserve system (DMRs) comprising of six small islands of Bongoyo, Pangavini, Mbudya, Makatube, Sinda, Kendwa and one sand bank of Funguyasini; Maziwe island located in Pangani district; Nyororo, Mbarakuni and Shungimbili located north of Mafia Island; and the four newly gazetted islands north of Tanga (i.e. Kwale, Mwewe, Kirui and Ulenge) near the border with Kenya.

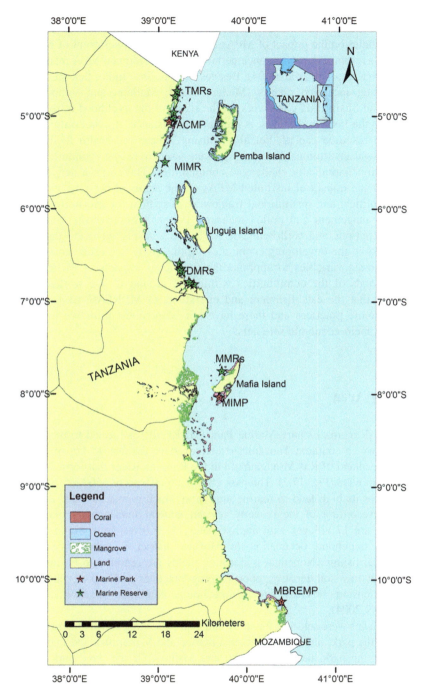

Fig. 17.1 Map of the coastal area of Tanzania showing MPAs managed under MPRU. *TMRs* Tanga Marine Reserves System (Kirui, Mwewe, Kwale and Ulenge), *TACMP* Tanga Coelacanth Marine Park, *MIMR* Maziwe Island Marine Reserve, *DMRs* Dar es Salaam Marine Reserves System (Bongoyo, Pangavini, Mbudya, Makatube, Sinda, Kendwa and Funguyasini), *MMRs* Mafia Marine Reserves System (Nyororo, Shungimbili and Mbarakuni), *MIMP* Mafia Island Marine Park, *MBREMP* Mnazi Bay-Ruvuma Estuary Marine Park

Often studies on the impact of MPAs have inclined to assessment of biological ecosystem's responsive performance pertaining to the enforced MPA management institutions (Kamukuru et al. 2004). Besides the baseline studies that are commissioned during inception phases (Malleret 2004; Malleret and Simbua 2004; Mangora and Shalli 2012), the work on the status of the livelihood trajectories in response to the instituted MPAs, is meager. For instance, it is not only access to the natural resource capital, but also housing, education, health facilities and access to legal institutions that are important assets to assure economic security to the socio-economically challenged communities. Therefore, a deeper understanding of the impact of instituted MPAs on the status of bio-physical, social, cultural, political and institutional framework is critical for decision support in resource management and policy measures that will improve the household's livelihood options and well-being, if we are to sensibly advocate the scaling up of MPAs in the developing economies. In this study, we used some of these socio-economic variables to appraise the functional impacts of MBREMP on the livelihoods of the communities within and around it. We worked on the hypothesis that the establishment and operations of MBREMP have impaired socio-economic practices and there have not been robust initiative to provide alternative socio-economic safe nets.

2 Study Area

Mnazi Bay-Ruvuma Estuary Marine Park (MBREMP) is located to the south of Mtwara town in southern Tanzania, stretching over the last 45 km of coastline from the headland of Ras Msangamkuu to the Ruvuma River that form the border with Mozambique (Fig. 17.2). The park covers a total area of 650 km^2. MBREMP is unique for its high land to marine area ratio which represents 33 % (220 km^2 of land). According to the recently revised general management plan of the park, there are 17 villages with approximately 44,000 residents within the park. The main livelihood occupation in Mtwara district is subsistence farming and artisanal fishing. Nonetheless, farming yields are reported to be low due to the inherent low soil fertility, poor farming practices and inputs, farm losses, limited extension services and compounding unreliable weather conditions (CONCERN 2004).

Five villages were selected for study and data collection, of which three villages are within the park, namely Msimbati, Litembe and Mahurunga and two villages are outside the park, namely Naumbu and Msijute. Of the three park villages, each represented either one of the three park eco-zones, i.e. seafront, mangrove surrounding and riverine habitat respectively. Non-park villages represented two major livelihood occupations, i.e. fishing and agriculture respectively.

17 Livelihoods of Coastal Communities in Mnazi Bay-Ruvuma Estuary Marine Park... 275

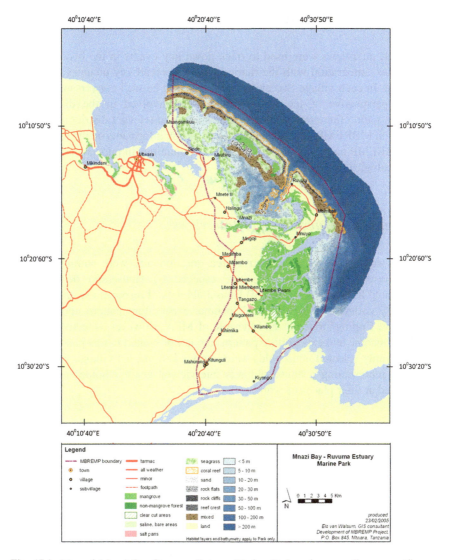

Fig. 17.2 Map of Mnazi Bay-Ruvuma Estuary Marine Park and surrounding areas (Source: MBREMP General Management Plan)

3 Study Methodology

3.1 Focus Group Discussions (FGDs)

Focus group discussions were used to rapidly visualize community profiles by eliciting primary information on livelihood assets and resource use patterns with reference to the existence and operations of the MPA. Separate groups of women, men and youth were formed. Group size varied from 6 to 10 individuals.

3.2 Key Informant Interviews

Open-ended interviews were used to track down key events in the history of the communities' integration with the marine park and recall key changes that have taken place. In each village two elders, considered to be well conversant with the historical perspective of the communities, were involved in the discussions. In addition, park, other conservation organizations and relevant district authorities working with MBREMP were also interviewed as key informants on the impact and role of MBREMP on the livelihoods of local communities in pursuance of biological and ecological integrity.

3.3 Household Questionnaire Interviews

Household surveys using a semi-structured questionnaire were administered to collect datasets on the following five aspects: (i) Household descriptions; (ii) Household livelihoods assets; (iii) Household natural resource use and productive activity patterns; (iv) Household income, expenditure pattern and material lifestyle; (v) Coping strategies and the role of MPA. However, selective data on household income and expenditure patterns are not presented in this paper. A total of 30 randomly selected households were interviewed in each village. Each of an individual representing a household was interviewed at their homes and where appropriated at their places of work.

3.4 Data Analysis and Presentation

Data from group discussions and in-depth interviews were subjected to content analysis paying an extra attention in filtering to avoid any possible misjudgment. Statistical Package for Social Sciences (SPSS) was used to process data from household questionnaires and present results in descriptive statistics, graphical presentations and cross-tabulations. For household income and expenditure data, regression analysis and ANOVA were used to test for statistical variations at $P = 0.05$.

4 Results and Discussion

4.1 Household Characteristics

4.1.1 Household Size and Age Structure

Average household size did not differ significantly between park and non-park villages (Table 17.1). Majority (44 %) of the households had 5–7 members

Table 17.1 Average household sizes in studied villages (N = 30 for each village)

Village	Minimum	Maximum	Mean ± SE
Msimbati	3	15	6.80 ± 0.558
Litembe	1	13	5.77 ± 0.467
Mahurunga	2	12	6.13 ± 0.516
Naumbu	2	14	6.53 ± 0.481
Msijute	1	10	5.63 ± 0.388

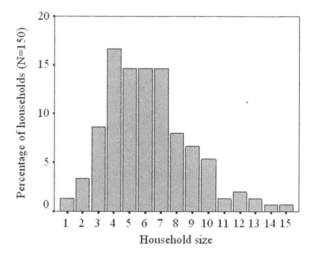

Fig. 17.3 Overall percentage distribution of household sizes in the study villages

(Fig. 17.3), making an overall average household size of 6. A considerable number of households (11 %) had sizes of 10 and above. In terms of the age distribution, over 80 % of the households had 1–3 adult members living at home. This formed the prime working force of the households. Over 50 % of the households comprised members of under 18, the age group which is composed of school children and babies. In Tanzania, 18 is the age legally recognized of being adult.

Household size and age structure are among indicative parameters of the level of household dependence and use of natural resources (Coad et al. 2008). These also have an implied relationship with the household wealth status. The observed household sizes in the study villages are all above the national, regional and district averages which are 4.9, 3.8 and 4.0 respectively as indicated in the population and housing census report of 2002 (URT 2002). Large households (high consumption levels) with less productive members (low level of human capital) and limited access to assets are prone to poverty and their livelihoods have much reliance on natural resource capital with accelerated use of inappropriate practices in order to maximize the output. People sacrifice their future livelihood opportunities to meet present needs. These observations are in conformity with reports by other workers, who suggest that poverty and dependence on marine and coastal resources is directly correlated in most villages of Mtwara (URT 1997). Furthermore, the poor have maintained relatively free access to the coastal and marine resources; and therefore, any activity that draws from natural resource bases, be it agriculture,

Table 17.2 Percentage of households which had at least one member who had attained a given level of education in the villages within study area (N = 30 for each village)

Village	Not attend school	Primary	Secondary	Tertiary/college	Other forms/informal
Msimbati	33.3	96.7	30.0	–	10.0
Litembe	83.3	90.0	6.7	–	3.3
Mahurunga	58.6	90.0	6.7	–	16.7
Naumbu	36.7	93.3	20.0	6.7	3.3
Msijute	69.2	78.6	28.0	–	–

fishing etc., has well remained to be an opportunity of last resort to make a living. Inherently therefore, the natural resources they depend upon, remain under intense pressure (Barbier 2007). Although the effects of demographic factors on the quality of natural resource may be indeterminate (Scherr 2000), many analysts argue that they are the major factors contributing to poverty especially in third world countries (Birdsall et al. 2001).

4.1.2 Household Education

Overall, 56 % of the surveyed households reported to have at least one member who did not attend school or have any formal education. In most cases, these comprised of the household heads and their spouses. For primary level of education, about 90 % of the households indicated to have at least one member who had attained primary education. Only 18 % of the households reported to have at least one member with secondary level of education and virtually below 2 % had one member with tertiary level of education. Table 17.2 summarizes the village specific percentage representation of households with at least a member in a given category of education attainment.

This trend of educational attainment indicated the prevalent illiteracy in the study areas showing no significant difference between park and non-park villages. Higher levels of illiteracy are often associated with stanchly limited livelihood opportunities and limited access to assets other than drawing from the existing natural resource corpus. Better education would have meant increased employment opportunities leading to better occupation and alternative livelihood opportunities assuring augmented income (Kideghesho et al. 2007). However, the inherent vicious cycle of poverty and the high costs of living reciprocate on the households' income security to pursue better education (Coad et al. 2008).

4.2 Livelihood Occupations and Security

There has been no major shift in household livelihood options and resource use patterns between pre- and post ante period of establishment of MBREMP. People had strived and continued to engage in similar activities, but there was a reduced

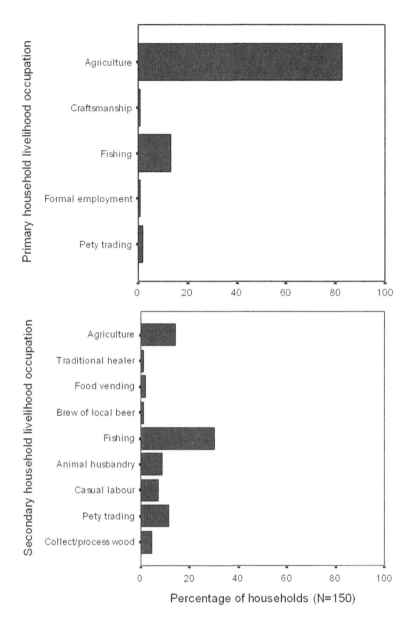

Fig. 17.4 Overall percentage distribution of the reported primary and secondary livelihood occupations in the study areas

effort particularly in fisheries in park villages due to the regulation and controlled access by the marine park. Agriculture remains the predominant primary activity from which households in both park and non-park villages (83 %) depended on to earn a living (Fig. 17.4). Of those, 66 % depended solely on agriculture for food and

Table 17.3 Village specific percentage distribution of the major household primary and secondary livelihood occupations

Occupations/villages	Msimbati	Litembe	Mahurunga	Naumbu	Msijute
Primary activities					
Agriculture	86.7 (73, 23)	76.7 (63, 20)	96.7 (83, 53)	60.0 (40, 10)	93.3 (70. 33)
Fishing	6.7 (-, 20)	23.3 (-, 60)	3.3 (-, -)	33.3 (7, 33)	- (-, -)
Petty trading	3.3	–	–	6.7	–
Secondary activities					
Agriculture	10.0	23.3	3.3	33.3	–
Fishing	46.7	63.3	13.3	26.7	–
Petty trading	10.0	–	16.7	20.0	10.0
Animal husbandry	3.3	–	13.3	–	23.3
Casual labor	6.7	–	3.3	–	26.7

In brackets are relative percentages of household reliance on primary activities of agriculture and fishing as sole sources of (food, income) (N = 30 for each village)

28 % earned some revenues from agriculture. Fishing accounted for only 13 % as a primary source of livelihood with only 20 % of them relying on it for food and another 23 % could secure little earnings from fishing. While agriculture is a common occupation, fishing is predominant in the seafront villages both within and outside the park. When reported as secondary activities, fishing was more important at 30 % than agriculture which was reported by 14 %. Other important secondary activities include petty business (11 %), animal husbandry (9 %) and casual labor (7 %). Table 17.3 presents an indication of the specific villages' percentage distribution of the household livelihood occupations. In terms of the overall livelihood security, over 50 % of the households indicated that most of the time they have little food or money and remained unemployed.

Though these occupations are all largely practiced at subsistence scales, agriculture is practiced for both food and income while fishing is mainly for cash income. These two primary livelihood activities have also remained characteristically unsecured and therefore for Mtwara rural communities, food and income security issues are inherently intrinsic phenomena that need concerted interventions that integrate both conservation and economic bases. These findings are comparable to those observed by Malleret and Simbua (2004) who also found that the most widely spread activity was farming (87 %) among the park households and that fishing was second with only 26 % of park households involved in active fishing. Resource use patterns significantly influence the communities' livelihood activities and their perception on need, role and acceptance of MPA. Park villages like Mahurunga for example, which leads in agriculture as the main stay source of livelihood may comparably be easy supporters if enhanced agriculture would be an agenda of the park.

The overriding concept of MPAs is an integrated community-ecosystem conservation paradigm for livelihoods; though the prevalence of the practice has remained biased to fishery management than the livelihood options of the coastal communities, as a whole. The perceived fishing communities are not really fishing villages (Fig. 17.4 and Table 17.3). This study has demonstrated that coastal communities in Mtwara are in essence agrarian and fishing plays just a secondary role to make a living. In this situation, and for the excellence of MBREMP, the empirical focus would really be to enhance agricultural production of both food and cash crops including animal husbandry, which would in turn relieve pressure on fishing. Contrary, MPAs do take matters at macro levels, where for example, the analysis of the relationships of livelihoods and dependency on the environment and natural resource capital, may be biased to proclaim that coastal communities are fishers (Chuenpagdee and Bundy 2005) and thus the focus of an MPA falls on fishery management under the precept of enhancing livelihoods. But often, the immediate shortfall of MPAs in developing countries as demonstrated in this study is the failure to highlight the specific choices that people make for their survival, the livelihood strategies that evolve, and the policy decision that are drawn in mediating these relationships at the micro level, i.e. at the household level where a living really matters.

4.3 Household Income and Expenditure Pattern

Fisheries and farming were variably the major sources of household income, with fishing at significantly the upper hand (Fig. 17.5a). Other reported sources of income included, small-business involving running of vending kiosks, food vending and livestock keeping. Overall, total mean annual income across villages was about 1.7 million Tanzania Shillings (TZS). Food accounted for the majority of expenses, taking up to over 66 % of overall household total mean annual expenditure (Fig. 17.5c). The expenditure on food referred to here combined estimates for both own-produced and purchased food. Apart from food, the items that had most widespread consumption by households were clothing, education, and housing. Nonetheless, analysis indicated that even if expenses on these non-food major household items were combined, they would still remain disproportionately less than the expenses on food. For expenses on health service, the observed less and meager expenditure represented the general experience among rural poor households, that they tend to spend less on medical care, a social behavior that is often related to low income and prevalent illiteracy, although this might have been counterchecked by the subsidence of, but yet poor medical services in government dispensaries and health centers. The majority of rural communities often tend to rely on traditional herbal remedies of which they can't put value on and account for (Mangora and Shalli 2012).

Specifically, Naumbu, a seafront non-park village was a better-off village with significantly higher household income to expenditure ratio of about 5 while Msimbati, a seafront park village had the lowest income to expenditure ratio of 1 (Fig. 17.5b, d). Other park villages of Litembe and Mahurunga had ratios of 1.4 and 1.8 respectively while Msijute, another non-park village had a ratio of 1.8. This income and expenditure pattern indicated that non-park villages are much better off with expressed household savings from income accruals for other household developmental activities, while in a village like Msimbati, households were more or

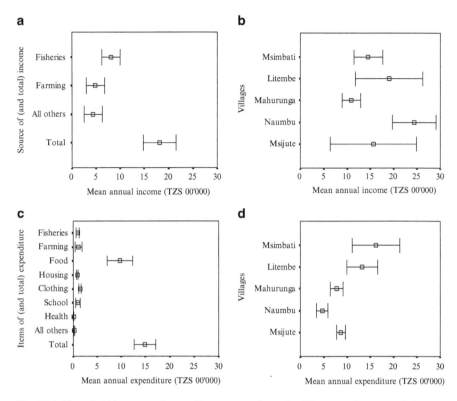

Fig. 17.5 Household income and expenditure patterns in study villages. a = income variation per different sources, b = income variation per villages, c = expenditure variation per different items, d = expenditure variation per villages

less proportionally spending much on recurrent commodities with no or little savings for household development endeavors like purchase of durable assets which often act as safety nets in events of household economic shocks. The lower the income to expenditure ratio, the higher a household would be at risk of income insecurity as households become more prone to not being able to self sustain in covering all household expenses, because household earnings are hardly sufficient. More expenditure over income is an indication of how much vulnerable are household budgets to deficits and economic shocks (Biostockpro 2012). Such households tend to be more prone to poverty (Lanjouw and Ravallion 1995). In a nutshell, households accrue income for survival, hardly accruing savings that would allow them to pursue extra opportunities of development. Comparable findings have recently been reported in other seafront villages adjacent to the newly established Tanga Marine Reserve system in the north coast of Tanzania (Mangora and Shalli 2012). The implication for this income sourcing pattern may mean that, disrupted artisanal fisheries operations exert considerable shocks in livelihoods dynamics of the communities in seafront villages if there will be no appropriate integrative safenet measures as access to fishing grounds within park boundaries become restricted.

4.4 Community Perceptions: Role and Acceptance of MPA

4.4.1 Participation and Empowerment

Though local communities reclined to the concepts of conservation and protection of natural resources, but, their overall attitude about conserving the marine park doesn't seem to be a serious concern. In Litembe village for example, a representative message was that "we are not willing to pay for the conservation and protection of resources, because the marine park is ruining our livelihood opportunities. Much importance is placed on the biodiversity excellence at the expense of our wellbeing". This observation demonstrated that community resistance to MBREMP is related to the perceived social costs, which are represented as a state of lack in community participation at the planning level of conservation - development initiatives. It is rather a clearer indication of a desire of the community for greater participation in managing marine and coastal resources. From the policy and management point of view it is unfortunate to note that although the need for community involvement in management of natural resources has been widely proposed as an important element in sustainable development (Lewis 1997; Sunderlin and Gorospe 1997) this has faced conflicts of interests while putting it into practice. The problem is on the management approach where the advocated participatory management is not practically in place. Even the established marine park village liaison committees for example, are blamed to have overshadowed the assumed responsibilities and jurisdictions of the village environmental committees. Perceptively, MBREMP is blamed further for taking issues on their own hands and the communities have eventually become powerless over their natural endowments and resources that they once commonly accessed.

4.4.2 Tenure and User Rights

Exclusively communities say, "We don't need the marine park to come and assume our right to conserve our own resources, they can only provide us with guidance and expertise to draw from". With regard to tenure and user rights on land and other resources, the community is accusing the marine park for seizing their rights to exercise ownership and use their customary and endowed resources. In Msimbati village for example, it was reported that one resident was allegedly denied from selling his portion of his land to one cellular phone company which wanted to erect a communication tower. The marine park intervened to block the deal on grounds that the procedures for having permanent development structures within the park boundaries were not adhered to and further that the mandate of authorizing any construction project is with the marine park and not the local authorities. It is for these such cases that Chatty and Colchester (2002) and Brockington et al. (2006) note that human rights campaigners do build on to accuse park authorities and their supporters in the conservation community of their illegal deprival of access and use rights of local people. Such kind of disagreements are typical of an increasing polarization of positions in which trade-offs are not just erupting between indigenous people versus conservationists, but also protection versus people, and parks versus development (McShane and O'Connor 2007). For MBREMP, issues of tenure rights and rights of occupancy are not yet explicitly addressed to the understanding of local communities and this has been a reference point to show how the marine park has betrayed the communities. "Marine park has become the land lord of our own customary land", people at Msimbati complained. Other studies have also demonstrated that clarity and congruence of rules governing resource use have much influence on MPA performance and other natural resource governance regimes (Mascia 2001; Pomeroy et al. 2006). Rules governing resource use that are explicitly linked to local conditions also tend to enhance reserve performance (Mascia 2000). MPAs tend to be enhanced when reserve resource use rights are consistent with existing informal or socio-cultural resource use rights (Fiske 1992; Mascia 2000).

4.4.3 Sharing the Costs and Benefits: A False Promise of Win-Win

While local communities are already carrying the opportunity costs of denied access to livelihood sources, the marine park has not adequately translated the accrued value and benefit of the improved resource capital in enhancing livelihood opportunities and ensuring welfare for the local populace. Instead, villagers are of the opinion that these conservation initiatives should have been mandated to the local government authorities like the Department of Natural Resources as they feel a sense of representation, ownership and respect towards these. These revelations demonstrated and confirmed the failure of promising a win-win scenario that has been widely advocated and applied by stakeholder organizations to rationalize simultaneous achievement of conservation and development (McShane and O'Connor 2007; Coad et al. 2008). But, conservation experience depicts that initiatives, which produce win-win outcomes are rare (Christensen 2004) and in practice, many such attempts to simultaneously meet the twin goals of conservation and human development have fallen short of expectations (Robinson 1993; McShane and Wells 2004). In Litembe village, which has the largest mangrove forest cover amongst all the park villages and also hosts turtle nesting grounds, it was alleged that already ecotourism is taking place there but the village authority has never been informed of any such developments. Further, it was enquired that if there are any revenues accrued how the same should be invested for the benefit of the villages. Likewise, at Msimbati, though there is a marine park control gate for visitors, the local authority is not well informed of the functions of the gate, and if there are any entry charges, how are they accounted for.

4.4.4 Social Development and Alternative Sources of Livelihoods

Deteriorating qualities of social services like water supply, health facilities, schools, roads etc. are asset based indices of poverty (Coad et al. 2008), during the field surveys we did not observe any commendable improvement which can be attributed to the presence of MBREMP as a social change agent. One noted assistance from the park was the construction of a classroom for a ward secondary school at Mahurunga village. Otherwise, the little improvement on water and health services is funded by other donors like Japanese International Cooperation Agency (JICA) which apparently had no mutual cognizance in interventional initiatives with MBREMP.

Promotion of alternative income generating activities has not been encouraging in terms of facilitation, inception, performance and sustainability. Only three pilot projects towards alternative livelihood were commissioned but none was in the seafront villages where there were strong resistance to the park. One of these projects was a poultry rearing project at Mahurunga led by women, and another was a mangrove crab fattening farm at Litembe. The projects were rendered unsustainable and local communities opined in despair that the scale of the promoted and supported activities is not worth the opportunity costs of the benefits foregone by the beneficiaries for the sake of conservation and protection of the park. This could easily be substantiated because within the course of 2 years of field work for this study, the two projects had collapsed owing to lack of competent management and appropriate extension guidance. The general public concern is that there have not been any significant changes to the livelihood and well being brought about by the establishment of the marine park. So, whenever there is lack of acceptable alternative sources of employment and livelihood opportunities, the insecurity of losing access to natural resource capital drives people to resist change (IUCN 2010) as it was also noted in MBREMP.

5 Conclusions

Despite the close inter-linkages between natural resource conservation and poverty alleviation, there is still considerable polarization between conservation and community development. Development agencies have often undervalued the potential role natural resource goods and services can play in poverty reduction. On the other hand conservation organizations have viewed poverty concerns to be outside their core business. The present study has demonstrated that there is contradictory evidence as to the efficacy of protected areas in conserving natural resources because they have not been without costs. Often, protected areas have been associated with forced loss of access to natural resources for the people living in and around them, with inadequate alternative opportunities. MPAs for example are ideally advocated as the solution for fisheries and ecosystem management problems, but in reality, they are not substitutes for fishery management, rather are one of several tools in the toolbox (Pomeroy et al. 2006).

MPAs can therefore either be beneficial or derogative to development for the local communities depending upon how they are designed and implemented. An often shortfall of MPAs to the coastal communities is that, strategic impact assessment of MPAs pertaining to community livelihood, and social auditing as well, have not been part of official planning processes. Planners are often surprised when communities resist the establishment or expansion of MPAs fearing that access to the naturally endowed resources will be restricted or cut off completely (McShane and O'Connor 2007), resulting in conflicts between people and state agencies or other government sponsored sector programs (Chatty and Colchester 2002; Brockington et al. 2006). It is from this rhetoric ideology that MPAs are skewed to the excellence of biodiversity at the expense of livelihoods of the people, cases in which their costs remain concentrated while benefits diffuse (Pomeroy et al. 2006). An MPA may attain a biological success by increased fish abundance, diversity and improved habitat (diffusing benefits), but derive a socio-economic failure in lack of broad stakeholder partnership and community participation in management, sharing of economic benefits and mechanisms of resolving conflicts of interest (concentrating costs). In this situation, biological gains may only be short-termed and likely to disappear unless the failing community livelihood issues are resolved.

References

Barbier EB (2007) Natural resources and economic development. Cambridge University Press, Cambridge, 428 pp

Biostockpro (2012) What is the right income expenditure ratio for your family? http://www.biostockspro.com/fuel/financial-planning/family-budget-planning/. Accessed 01 June 2012

Birdsall N, Pinckney T, Sabot R (2001) Natural resources, human capital, and growth. In: Auty RM (ed) Resource abundance and economic growth. Oxford University Press, Oxford, pp 57–75

Brockington D, Igoe J, Schmidt-Soltau K (2006) Conservation, human rights and poverty reduction. Conserv Biol 20:424–470

Chatty D, Colchester M (eds) (2002) Conservation and mobile indigenous peoples: displacement, forced settlement, and sustainable development. Berghahn Books, Oxford, 420 pp

Christensen J (2004) Win-win illusions: facing the rift between people and protected areas. Conserv Pract 5(1):12–19

Chuenpagdee R, Bundy A (eds) (2005) Innovation and outlook in fisheries: an assessment of research presented at the 4th World Fisheries Congress. Fisheries Centre Research Report, vol 13(2). University of British Columbia, Vancouver, 113 pp

Coad L, Campbell A, Miles L, Humphries K (2008) The costs and benefits of protected areas for local livelihoods: a review of the current literature. Working paper. UNEP World Conservation Monitoring Centre, Cambridge, UK

CONCERN (2004) Community empowerment for livelihood security programme. Baseline Survey Report, CONCERN Worldwide Tanzania, Dar es Salaam

Crossland CJ, Kremer HH, Lindeboom HJ, Marshall Crossland JI, Le Tissier MDA (eds) (2005) Coastal fluxes in the anthropocene – the land-ocean interactions in the coastal zone project of the international geosphere-biosphere programme, Global change – the international geosphere-biosphere program series. Springer, Berlin, 231 pp

Fiske SJ (1992) Sociocultural aspects of establishing marine protected areas. Ocean Coast Manag 18:25–46

Halpern B (2003) The impact of marine reserves: do reserves work and does reserve size matter? Ecol Appl 13(1):117–137

IUCN (2010) Building policy for supporting livelihoods through conservation: review of bottlenecks to influencing environmental and development policy. IUCN ESARO Office, Nairobi, iv + 18 pp

Kamukuru AT, Mgaya YD, Öhman MC (2004) Evaluating a marine protected area in a developing country: Mafia Island Marine Park, Tanzania. Ocean Coast Manag 47:321–337

Kideghesho JR, Roskat E, Kaltenborn BP (2007) Factors influencing conservation attitudes of local people in Western Serengeti, Tanzania. Biodivers Conserv 16(7):2213–2230

Lanjouw P, Ravallion M (1995) Poverty and household size. Econ J 105(433):1415–1434

Lewis D (1997) Rethinking aquaculture for resource-poor farmers: perspective from Bangladesh. Food Policy 22(6):533–546

Malleret D (2004) A socio-economic baseline assessment of the Mnazi Bay – Ruvuma Estuary Marine Park. IUCN EARO Publication Service Unit, Nairobi, x + 126 pp

Malleret D, Simbua J (2004) The occupational structure of the Mnazi Bay – Ruvuma Estuary Marine Park communities. IUCN EARO Publication Service Unit, Nairobi, vi + 42 pp

Mangora MM, Shalli MS (2012) Socio-economic profiles of communities adjacent to Tanga Marine Reserve Systems, Tanzania: key ingredients to general management planning. Curr Res J Soc Sci 4(2):141–149

Mascia MB (2000) Institutional emergence, evolution, and performance in complex common pool resource systems: marine protected areas in the Wider Caribbean. PhD thesis, Department of the Environment, Duke University, Durham, NC

Mascia MB (2001) Designing effective coral reef marine protected areas: a synthesis report based on presentations at the 9th International Coral Reef Symposium. IUCN World Commission on Protected Areas-Marine, Washington, DC

Mascia MB (2004) Social dimensions of marine reserves. In: Dahlgren C, Sobel J (eds) Marine reserves: a guide to science, design, and use. Island Press, Washington, DC, pp 164–186

McShane TO, O'Connor S (2007) Hard choices: understanding the trade-offs between conservation and development. In: Redford KH, Fearn E (eds) Protected areas and human livelihoods, WCS working paper no. 32. Wildlife Conservation Society, Bronx, pp 145–152

McShane TO, Wells MP (2004) Integrated conservation and development? In: McShane TO, Wells MP (eds) Getting biodiversity projects to work: towards better conservation and development. Columbia University Press, New York, pp 3–9

Pollnac R, Christie P, Cinner JE, Dalton T, Daw TM, Forrester GE, Graham NAJ, McClanahan TR (2010) Marine reserves as linked social–ecological systems. http://www.pnas.org/content/107/43/18262.full

Pomeroy RS, Oracion EG, Pollnac RB, Caballes DA (2004) Perceived economic factors influencing the sustainability of integrated coastal management projects in the Philippines. Ocean Coast Manag 48:360–377

Pomeroy RS, Mascia MB, Pollnac RB (2006) Marine protected areas: the social dimension. FAO expert workshop on marine protected areas and fisheries management: review of issues and considerations, Background paper 3, FAO Rome, Italy

Robinson JG (1993) The limits to caring: sustainable living and the loss of biodiversity. Conserv Biol 7:20–28

Ruitenbeek J, Hewawasam I, Ngoile M (2005) Blueprint 2050: sustaining the marine environment in mainland Tanzania and Zanzibar. The World Bank, Washington, DC, 144 pp

Scherr S (2000) A downward spiral? Research evidence on the relationship between poverty and natural resource degradation. Food Policy 25:479–498

Sesabo JK, Lang H, Tol RSJ (2006) Perceived attitude and marine protected areas (MPAs) establishment: why households' characteristics matters in coastal resources conservation initiatives in Tanzania. Working paper FNU-99. Research Unit Sustainability and Global Change and Centre for Marine and Atmospheric Science, Hamburg University, Hamburg

Sunderlin WD, Gorospe MLG (1997) Fishers organizations and modes of co-management; the case of San Miguel Bay, Philippines. Hum Organ 56(3):333–343

Tobey J, Torrel E (2006) Coastal poverty and MPA management in mainland Tanzania and Zanzibar. Ocean Coast Manag 49(11):834–854

URT (1997) Mtwara region socio-economic profile. United Republic of Tanzania, The Planning Commission, Dar es Salaam and Regional Commissioner's Office, Mtwara

URT (2002) Population and housing census report 2002. United Republic of Tanzania, National Bureau of Statistics, Dar es Salaam

Chapter 18
Livelihood Strategy in Indonesian Coastal Villages: Case Study on Seaweed Farming in Laikang Bay, South Sulawesi Province

Achmad Zamroni and Masahiro Yamao

Abstract Seaweed farming has been done by fishermen for many years and contributing to the local economy. This paper aims to analyze the sustainable coastal livelihoods in a coastal community in Indonesia. It is noteworthy that some alternative livelihoods are important ways to raise local economy, which are affected by decreased production of capture fisheries. This paper would also assess the roles of marine culture in the local economy of coastal villages, to investigate challenges and opportunities of coastal livelihoods, and to analyze fishers' perception and stakeholder participation in developing alternative livelihoods. Structured and semi-structured questionnaires as qualitative approach, and interviews were conducted with 200 fishermen in Takalar and Jeneponto Districts, South Sulawesi, Indonesia. Secondary data included a series of reports and other statistical data. SWOT analyses were used to assess the problems and opportunities of coastal livelihoods development. The finding showed that there are two livelihood activities namely seaweed farming (*Eucheuma cottonii*) and fishing activity, where in seaweed farming having bigger contributions to household income. Culture of seaweed has given a positive impact to local economic condition of coastal villages in South Sulawesi in the form of increased household income of fishermen. However, changes in monsoon periods, environmental awareness of

---

Achmad Zamroni (✉)
Research Center for Marine and Fisheries Socio Economics, Ministry for Marine Affairs and Fisheries Republic of Indonesia, Jl. K.S. Tubun Petamburan VI, 10260 Jakarta, Indonesia
e-mail: roni_socio@yahoo.com

M. Yamao
Graduate School of Biosphere Science, Laboratory of Food Production Management, Department of Bioresource Science, Hiroshima University, 1-4-4 Kagamiyama, Higashi Hiroshima-City 739-8528, Japan
e-mail: yamao@hiroshima-u.ac.jp

fishermen and post-harvest technology are major constraints in developing seaweed farming. Finally, fishermen's livelihood are transformed from the fishing activity to seaweed (*Eucheuma cottonii*) as main income source.

Keywords Livelihoods • Sustainable • Seaweed farming • Coastal Village and Indonesia

1 Introduction

Indonesian national fisheries production grows at an average annual rate of 9.92 % from 6.12 million tonnes to 9.82 million tonnes during 2004–2009. In terms of capture fisheries, production increased annually at a mean rate of 1.91 % from 4.65 million tonnes in 2004 to 5.11 million tonnes in 2009. As for aquaculture, a sharp increase during those period was seen about at 26.64 %, from 1.47 million tonnes to 4.71 million tonnes (MMAF and JICA 2011). However, some fisheries management areas (FMA) have already experienced overfishing. The production of fish species has decreased by around 20 % in some fisheries management areas in recent years. As a consequence, coastal poverty, population growth, heavy dependence on fisheries resource and resource degradation continue to affect coastal community livelihood, where approximately 16.4 million people live in coastal areas, and 32 % of whom are living under the poverty line (Kusnadi et al. 2006).

During the last two decades, The Government of Indonesia (GoI) has promoted and implemented a series of coastal projects for poverty alleviation, focusing on sustainable use of coastal resources and enhancement of fisheries livelihood. Marine and Coastal Resource Management Project (MCRMP) is one of coastal projects aimed to sustain livelihood, improve management, conservation of the environment by developing seaweed farming as main livelihood activity in fishery communities and by supported international donor agencies. As part of MCRMP, Small-Scale Natural Resource Management (SNRM) has been promoted to assist fishermen improve their incomes, mainly to extend financial support for the economically challenged coastal communities. Besides these other programs for environmental restoration such as marine protected area (MPA), mangrove rehabilitation, infrastructure development and strengthening of regulation framework have also been promoted.

Seaweed farming has been done by fishermen for many years, contributing to the revitalization of the local economy. It is noteworthy that some alternative livelihoods have become important ways to improve local economy, which is negatively affected by decreased production of capture fisheries. This paper aims to assess the socio-economic status of fishermen in the study sites, to investigate challenges and opportunities of coastal livelihoods, to analyze fishers' perception and stakeholder participation in developing alternative livelihoods.

2 Conceptual Framework

Community based resource management (CBRM) involves the concept of co-management (Hanna 1998). Co-management is the sharing of authority and responsibility among government and stakeholders, a de-centralized approach to decision-making that involves user groups as consultants, advisors, or co-equal decision-makers with government (Berkes 1991, Jentoft and Kristoffersen 1989). Charter (1996) gives the definition of CBRM as a strategy to achieve development which centers on humans, where the centers of decision making about continuous utilization of resource in an area depends on people's organizations (organization refers to community group or informal organization) in that area. People have responsibility to manage their resource. They define need, aim and decision-making by themselves. In Indonesia, a wide variety of community based management in marine and coastal resources have developed differently from long time experiences between one region and another. Capacity building and decentralization are one of the ways to develop partnership with the community in coastal management. Improved capacity can help coastal communities tackle adverse socio-economic pressures. Thus, the central government decentralized the authority to local government (provincial) and district/city to participate in coastal zone management, particularly in improving economic condition of small-scale fishermen.

3 Study Area

Laikang Bay is located in South Sulawesi Province, Indonesia, being on the southernmost section of the island of Sulawesi (old name is Celebes). South Sulawesi Province with Makassar as its capital city is located between zero and $1°52'33''$ South Latitude and $116°48'$ up to $122°36'$ East Longitude. The average temperature in Makassar is between 22.8 and 34.7 °C (https://www.google.com/maps/) (Fig. 18.1).

This study includes two districts; (1) Takalar District which is located on the southern side of capital city of South Sulawesi Province. Land area covers 566.51 km^2 located between $5°3'-5°38'$ S and $119°22'$ up to $119°39'$ E. Laikang Village was chosen for data sampling. (2) Jeneponto District is located between $5°16'13''-5°39'35''$ S and $12°40'19''-12°7'51''$ E. This district is bounded by Gowa District in north side, Flores Sea in south side, Takalar District in western side and Bantaeng in eastern side. It is located on the west side Ujung Pandang. Garassikang, LP. Bahari and Ujunga Villages was used for data collection.

4 Study Methodology

The current investigation involved survey, observation and face-to-face interview. Total number of respondents is 200 seaweed farmers selected by the appropriate sampling method. Representative of 11 different stakeholders was interviewed too.

Fig. 18.1 Map of Laikang Bay, South Sulawesi Province

Data were obtained through direct interview and focus group discussion. The structured questionnaire was used for direct interview, while semi structured questionnaires were used to guide in focus group discussion. The topics of questionnaire covered socio-economic information of seaweed farmers. Key informants were selected purposively. They included the government officers and local government officers, researchers at research centers and universities, social leaders (*tokoh masyarakat*), the head of the village (*kepala desa*) who understand well about the social and economic condition of the village. Secondary data were included from statistics, published books, scientific journals and other resources. Data collected were analyzed by using simple statistics, such as descriptive statistics to analyze percentage, arithmetic mean, number and standard deviation. A Likert-type scale was used when the respondents were asked to point out their perceptions.

5 Results and Discussion

5.1 Socio-economic Characteristics of Fishermen

Livelihood activities of fishermen at Laikang Bay can be divided into two types: capture fisheries and seaweed farming. One day fishing was mostly adopted by fishermen there prior to the expansion of seaweed farming. Then, they began to implement seaweed farming within floating long line method, together with fishing activity that adopting fishing net around seaweed farm.

Table 18.1 Socioeconomic data of seaweed farmers in Laikang Bay

Variable	Frequency N = 200	%	Mean	S.D
Age (years)			37.04	9.6
≤25	17	8.5		
26–40	115	57.6		
41–60	68	34		
Gender (male/female)			1	0
Male	200	100		
Female	0	0		
Education			2.42	2
Elementary school	105	52.5		
Junior high school	41	20.5		
Senior high school	9	4.5		
None	45	22.5		
Marital status			2	0
Single	0	0		
Married	200	100		
Widow	0	0		
Ethnicity			2	0
Bugis	0	0		
Makassar	200	100		
Javanese	0	0		
Main income generating activity			4.02	2.9
Seaweed culture	92	46		
Seaweed culture + capture fishing	74	37		
Seaweed culture + public officer	4	2		
Seaweed culture + non-fishing	30	15		
Number of family member (persons)			1.86	0.34
≤2	27	13.5		
3–5	173	86.5		
Income value per month			1.44	0.5
≤500,000	115	57.5		
501,000–1,000,000	83	41.5		
>1,000,000	2	1		

Source: Primary data analyzed, 2010

The survey showed that most seaweed farmers in Laikang Village were 26–40 years old. They have between 2 and 5 persons in each family. Most of them have poor level of education, having graduated only from elementary school. The income of respondents came from two main activities, capture fisheries and seaweed farming. Both activities were conducted by people in Laikang, Garassikang, LP. Bahari and Ujunga Villages. The fishers who get more profit from seaweed farming compared with culture fisheries naturally preferred to give a higher priority to seaweed culture as their main income source. Most of respondents (70.5 %) have income less than one million Indonesian Rupiah (IDR) per month (Table 18.1).

Nowadays, fishermen at Laikang Bay do both livelihood activities. Local people become interested in seaweed farming on an individual basis. Because of this, the

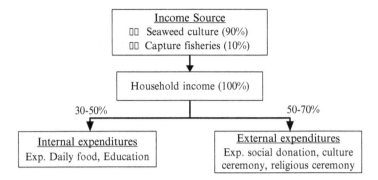

Fig. 18.2 Income utilization of fishermen at Laikang Bay

number of farmers and farms of seaweed have increased sharply. The fishing activities also are done by using simple fishing gears, such as fish net and crab net. Fishers went to the beach around seaweed farm to set up the fishing nets at evening and they took-up these fishing nets in the morning. There are leisure times between set-up and take-up time of fishing nets. Fishermen use this leisure time to do activities related to seaweed farming.

There are many reasons of fishermen for doing seaweed farming as the income source. Firstly, the government introduced it through government projects that aim to improve the economy of coastal communities. Secondly, relatively small operational cost. Thirdly, it was easy to maintain. Finally, they can get more profit from this farming activity. Satria (2009) argued about double strategy to solve fisher's household income problem. That strategy means, fisher together with family should do both fisheries activity and alternative jobs outside fisheries. In this study, almost all the seaweed farmers (97.5 %) agreed that the benefit of seaweed farming is better than catching fish. The indication was 77.5 % farmers stated that the number of seaweed farm has increased. However, 71.5 % farmers still used old seed, which has long strain. Therefore, 77 % of farmers expressed to make breeding hatchery to create a new strain of seaweed.

Fishermen have various livelihood activities such as capture fisheries, seaweed farming, and seaweed cultivation combined with fishing, seaweed farming with a combination of public services. There are some seaweed farmers (46 %) who conducted seaweed farming as a single activity. Meanwhile, others (37 %) combined seaweed farming and fishing activities. In this study, most fishermen still have low income. Some fishermen (57.5 %) had income below 500,000 rupiah (IDR) per month, while others (41.5 %) had incomes between 501,000 and 1,000,000 rupiah (IDR) per month. This amount represents the total income derived from all livelihood activities of fishermen. This income is used to support the family needs where each respondent had 1.86 child on average (Table 18.1).

Fishermen use their incomes mostly for social and cultural ceremonies and other ritual, with the remainder for foods and the school of their children. This condition affected fishermen facing the problem with lack of financial capital when they re-start to do seaweed farming. Because of this, fishermen really need to change the priority to the school, foods and capitals at first, then others (Fig. 18.2).

5.2 Seaweed Farming as Main Livelihood Activity

Seaweed farming in South Sulawesi is spread throughout the west coast (Makassar Strait) and the east coast (Gulf of Bone). Small financial, small production and traditional technology are characteristics of farmers. In this study, the perception of fishermen/farmers to seaweed farming development in Laikang Bay is divided into four parts: (1) perception of farming activity, (2) perception of environmental management, (3) perceptions of harvesting and (4) marketing activity.

The state of coastal resources is a significant factor affecting livelihood prospects for poor coastal communities (Glavovic and Boonzaier 2007). At present, seaweed cultivation has become a major source of livelihood for fishing communities along the coast of Laikang Bay. Capture fisheries cannot be done throughout the year, because it depends on the condition of local waters where fishers do fishing activity (Karubaba et al. 2001). This activity had improved the household economy of fishermen at least for the last 10 years. In this study, the planting process, maintenance and harvesting were done by the husband (head of household) and sometimes assisted by the children. Meanwhile, wife and daughter could support in the process of breeding. Seedling and planting activities is done by the fishermen (76 %) for 45 days. Most of them (71 %) used the services of labor in the planting process. After the planting is finished, the next step is maintenance. They (91 %) were to check the farm plots 2–4 times in 1 week. Almost all the seaweed farmers (97.5 %) agreed that the benefit of seaweed farming is better than catching fish. The indication was 77.5 % farmers said that seaweed farm was increase. However, 71.5 % farmers is still use old seed which had long strain. Therefore, 77 % of farmers expressed to make breeding hatchery to create a new strain of seaweed.

In Takalar, high productivity occur during December to April, because this area is protected from big waves and has a low salinity. In Jeneponto, highest productivity occurs during May to November. At that time, the area is protected from big waves and has a supply of fresh water from the Allu River. According to this condition, some fishermen moved to those places following the environmental condition and productivity. During May to November, farming activity of some fishermen moved from Takalar to Jeneponto. Meanwhile, during December to April, they moved to Takalar to do the same activity. However, not all the fishermen transfer their farming activity to highest productivity area.

According to respondents (81.5 %), environmental conditions in Laikang Bay is still to be developed for cultivating the seaweed. At present, fishermen/farmers (82.5 %) argued that the farm area is already crowded. Therefore, most of them (89 %) stated they need re-arrangement of farm/plot of seaweed and identify farm ownership. It causes some farm/plot not to be used for long time, and on the other hand, there are some farmers who want to use these plots. Public awareness on environmental quality is still low. Some farmers stated that the environmental condition is still suitable for planting the seaweed, but some other argued that there is need for rearranging the farm. Inconsistency of fishermen's answer because of the simple minds of fishermen believing as long as the environment is capable of producing seaweed, then they will continue to expanding the farm.

Utilized seaweed farm over several years is expected to add to the problems in the environment. It can be seen with equipment such as main ropes and anchors in the farm structure becoming dirty and disorganized. Many cases like this usually occur if the owners of farm is no longer cultivating seaweed as out of the village or went to the city to look for other jobs or they do not have enough capital to do the cultivation of seaweed. On the other hand, there are some fishermen who want to use the capital/money to start planting the seaweed.

Traditionally, fishermen dried the seaweed under sunshine. They (62 %) people used the bamboo racks for drying the wet seaweed. The problems were seaweed could not dry well on the rainy season and it takes longer times. Fishermen (95 %) stated that they have suffered many losses during the rainy season. Seaweed is not commercially produced yet to become value added product. In Laikang Bay, 25 % farmers stated that some seaweed made into traditional products such as a toffee (lunkhead), candy, jelly etc. Some fishermen (83.5 %) people want the existence of appropriate technology to process the dried seaweed into value added goods. Seaweed farmers in Laikang Bay sold them in the form of dried seaweed. Some (81.5 %) people kept dried seaweed before selling to middlemen. The farmers do not directly sell seaweed after each harvest. They sell dried seaweed after harvesting 2–3 times. The farmers feel that seaweed market channels are still long and respondents (69.5 %) stated the price can fluctuate. However, they (65 %) were still able to accept as long as they still produce the seaweed.

5.3 Stakeholders and Their Role in Developing Coastal Livelihoods

The increased attention on CBRM is partly a product of the emphasis on participatory and democratic governance and civil society participation within international institutions since the early 1980s (Wilson 2003). There are many failed experiences of livelihood projects in coastal areas of Indonesia. The failures were caused by lack of internal coordination, cooperation and communication between participants, local and central government, local government and the peoples, etc. Cooperation with all stakeholders is needed. Cooperation of local stakeholders should provide greater role, real actions in developing coastal areas. Central government will play a role as a partner for local government in developing any livelihood activities. Local government is expected to gradually reduce the dependence on the central government particularly in funding sources.

The descriptive statistics within the Likert-scales statements are presented in Table 18.2. The stakeholders here consists of 11 different categories who come from governmental sectors, non government organizations (NGOs), university, small business groups, fishermen group and research centers. They have answered six questions that were basically related to management, roles and communication. Perception of stakeholders regarding roles and responsibility, they stated that

Table 18.2 Participation of stakeholder in local-level

Likert-scale statements N = 11	Responses (%) Agree	Neutral	Disagree	Mean (± SD)
Current farming area Management prone to horizontal conflict	45.5	0	54.5	1.55 (0.522)
Respondents can communicate with other stakeholders	63.6	9.1	27.3	1.45 (0.688)
Some fishermen still doing illegal fishing practice	36.3	18.2	45.5	1.82 (0.751)
Stakeholder do their responsibility	90.9	0	9.1	1.09 (0.302)
Stakeholders need communication between each other	90.9	9.1	0	1.18 (0.603)
No integrated management concept between two districts	63.6	27.3	9.1	1.64 (0.924)

Source: Primary data analyzed, 2011

90.9 % stakeholders did roles and responsibilities to support fishermen to develop alternative livelihoods. Perceptions regarding to farming area management showed that 45.5 % perceived current management scheme as prone to horizontal conflict and 54.5 % perceiving not to be prone about the conflict. In addition, 63.6 % stakeholders stated that there is no integrated management concept established in the two districts for regulating the seaweed farm in Laikang Bay, 27.3 % said neutral because they were not sure if there was any or not. Coastal communities and stakeholder change over time and this requires fisheries management approach to be adaptive, not only to ecological fluctuations but also to shifts in social values, perceptions and to interests (Alpizar 2006).

Some stakeholders (36.3 %) perceived illegal fishing practiced by fishermen, but 45.5 % stated there were no illegal fishing practices. Therefore, they set up informal agreements among all seaweed stakeholders includes fishermen who as the seaweed farmers and seaweed farmers alone. This agreement aims to prevent conflicts, to control coastal utilization around the beach, to optimize the result of farming and fishing. Perceptions regarding the communications process showed 63.6 % agreed that fishermen can communicate well with other stakeholders, 27.3 % perceiving not to be good communicators and 9.1 % neutral. Therefore, almost all stakeholders 90.1 % want to arrange more intense communication between existing stakeholders. Public participation at the local level supported to the empowerment of coastal communities or stakeholders in the active management and implementation of policies (Berkes et al. 2001, Jentoft et al. 1998). Participation should be understood as a process which not only includes the opportunity of different sectors (resource users, stakeholders) to have something to say (share information), but also the opportunity to be listened to and considered in the decision making process (empowerment) in order to achieve a consensus that leads to effective resource management (Alpizar 2006).

5.4 Current Problems and Opportunities for Future Development

There were a number of problems faced by fishermen doing livelihood activities. Finding shows that changes in monsoon season and its cycles are the major problem in the development of seaweed cultivation at this time. Long market channel and income distribution are still major problems both in the business seaweed in Laikang Bay. Financial capital is the next problem most often experienced by fishermen especially when they start planting. In addition, the availability of seaweed seedlings, quality of seed, land tenure seaweed, a disease that attacks the seaweed plant and the post harvest process. Fluctuation of seaweed price is as a minor problem.

At present, fishermen have limited capabilities to improve their livelihood activities. This study shows that there are contributing factors that can be strengths to developing fishermen's livelihood activities. These are consist of decreasing the amount of fish catch, promotion of seaweed culture as alternative livelihood, beneficiaries from seaweed culture, supporting from local government, good market opportunity. These factors are the basic interests that encourage fishermen to do seaweed farming as alternative livelihoods.

In the future, at least some of these factors can be opportunities. First, the demand of raw material has been increasing year by year of these from domestic and foreign markets. Second, the policies of national government support developing seaweed farming. Third, Government of Indonesia (GoI) encourages private companies and national business agencies to develop seaweed processing. However, climate change, profit-taking action, decreasing the environmental quality and no standard price of dried seaweed would be the threats for fishermen to develop their livelihood activities.

Government of Indonesia (GoI) should encourage all stakeholders, particularly local government to realize greater roles. Private sectors can play the roles to diversify the product of the seaweed. Informal leaders can use their power to encourage the local people to conduct self-management for local resources. Kay and Alder (1999) stated capacity building as the process of increasing the capacity of those charged with managing the coast to make sound planning and management decisions. In addition, community based capacity-building serves to enhance a moral sense of duty (Fletcher 2003).

6 Conclusions

Seaweed farming *Eucheuma cottonii* is the main livelihood activity and major household income source in Laikang Bay. Fishermen could not usually divert household income for reinvestment in seaweed farming or other livelihood activities. They had used/have been using their income for social matters such as cultural events and ceremonies etc. Fishermen have high expectation in seaweed farming development at Laikang Bay within suitable environmental condition though they have various constraints like lack of post harvest technology, low price of dried seaweed, and complicated market channel. Integrated management is still a problem in managing Laikang Bay, although the stakeholders can communicate among themselves in Takalar and Jeneponto Districts. Finally, this paper infers that to improve livelihood activities in Laikang Bay, some aspects should be addressed, particularly changes in monsoon seasons, marketing channel, quality of seaweed seed, farm ownerships and commercial price. High demand of raw material from domestic and export market, and national policy can be opportunities for future development of seaweed farming and it can expectedly improve livelihood of fishermen in Indonesia.

Acknowledgment The authors would like to express their thanks and appreciation to Dr. Lawrence Liao, associate professor of Graduate School of Biosphere Science, Hiroshima University for his academic suggestions.

References

Alpizar MAQ (2006) Participation and fisheries management in Costa Rica: from theory to practice. Mar Policy 30(6):641–650
Berkes F (1991) Co-management: the evolution in theory and practice of the joint administration of living resources. Alternatives 18(2):12–18
Berkes F, Mahon R, McConney P, Pollnac R, Pomeroy R (2001) Managing small scale fisheries: alternative direction and methods. IDRC, Canada, 310 pp
Charter JA (1996) Introductory course on integrated coastal zone management (training manual). Pusat penelitian sumberdaya alam dan lingkungan Universitas Sumatera Utara, Medan dan Pusat Penelitian Sumberdaya Manusia dan Lingkungan Universitas Indonesia, Dalhousie University, Environmental Studies Centres Development in Indonesia Project, Jakarta
Fletcher S (2003) Stakeholder representation and the democratic basis of coastal partnerships in the UK. Mar Policy 27(3):229–240
Glavovic BC, Boonzaier S (2007) Confronting coastal poverty: building sustainable coastal livelihoods in South Africa. Ocean Coast Manag 50(1–2):1–23
Hanna S (1998) Co-management in small-scale fisheries: creating effective links among stakeholders. Plenary presentation, international CBNRM workshop, Washington, DC, USA, 10–14 May 1998. http://citeseerx.ist.psu.edu/viewdoc/summary?doi=10.1.1.194.7244
Jentoft S, Kristoffersen T (1989) Fisheries co-management: the case of the Lofoten fishery. Hum Organ 48(4):355–365
Jentoft S, McCay B, Wilson DC (1998) Social theory and fisheries co-management. Mar Policy 22 (4–5):423–436
Karubaba CT, Dietriech GB, Nikijuluw VPH (2001) *Kajian Pemenuhan Kebutuhan Pangan Nelayan pada Musim Timur dan Musim Barat Kaitannya dengan Pemanfaatan Sumberdaya Pesisir* – study of needs assessment of fishermen food on two monsoon seasons in relation with coastal resource uses. Indones J Coast Mar Resour 3(3):1–11
Kay R, Alder J (1999) Coastal planning management. SPON, New York, 380 pp

Kusnadi et al (2006) 6 tahun program PEMP "sebuah refleksi". Directorate General of Coastal Society Empowerment. Ministry for Marine Affairs and Fisheries, Jakarta, Republic of Indonesia

MMAF and JICA (2011) Indonesian fisheries book 2011, Jakarta. Available in: http://www.kkp.go.id/

Satria A (2009) *Pesisir dan Laut untuk Rakyat*-Marine and coast for people. IPB Press, Bogor, 178 pp

Wilson DC (2003) The community development tradition and fisheries co-management. In: Wilson DC, Nielsen JR, Degnbol P (eds) The fisheries co-management experience: accomplishments, challenges and prospects. Kluwer Academic Publishers, Dordrecht

Chapter 19
Monosex Fish Production in Fisheries Management and Its Potentials for Catfish Aquaculture in Nigeria

Wasiu Adekunle Olaniyi and Ofelia Galman Omitogun

Abstract Sustainable fisheries and aquaculture production serve as means of achieving nutritional security and sources of employment for large number of people in the world. Fisheries play an important role in enhancing foreign earnings and solving malnutrition problems. However the state of global food fish insecurity due to continuous decline in capture fisheries needs serious attention. This is as a result of variability in climate, anthropogenic pressures, ecosystem degradation and increasing demand for fish and its products by increasing human population. Developments of biotechnology strategy such as genetic manipulations in fisheries management have been recently employed in developed economies to increase fish production. In this study, monosex larvae of African catfish (*Clarias gariepinus*) were produced using biotechnology techniques such as androgenesis and gynogenesis where catfish gametes were treated with UV irradiation at 30,000 µWcm^{-2} for 15 min. The fertilized treatment were then subjected to cold shock at 2 °C for 20 min. Eggs numbering 100 ± 10 each quadruplicates were induced for gynogenesis through activation with UV irradiated sperm and then cold shock. For androgenesis, irradiated eggs were fertilized with normal sperm followed by cold shock. For the control experiments, 100 ± 10 normal eggs in quadruplicates were fertilized with irradiated sperm to produce haploid embryos and normal milt to produce normal diploid embryos. Fertility, hatchability, and survival were monitored and recorded. Androgeneic, gynogeneic, haploid and normal diploid treatments gave fertility of 80 %, 72.5 %, 100 % and 100 % with standard error of mean (SEM) of 3.19 respectively. Hatchability (number of hatched embryos) was 5.5 %, 22 %, 15 % and 93 % with SEM of 9.05 for androgeneic, gynogeneic, haploid and normal diploid embryos. No survived larvae

W.A. Olaniyi (✉) • O.G. Omitogun
Biotechnology and Wet Laboratories, Department of Animal Sciences,
Obafemi Awolowo University, Ile-Ife 220005, Nigeria
e-mail: tawakkull@yahoo.com; waolaniyi@gmail.com; aomitog@oauife.edu.ng

was recorded at yolk absorption for the haploid group while survival after 1 week for androgeneic, gynogeneic and normal diploid embryos gave 5 %, 13.25 % and 91 % (SEM = 9.65) respectively. Ploidy levels of the embryos were determined in 1-day old post-hatched embryos following modified protocol of Don J, Avtalion RR (Theor Appl Genet 72:186–192, 1986). The mean chromosome number (n) of 28 was obtained for haploid and 56 for normal diploid, androgeneic and gynogeneic larvae. However, sex specific DNA analysis is necessary to further confirm their sex determination, hence its potential applications in enhancing breeding strategies for research and commercial catfish aquaculture in Nigeria are discussed.

Keywords Biotechnology • Androgenesis • Gynogenesis • *Clarias gariepinus*

1 Introduction

Global fish production and its products contribute significantly to food and nutrition security. Fish productions serve not only as means of achieving individuals and households food and nutrition security objectives to meet up minimum protein requirements but also a source of employment for large number of people in the world. It plays important roles in enhancing foreign earnings, poverty alleviation and solving health and malnutrition problems. However the state of global food fish security is being threatened. This is due to continuous decline in capture fisheries, fish stock genetic resources depletion as a result of climate change or climate variability, over-fishing, environmental degradation with increasing demand for fish and its products by ever-increasing world population.

Fish production in global food supply contributes about one-fifth (20 %) of all animal protein in human diet, and which is comparable to other animal protein sources. In this light, production needs to be increased tremendously to meet up the speculated requirement of additional 40Mt of fish per year by 2030 (FAO 2009); and even more by 2050 when global population projection is expected to grow to 9.2 billion (UNPP 2008). China and other developing countries contributed 95 % of global inland capture production but sustainable increment will be of high significance to these countries that supply most of the World's capture fisheries (FAO 2009). Moreover, fisheries potentially provide benefits of vital source of food security and employment for the rural and urban-poor but these benefits are severely threatened. This is as a result of many challenges like change in climate, fish stock depletion, ecosystem degradation, overfishing, illegal fishing and ineffective fisheries governance which resulted in fisheries being over-exploited economically and well beyond biologically sustainable limits. These threats in capture fisheries have been compounded by challenges in rising demand for fish and its products. Hence the urgent need for innovative and high-tech management implementation strategy such as genetic manipulations among others that can be employed in aquaculture practices.

Genetic manipulation techniques in fisheries and aquaculture present potential opportunities in fisheries management. It falls into two categories, viz: cytogenetics and molecular (Standish 1986); and can be used mainly for two different purposes in fishery management: (i) to reproduce fish by uniparental chromosome inheritance to obtain gynogeneic and androgeneic fish, that is monosex population; and, (ii) to induce polyploidy, essentially to obtain triploid and tetraploid fish even to heptaploid (Pandian and Varadaraj 1990). Genetic manipulation techniques of Androgenesis – that is all-male production and Gynogenesis – that is all-female production, are being used to produce rapid inbreeding, clonal or monosex populations (Stanley and Sneed 1974; Purdom 1976, 1983; Cherfas 1981; Chourrout 1982).

1.1 Gynogenesis

Gynogenesis is a naturally occurring phenomenon in lower vertebrates in which offspring receives two sets of chromosome from the female parent (Dawley 1989). It is an inbreeding technique in artificial reproduction whereby eggs are activated by irradiated sperm then physical or chemical shock is applied to restore diploid to the embryo. The shocks destroy the aster formation or the microtubules of the spindle figure and inhibit nuclear division (Diter et al. 1993), thereby resulting in a diploid embryo containing maternal genetic material only. Purdom (1969) proposed this technique as a rapid means of developing inbred lines of maternal traits, and monosex population in fish. Gynogenesis has been widely investigated for its potential applications in numerous commercial species (Cherfas 1981; Purdom 1983; Thorgaard 1983, 1986; Ihssen et al. 1990; Tave 1993); in general, it has two major potential applications: (i) production of isogenic or clonal lines depending on the timing of initial diploidy restoration; and (ii) rapid inbreeding for the generation of inbred lines (Stanley and Sneed 1974; Purdom 1976, 1983; Cherfas 1981; Chourrout 1982).

Two kinds of gynogens can be produced: (a) Meiogynogenesis is achieved by inhibiting the extrusion of the second polar body (Volckaert et al. 1994, 1997; Aluko 1998). The resulting gynogens are homozygous at a locus only if no recombination occurred. Inbreeding is lower for those traits showing much residual heterozygosity. For some species, like rainbow trout, this residual heterozygosity is so high up to 100 % at some loci that meiogynogenesis cannot be considered as an efficient inbreeding tool (Thorgaard et al. 1983; Guyomard 1984); (b) Mitogynogenesis or endomitosis is achieved by inhibiting the first mitotic cleavage after duplication of the (haploid) genome and this result in fully homozygous offspring (Galbusera et al. 2000). Homozygous inbred strains of genetically identical fish (clonal lines) may be obtained after two generations using this reproduction method. It has been achieved in zebrafish *Danio rerio* (Streisinger et al. 1981), medaka *Oryzia latipes* (Naruse et al. 1985), common carp *Cyprinus carpio* (Komen et al. 1991), ayu *Plecoglossus altivelis* (Han et al. 1992), rainbow trout *Oncorhynchus mykiss* (Quillet et al. 1991; Scheerer et al. 1991), hirame (Yamamoto 1999) and Nile tilapia *Oreochromis niloticus* (Hussain et al. 1993).

1.2 Androgenesis

This is the induction of all-paternal inheritance; that is the irradiation process of the eggs (at sub-lethal dose) to make it genetically inactivated and then fertilizing with normal (unirradiated) milt with subsequent restoration of diploidy by physical shock or chemical induction (Bongers et al. 1995). It can further be described as the diploidization of the paternal genome (androgenote) which is accomplished by interference with first mitosis for eggs that have been genome-neutralized before activation. Moreover, it is the selection of individuals for sex inheritance characteristics that will be used in direct production of YY-males to be used as broodstock to produce all-male progeny. Androgenesis is a very difficult technique which has not been widely investigated and reported in fish as gynogenesis and triploidy (Pandian and Varadaraj 1990; Bromage and Roberts 1995). This is due to genetic inactivation of the eggs through irradiation which destroys structural organization of the cytoplasmic organelles, that is, not only nuclear DNA are damaged but also mitochondrial DNA and ooplasmic RNAs which are essential for embryonic development; thereby resulting in low survival of the androgeneic larvae; as it has been reported in various species (Scheerer et al. 1991; Bongers et al. 1995).

1.3 Merits of Monosex Population

The essence of monosex production in fisheries management is that in some fish species the males grow faster than the females while in others its vice–versa, therefore production of the faster growing sex will be better for high production purposes. This is very easy in cultured fish because they exhibit sexual dimorphism in growth (Dunham 2004). Other reasons are that some species do mature at small size or very young age prior to the desired harvesting time, and this may lead to precocious mating and unwanted reproduction that will affect management of pond due to over-crowding – for example, in the strain *Rhamdia quelen* (Siluriformes) that is sexually precocious (da Silva et al. 2007) and tilapia culture (Pandian and Varadaraj 1990) that quickly overpopulate ponds, thereby monosex production (by gynogenesis and androgenesis) can be employed to eliminate undesirable reproduction. Differences in sex may also affect fillet quality and its yield. Dunham (2004) raised that there is close relationship between sexual maturity with both carcass yield and growth rate – that is, as the fish becomes sexually mature, growth rate slows and carcass yield decreases. He further stated that sexual dimorphism could exist for economic traits like tolerance to thermal as in the case of climate change or climate variability, disease and poor water quality etc. Hence, aquacultural scientists have being able to transfer elegant developments from the laboratory to the culture tanks, which have been feasible and are viable commercially. Other genetic manipulation techniques are hybridization, sterilization, polyploidy and sex reversal.

2 Study

The present study employed genetic manipulation techniques of gynogenesis and androgenesis to produce monosex larvae of African catfish (*Clarias gariepinus*) using simple and safe biotechnology techniques where catfish gametes were treated with UV irradiation at 30,000 μWcm^{-2}. The African catfish, *C. gariepinus* Burchell 1822, also referred to as sharp-tooth catfish because of fine, pointed bands of teeth (Skelton 1993), are favorite food source for humans and contributed significantly to annual fish production in many countries in the world. They are mostly cultured in tropical and subtropical Africa and Asia (Hecht et al. 1996); and are farmed extensively all over the world. This is due to their ease of culturing, high growth performance, resistance to diseases, tolerance to wide range of temperatures, low dissolved oxygen levels and high stocking density, and most importantly low production input requirements and yet yield high returns (Bovendeur et al. 1987; Omitogun and Aluko 2002). With these potentials, *C. gariepinus* species among other farmed species contributed greatly to the total world fish supply (Bromage and Roberts 1995; FAO 2001). However in developing countries like Nigeria there is need to optimize production tremendously to meet rising demand of fish.

3 Materials and Methods

3.1 Fish Broodstock Collection and Management

The study was carried out in the Biotechnology and Wet Laboratories of the Department of Animal Sciences, Obafemi Awolowo University, Ile-Ife, Nigeria. Fish broodstock (600 ± 100 g) were obtained from a reliable commercial fish farm in Ile-Ife, Southwestern part of Nigeria, and acclimatized to full sexual maturation.

3.2 Induced Breeding

Sexually matured *C. gariepinus* broodstock, male and female (600 ± 100 g) were injected intramuscularly with 0.5 ml of Ovaprim® (Syndel) per Kg of the body weight, and left for about 12 h latency period (Olufeagba et al. 1997; Fafioye and Adeogun 2005). The testes of male broodstock were dissected out and then macerated to get the milt while the eggs from the female fish were stripped into a dry bowl.

3.3 Gametes Inactivation

The milt for gynogenesis and the eggs for androgenesis were subjected to irradiation by placing the Petri dishes containing them in a UV-sterilization chamber at 8 cm distance under the UV lamp providing 30,000 µW/cm^2 at 254 nM wavelength for 15 min duration.

3.4 Experimental Procedure

Good quality greenish olive eggs were randomly chosen for fertilization while white eggs and those contaminated with blood or excreta were removed from the batch. Eggs (100 ± 10) were randomly distributed into Petri dishes in quadruplicates. Each batch was either fertilized with normal (non-irradiated) or irradiated 0.5–1 ml milt (diluted with 0.9 % normal sterile saline solution), and were set in plastic hatching tanks (20″ × 15″ × 12.5″) for incubation. The fertility, hatchability, and survival were monitored and recorded for a week.

3.5 Control Group, Androgeneic and Gynogeneic Productions

Normal milt was used to fertilize normal eggs for the diploid control, while the irradiated milt was used to activate normal eggs for the haploid control. Both groups were not subjected to any cold shock but immediately set in aerated plastic hatching tanks at 27 ± 1 °C, after 3–4 min of activation. The normal milt was used to activate the irradiated eggs in a Petri dish for androgeneic production while the irradiated milt was used to activate the normal eggs for the gynogeneic production; and after 3–4 min of activation, each was transferred into a thermoregulated refrigerator and maintained at 2 ± 1 °C for 20 min to suppress the cell divisions. These were immediately set in aerated plastic hatching tanks at 27 ± 1 °C. pH, alkalinity, dissolved oxygen and temperature parameters were monitored.

3.6 Fertilization, Hatchability and Survival

Fertility was assessed by observation of the morula stage of cell division at 2–4 h after fertilization. The fertilization, hatchability and survival were calculated following Don and Avtalion (1986). The percentage hatchability (Eq. 19.1) was calculated by counting the number of larvae hatched (N_H) with respect to the number of eggs fertilized (N_F). Seventy-two (72) hours after hatching, that is, after the yolk absorption, the fry were fed with *Artemia naupli* shell free feed (54 % CP) thrice daily *ad libitum*. The survival (Eq. 19.2) was calculated at the end of 7th day for each experiment as the relative percentage of survivals multiplied (×) at a particular stage [Day 7 (z)].

$$\% \text{hatch} = \frac{N_H \times 100}{N_F} \qquad (19.1)$$

$$\% \text{survival}(X_z) = \frac{n \times 100}{i \times c/100} = \frac{n \times 10^4}{i \times c} \qquad (19.2)$$

where; c is the absolute percent of fertile eggs (% of morula at 2–4 h post-fertilization); i is the initial number of the eggs and n is the number of eggs which survived up to a given developmental stage (i.e. Day 7) set for this experiment.

The survival were monitored and assessed daily but were evaluated on the last day, that is, Day 7 with respect to hatchability (Day 1).

3.7 Ploidy Evaluation and Karyotyping

Ploidy conditions of the embryos were determined in 1-day old post hatched embryos following modified protocol of Don and Avtalion (1986). The hatched larvae immersed in 0.01–0.02 % Colchicine (BDH) for 2–4 h and later treated with diluted catfish serum 1:3 with 0.9 % NaCl solution for 25 min, then 5 min in distilled water. The fragments were later fixed in acetic acid – methanol mixture (1:4 v/v) and the fixed cells were aged and then stained with 20 % Giemsa (Scharlau). The coded slides were screened under the high power objective (40x) microscope 3–4 days after the chromosome preparation and the chromosome counted in good metaphasic spread.

3.8 Statistical Analysis

Data recorded for fertility, hatchability, and survival was subjected to statistical testing to verify differences between the groups by General Linear Models procedure which runs the two-way analysis of variance (ANOVA) using software SAS®, 2003.

4 Results

The pH, alkalinity, dissolved oxygen and temperature were monitored and measured to be 7.1 ± 1, 112.31 ± 1.14 mg/l, 24.5 ± 0.5 mg/l and 27 ± 2 °C respectively. Figure 19.1 and Table 19.1 showed the effects of genetic manipulation techniques

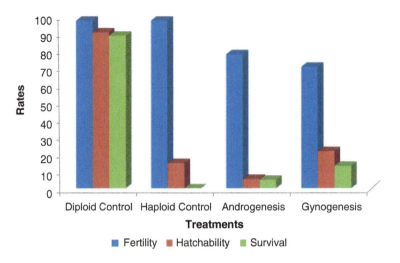

Fig. 19.1 Effects of genetic manipulation techniques on fertility, hatchability and survival in African catfish (*Clarias gariepinus*)

Table 19.1 Effects of genetic manipulation techniques on fertility, hatchability and survival in African catfish (*Clarias gariepinus*)

Parameters	Diploid control	Haploid control	Androgenesis	Gynogenesis	SEM	p value
		%				
Fertility	100^a	100^a	80^b	72.5^c	3.19	0.0001
Hatchability	93^a	15.0^{bc}	5.5^c	22.0^b	9.05	0.0001
Survival	91^a	0.0^c	5.0^{bc}	13.25^b	9.65	0.0001

[abc]Means within a row with different superscripts are significantly different ($p < 0.05$)
General linear model procedure (SAS®, 2003)

on fertility, hatchability and survival in *C. gariepinus*. The mean fertility of 100 ± 3.19 % was recorded for both diploid and haploid controls while the androgeneic and gynogeneic treatments gave 80 ± 3.19 % and 72.5 ± 3.19 % (p < 0.05) respectively. Therefore, there were significant differences between the monosex group and the control group (p < 0.05), while the controls were not statistically different from each other (p > 0.05). The hatching of larvae began at 26 h post-fertilization (pf) in the diploid and haploid controls while for androgeneic and gynogeneic treatments started at 27 h pf. The diploid control and the monosex treatments gave viable larvae that were wriggling about in the breeding tanks. The haploid control produced larvae of curved spine, rounded yolk sacs and exhibited other morphological deformities. Moreover, the hatched larvae of haploid control did not live up to yolk-absorption stage. The hatchability data (Table 19.1) for diploid control, haploid control, androgenesis and gynogenesis were 93 %, 15.0 %, 5.5 %, and 22.0 % respectively with 9.05 SEM. The percent hatchability in all the treatments were all statistically different (p < 0.05) except the haploid control that was not statistically different (p > 0.05) from the monosex treatments.

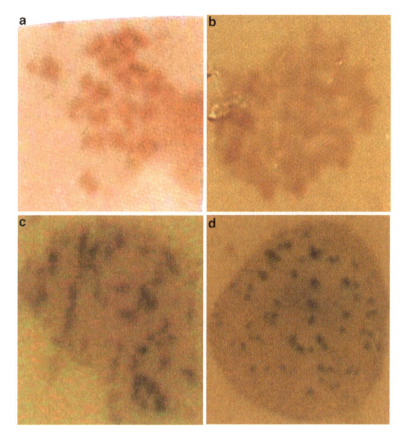

Fig. 19.2 Chromosome spreads (×40) from 24 h larvae of African catfish (*Clarias gariepinus*). (**a**) Haploid control = 28; (**b**) Diploid control = 56; (**c**) Androgenesis = 56; and (**d**) Gynogenesis = 56

The survival value for diploid control gave 91 ± 9.65 % and was statistically different from haploid control, androgenesis and gynogenesis which gave 0.00 %, 5.00 % and 13.25 % respectively ($p < 0.05$). There were no statistical differences in the androgeneic treatment compared to gynogeneic treatment and haploid control ($p > 0.05$), while gynogeneic treatment was different ($p < 0.05$) from the haploid control.

Ploidy conditions of the embryos, Fig. 19.2a–d, revealed mean chromosome number (n) of 28 for haploid, 56 for normal diploid, androgeneic and gynogeneic larvae respectively.

5 Discussion

The differences in fertility primarily rest on fecundity, egg size and quality, though the egg size determines the quality of eggs (Bromage and Roberts 1995). The differences in hatchability can be as a result of the developmental process from fertilized egg to hatching, like all other biological processes is dependent upon water temperature. The higher the water temperature the faster the eggs hatched and the better the survival (de Graaf and Janssen 1996). Moreover, it can also be explained that the kind of inductors used have significant effects on the hatchability and most importantly can cause mutation (Pandian and Koteeswaran 1998).

The haploid control that recorded zero survival confirmed the effective genetic inactivation of the spermatozoa. The highest survival data of the diploid control can be attributed to its non-treatment with the UV irradiation and physical or chemical shock. This study showed low survival for both gynogeneic and androgeneic individuals with gynogeneic (13.25 %) having higher value than androgeneic (5 %) larvae. This can be attributed to the expression of lethal mutant genes as a result of UV irradiation, though sublethal doses were administered for a short period during the experimentation that enabled the successful production of the embryos. However, the application of this technique is important in fisheries and aquaculture management in solving some of its major drawbacks. For example, in the strain *Rhamdia quelen* (Siluriformes) that is sexually precocious (da Silva et al. 2007) and tilapia culture (Pandian and Varadaraj 1990) that quickly overpopulate the aquatic systems. The technique is also employed to produce inbred strains (Streisinger et al. 1981; Váradi et al. 1999), and to produce fish species in which males or females are capable of maturing earlier than their opposite sex (Parson and Thorgaard 1985). However, sex specific DNA analysis is necessary to further confirm their sex determination.

6 Conclusion

Improvement on fish breeding strategies through biotechnological sex control techniques of gynogenesis and androgenesis will have great contributions to fisheries and aquaculture management in the identification of good breeders (for thermo tolerance with respect to climate change or climate variability, disease resistance and poor water quality etc.) by the farmers to use for culture. The techniques of gynogenesis and androgenesis in this study are of high value due to chemical-free methodologies, consumer safety and offer unique solutions to specific problems in fisheries and aquaculture management.

References

Aluko PO (1998) Induction of meiotic gynogenesis with ultra-violet ray, in the African catfish, *Clarias anguillaris*. Nig J Biotechnol 1:76–81

Bongers ABJ, Ngueriga D, Eding EH, Richter CJJ (1995) Androgenesis in the catfish *Clarias gariepinus*. Aquat Living Resour 8:329–332

Bovendeur J, Eding EH, Henken AM (1987) Design and performance of water recirculation system for high-density culture of the African catfish, *C. gariepinus* (Burchell 1822). Aquaculture 63:329–355

Bromage NR, Roberts RJ (1995) Broodstock management and egg and larval quality. Iowa State University Press, A Blackwell Science Companion. 2121 S. State Avenue, Ames, Iowa, pp 6–24, 76–93

Cherfas NB (1981) Gynogenesis in fishes. In: Kirpichnikov VS (ed) Genetic bases of fish selection. Springer, Berlin, pp 255–273

Chourrout D (1982) Gynogenesis caused by ultraviolet irradiation of salmonid sperm. J Exp Zool 223:175–181

da Silva FSD, Moreira RG, Orozco-Zapata CR, Hilsdorf AWS (2007) Triploidy induction by cold shock in the South American catfish, *Rhamdia quelen* (Siluriformes) (Quoy & Gaimard, 1824). Aquaculture 272:S110–S114

Dawley RM (1989) An introduction to unisexual vertebrates. In: Evolution and ecology of unisexual vertebrates. University of the State of New York, Albany, New York, USA, pp 1–18

de Graaf GJ, Janssen J (1996) Handbook on the artificial reproduction and pond rearing of the African catfish *Clarias gariepinus* in sub-Saharan Africa, FAO fisheries technical paper 362. Nefisco Foundation, Rome/Amsterdam, pp 13–50

Diter A, Quillet E, Chourrout D (1993) Suppression of first egg mitosis induced by heat shocks in the rainbow trout. J Fish Biol 42:777–786

Don J, Avtalion RR (1986) The induction of triploidy in *Oreochromis aureus* by heat shock. Theor Appl Genet 72:186–192

Dunham RA (2004) Aquaculture and fisheries biotechnology: genetic approaches. CABI Publishing, Wallingford, p 385

Fafioye OO, Adeogun OA (2005) Homoplastic hypophysation of African Catfish, *Clarias gariepinus*. Niger J Anim Prod 32(1):153–157

FAO (2001) FAO Corporate Document Repository. What is the code of responsible Fisheries? Fisheries and Aquaculture Department. Food and Agriculture Organization of the United Nations' Code of Conduct for Responsible Fisheries

FAO (2009) The state of world fisheries and aquaculture, 2008. FAO Fisheries and Aquaculture Department, Food and Agricultural Organization of the United Nations, Rome, Italy

Galbusera P, Volckaert FAM, Ollevier F (2000) Gynogenesis in the African catfish *Clarias gariepinus* (Burchell, 1822) III. Induction of endomitosis and the presence of residual genetic variation. Aquaculture 185:25–42

Guyomard R (1984) High levels of residual heterozygosity in gynogenetic rainbow trout, *Salmo gairdneri*, Richardson. Theor Appl Genet 67:307–316

Han H, Mannen H, Tsujimura A, Taniguchi N (1992) Application of DNA fingerprinting to confirmation of clone in ayu. Nippon Suisan Gakkaishi 58:2027–2031

Hecht T, Oellermann L, Verheust L (1996) Perspectives on clariid catfish culture in Africa. Aquat Living Resour 9:197–602

Hussain MG, Penman DJ, Mcandrew BJ, Johnstone R (1993) Suppression of first cleavage in the Nile tilapia, *Oreochromis niloticus* L. – A comparison of the relative effectiveness of pressure and heat shocks. Aquaculture 111:263–270

Ihssen PE, McKay LR, McMillan I, Phillips RB (1990) Ploidy manipulations and gynogenesis in fishes: cytogenetic and fisheries applications. Trans Am Fish Soc 119:698–717

Komen J, Bongers ABJ, Richter CJJ, Van Muiswinkel WB, Huisman EA (1991) Gynogenesis in common carp (*Cyprinus carpio*): II. The production of homozygous gynogenetic clones and F1 hybrids. Aquaculture 92:127–142

Naruse K, Ijiri K, Egami N (1985) The production of cloned fish in the medaka (*Oryzias latipes*). J Exp Zool 236:335–341

Olufeagba SO, Aluko PO, Omotosho JS, Oyewole SO, Raji A (1997) Early growth and survival of cold shocked *Heterobranchus longifilis Valenciennes,* 1840 (Pisces: Clariidae). In:

Proceedings of the 10th annual conference of the biotechnology society of Nigeria. Biotechnology and Sustainable Development in Nigeria, pp 213–220

Omitogun OG, Aluko PO (2002) Application of molecular and chromosomal tools in Tilapiine and Clariid Species for genetic improvement and stock management. Niger J Genet 17:15–22

Pandian TJ, Koteeswaran R (1998) Ploidy induction and sex control in fish. Hydrobiologia 384(1–3):167–243

Pandian TJ, Varadaraj K (1990) Techniques to produce 100 % male Tilapia. ICLARM, Makati, Metro Manila, Philippines 13(3):3–5

Parson JE, Thorgaard GH (1985) Production of androgenetic rainbow trout. J Hered 76:177–181

Purdom CE (1969) Radiation-induced gynogenesis and androgenesis in fish. Heredity 24:431–444

Purdom CE (1976) Genetic techniques in flatfish culture. J Fish Res Board Can 33:1088–1093

Purdom CE (1983) Genetic engineering by the manipulation of chromosomes. Aquaculture 33:287–300

Quillet E, Garcia P, Guyomard R (1991) Analysis of the production of all homozygous lines of rainbow trout by gynogenesis. J Exp Zool 257:815–819

Scheerer PD, Thorgaard GH, Allendorf FW (1991) Genetic analysis of androgenetic rainbow trout. J Exp Zool 260:382–390

Skelton P (1993) A complete guide to the freshwater fishes of Southern Africa. Southern Book Publishers Ltd., Halfway House

Standish KA Jr (1986) Genetic manipulations: critical review of methods and performances, shellfish. EIFAC/FAO symposium on selection, hybridization and genetic engineering in aquaculture of fish and shellfish for consumption and stocking, Bordeaux, France, 27–30 May

Stanley JG, Sneed KE (1974) Artificial gynogenesis and its application in genetics and selective breeding of fishes. In: Blaxter JHS (ed) The early life history of fish. Springer, New York, pp 527–536

Streisinger G, Walker C, Dower N, Knauber D, Singer F (1981) Production of clones of homozygous diploid zebrafish (*Brachydanio rerio*). Nature 291:293–296

Tave D (1993) Growth of triploid and diploid bighead carp, *Hypophthalmichthys nobilis*. J Appl Aquac 2:13–25

Thorgaard GH (1983) Chromosome set manipulation and sex control in fish. In: Hoar WS, Randall DJ, Donaldson EM (eds) Fish physiology, vol 9B. Academic, New York, pp 405–434

Thorgaard GH, Allendorf FW, Knudsen KL (1983) Gene-centromere mapping in rainbow trout: high interference over long map distances. Genetics 103:771–783

Thorgaard GH (1986) Ploidy manipulation and performance. Aquaculture 57:57–74

UNPP (2008) Population Division of the Department of Economic and Social Affairs of the United Nations Secretariat, World Population Prospects: the 2006 revision and world urbanization prospects: the 2005 revision. http://www.un.org/esa/population/publications/WPP2006Rev Vol_III/WPP2006RevVol_III_final.pdf. Accessed 26 Mar 2008

Váradi L, Benkó I, Varga J, Horváth L (1999) Induction of diploid gynogenesis using interspecific sperm and production of tetraploids in African catfish, *Clarias gariepinus* Burchell (1822). Aquaculture 173:401–411

Volckaert FAM, Galbusera PHA, Hellemans BAS, Van Den Haute C, Vanstaen D, Ollevier F (1994) Gynogenesis in the African catfish (*Clarias gariepinus*): III. Induction of meiogynogenesis with thermal and pressure shocks. Aquaculture 128:221–233

Volckaert FA, Van Den Haute C, Galbusera PH, Ollevier F (1997) Gynogenesis in the African catfish (*Clarias gariepinus*) (Burchell, 1822). Optimising the induction of polar-body gynogenesis with combined pressure and temperature shocks. Aquac Res 28:329–334

Yamamoto E (1999) Studies on sex-manipulation and production of cloned populations in hirame, *Paralichthys olilaceus* (Temminck et Schlegel). Aquaculture 173:235–246

Chapter 20
Stock Assessment of Bogue, *Boops Boops* (Linnaeus, 1758) from the Egyptian Mediterranean Waters

Sahar Fahmy Mehanna

Abstract The trawl fishery off Egyptian Mediterranean waters consists of 1,400 vessels, yielding a total annual landing of approximately 16,000 t. The bogue (*Boops boops*) is a target species for this fishery, with annual landings oscillating between 1,222 and 3,980 t during the last 18 years. The stock of *Boops boops* in the Egyptian Mediterranean waters has been assessed using a 3 years length frequency data (2007–2009) from the trawl fishery. The vector of natural mortality by age was calculated from Caddy's formula, using the PROBIOM Excel spreadsheet Abella AJ, Caddy JF, Serena F (Aquat Living Resourc 10:257–269, 1997). The method applied was a tuned virtual population analysis (VPA), applying the Extended Survivor Analysis (XSA) method on the period 2007–2009 and Y/R analysis on the pseudo-cohort 2007–2009. The software used was the VIT program. Results showed that the stock is over exploited, being the fishery operating below the optimal yield level since the current Y/R is lower than the maximum. Results suggest a decreasing trend in the average fishing mortality during the study period by about 40 %.

Keywords Age and growth • Population dynamics • Stock assessment • Management

1 Introduction

The Egyptian Mediterranean coast (Fig. 20.1) is about 1,100 km with a mean annual fish production of 55,000 tonnes between 1990 and 2009. The main fishing gears are trawling, purse-seining and lining especially long and hand lining. The fishing grounds along the Egyptian Mediterranean coast are divided into four regions;

S. Fahmy Mehanna (✉)
Fish Population Dynamics Laboratory, Fisheries Division, National Institute of Oceanography and Fisheries (NIOF), P. O. Box 182, Suez, Egypt
e-mail: sahar_mehanna@yahoo.com

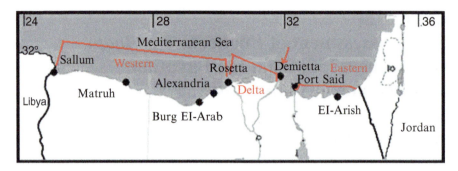

Fig. 20.1 Egyptian Mediterranean coast

Western region (Alexandria and El-Mex, Abu-Qir, Rasheed, El-Maadiya and Mersa Matrouh), Eastern region (Port Said and El-Arish), Demietta region and Nile Delta region (GAFRD 2009). The trawl fishery contributed about 33 % of the total fish production from Egyptian Mediterranean. It is a multi-species fishery targeting a number of commercial important fish species; red mullet, soles, triglid fish, breams, hakes, lizardfish, snappers, elasmobranches and barracuda. Invertebrates are represented by shrimp, cuttlefish, squid, crab and bivalves.

The number of trawlers in the Egyptian Mediterranean ranged between 1,100 and 1,500 during the period from 1990 to 2009. The vessel length varied between 18 and 22 m and its width varied from 4 to 6 m. Each vessel is powered by main engine of 150–600 hp but the majority of 250 hp engines. The fishing trip is about 7–10 days and the number of crew is about 6–15 persons.

Porgies (family: Sparidae) are among the most abundant demersal fishes inhabiting the Egyptian Mediterranean. They represented by more than 10 species and exploited by a variety of fishing gears. *Boops boops* is the most abundant sparid species in the area contributing a mean annual catch of 2,330 t (1991–2009) and exploited by bottom trawlers.

The present work was undertaken in order to provide fishery management advice for the *B. boops* stock in the Egyptian Mediterranean waters. It is aimed at estimating the biological and population parameters required for proposing future plans for sustainable development and management this valuable fish resource.

2 Material and Methods

Monthly random samples of *Boops boops* were collected from Alexandria landing site during the period from June 2007 until April 2009. Each fish was measured to the nearest mm for total length, weighed to the nearest 0.1 g, and its sex and maturity stage were determined macroscopically. Gonads were weighed to the nearest 0.01 g.

Otoliths were removed, rinsed of any adhering tissues, and sorted dry in labeled vials until processing. Annual rings on the otoliths were counted using an optical system consisting of Nikon Zoom – Stereomicroscope focusing block and Heidenhain's electronic bidirectional read out system V R X 182, under transmitted light. The total otolith radius and the radius of each annulus were measured to the nearest 0.001 mm. Regression analyses of otolith maximum radius on total length was calculated by the method of least squares. Back-calculated lengths-at age were computed by using the Lee method (Lagler 1956).

To estimate the relation between length (L) and weight (W), the variables were log transformed to meet the assumptions of normality and homogeneous variance. A linear version of the power function: $W = a*L^b$ was fitted to the data. Confidence intervals (CI) were calculated for the slope to see if it was statistically different from 3.

The growth parameters of the von Bertalanffy (1938) growth model (L_∞ and K) were computed by fitting the Gulland and Holt (1959) plot while the growth performance index was computed according to the formula of Pauly and Munro (1984) as $\emptyset = \text{Log } K + 2 \text{ Log } L_\infty$: where K and L_∞ are the parameters of von Bertalanffy growth model.

The length at first capture L_c (the length at which 50 % of the fish at that size are vulnerable to capture) was estimated by the method of Pauly (1984) in which the ascending limb of the linearized catch curve is considered to estimate the selection ogive.

The length at first sexual maturity L_{50} (the length at which 50 % of fish reach their sexual maturity) was estimated by fitting the maturation curve between the observed points of mid-class interval and the percentage maturity of fish corresponding to each length interval. Then L_{50} was estimated as the point on X-axis corresponding to 50 % point on Y-axis.

Natural mortality by age was estimated according to the Caddy's formula using the PROBIOM Excel spreadsheet (Abella et al. 1997). Fishing mortalities by age were estimated from VPA. Yield per Recruit analysis (Y/R) was estimated based on the exploitation pattern resulting from the XSA model and population parameters.

The assessment of this stock has been carried out by means of VPA using VIT software (Lleonart and Salat 1997) on a mean pseudo-cohort for the period 2007–2009. VIT model was applied to get an approximate estimate of the fishing mortality, the level of magnitude of recruitment and an indicative estimate $F_{0.1}$.

3 Results and Discussion

3.1 Length-Weight Relationship

Analysis of residual sums of squares indicated no significant difference between the sex-specific length-weight relationships of *B. boops* in the Egyptian Mediterranean

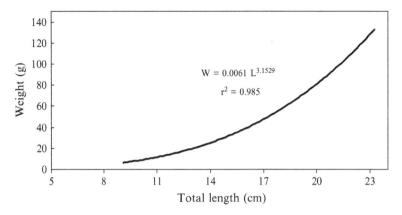

Fig. 20.2 Length-weight relationship of *Boops boops* from Egyptian Mediterranean waters

waters, consequently a power regression was applied to the length-weight data of all individuals combined (Fig. 20.2). The total length varied from 9 to 23.1 cm, while the total weight ranged between 5 and 140 g. The results of the length-weight analyses indicated that this relationship for *B. boops* is highly significant ($p < 0.05$) with high r^2 value (0.985) which indicate increase in length with increase in weight. The slope (b value) of the length weight relationship was 3.1529 which is not significantly different from 3 (95 % CI = 3.0784–3.2273) indicating isometric growth. Also, the value of b (growth exponent) for *B. boops* in the Egyptian Mediterranean waters is within the limits (two and four) reported by Tesch (1971) for most fishes. The obtained value is close to those recorded by Hernandez (1989) who gave b = 3.088 for females and 3 for males in Adriatic Sea, Abdallah (2002) gave b = 3.130 for sexes combined in Alexandria waters, Karakulak et al. (2006) estimated b as 3.258 for sexes combined in Turkey, Kara and Bayhan (2008) gave b = 3.272 for females and 3.522 for males in Turkey.

3.2 Age and Growth

Reliable age determinations are essential for almost all aspects of fishery assessment but especially for studies of growth, production, population structure and dynamics. Accurate estimates of age and growth are required for the management of Bogue fisheries (Mills and Beamish 1980; Girardin and Quignard 1986; Anato and Ktari 1986; Gordo 1996; Panfili et al. 2002; Sana et al. 2005). A total of 900 otoliths of fishes ranging from 9 to 23.1 cm total length were read and used in the estimation of the parameters of the growth models. The results revealed that the maximum life span of *B. boops* was 3 years (Fig. 20.3) and age group one was the most frequent group in the samples and constituted 75 % while the age group three was the least one in the samples (1 %).

Fig. 20.3 Growth curve of *Boops boops* from Egyptian Mediterranean waters

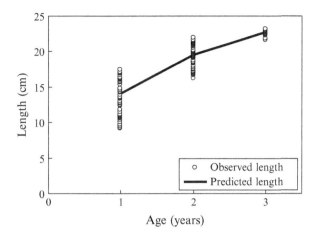

Table 20.1 Growth parameters (L_∞, K and t_o) and the Phi index of *Boops boops*

Locality	Sex	L_∞	K	t_o	Ø	Author
Adriatic Sea	M+F	33.9	0.17	−1.3	2.28	Hernandez (1989)
Cyprus	M+F	24.0	0.53	−0.45		Livadas (1989)
Algeria	M	26.6	0.21	−2.6	2.17	Djabali et al. (1990)
	F	27.3	0.22	−1.94	2.21	
Egypt	M+F	29.8	0.18	−1.33	2.2	Hassan (1990)
Greece	M+F	36.0	0.40		2.71	Tsangridis and Filippousis (1991)
Morocco	M+F	31.5	0.28	−0.96		Zoubi (2001)
Egypt	M+F	28.1	0.18	−1.13	2.15	El-Haweet et al. (2005)
		29.7	0.25	−0.70	2.34	
Portugal	M+F	28.06	0.22	−1.42		Monteiro et al. (2006)
Egypt	M+F	27.24	0.54	−0.33	2.6	Present study

3.2.1 Growth Parameters

Back-calculated lengths of pooled data were applied according to Gulland and Holt (1959) plot to estimate the von Bertalanffy growth parameters (L_∞, K and t_o). The obtained equations were as follows:

For growth in length $\quad L_t = 27.27\left(1 - e^{-0.54(t+0.33)}\right)$

For growth in weight $\quad W_t = 204.36\left(1 - e^{-0.54(t+0.33)}\right)^{3.1529}$

The growth parameters estimated for *B. boops* in different water bodies were summarized in Table 20.1. The difference in growth parameters between different locations could be attributed to the difference in maximum fish size in the used samples or due to the difference in ecological parameters.

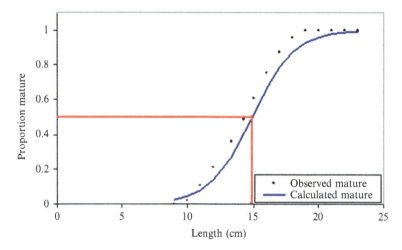

Fig. 20.4 Maturation curve of *Boops boops* from Egyptian Mediterranean waters

3.2.2 Growth Performance Index

Pauly and Munro (1984) have outlined a method to compare the growth performance of various stocks by computing the Phi index ø. The obtained results indicated that the growth performance index (ø) of *B. boops* in the Egyptian Mediterranean waters was 2.6 (Table 20.1).

3.3 Length and Age at First Sexual Maturity

The smallest length recorded in the catch was 9 cm, all fishes of lengths 9–11 cm were immature and those of lengths ≥ 17 cm were mature. The estimated L_{50} was 14.9 cm (1.17 year old) (Fig. 20.4). It was obvious that the length at first capture was smaller than the L_{50}. This means that the exploited *B. boops* must be protected till in order to share at least once in the spawning activity. Therefore, mesh sizes used should be increased to catch fish of about at least 15 cm length. El-Agamy et al. (2004) reported that for *B. boops* in Egyptian Mediterranean all fishes smaller than 12 cm TL were immature while the fishes above 17 cm were mature and the length at first maturity was about 13 cm TL. Monteiro et al. (2006) found that the length at first maturity in Portugal was similar for males and females and the value for both sexes combined was estimated to be 15.22 cm, corresponding to an age range of 1–3 year.

3.4 Length at First Capture L_c

The estimated length at first capture L_c in the present study was 11.3 cm TL (Fig. 20.5). Both the estimated L_c and the observed lengths of fish captured

Fig. 20.5 L_c estimation for *Boops boops* from Egyptian Mediterranean waters

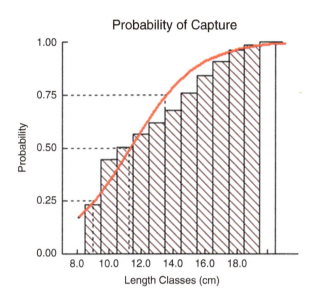

indicated growth and recruitment overfishing ($L_{50} = 14.9 > L_c = 11.3$ cm). In the light of these results, a minimum size limit should be implemented for *B. boops* in the Egyptian Mediterranean waters.

3.5 Virtual Population Analysis VPA

The assessment of bogue stock has been carried out by means of VPA (VIT) on a mean pseudo-cohort for the period 2007–2009. A natural mortality at age vector was used within the assessment, based upon the PROBIOM Excel spreadsheet (Abella et al. 1997). The mortality rates for the age groups were 1.67 total mortality, 1.18 fishing mortality and 0.49 natural mortality per year. Also, VIT gave an estimate for total and fishing mortality rates by length groups. The mean total mortality was 1.74 while the mean and global fishing mortality were 1.14 and 0.87, respectively.

Yield per Recruit Y/R analysis shows a clear status of growth overexploitation, due both to a high fishing mortality and an exploitation of the fishery based on juveniles under the minimum legal size. Also, the stock is in danger of recruitment overexploitation due to the decreasing trend in recruitment and very low levels of the spawning stock. Y/R analysis using VIT model (Fig. 20.6) provided the following values of reference points (Table 20.2):

To achieve the $F_{0.1}$ as a reference point, the current F should be reduced by about 38 % if we consider global F as the current F or 53 % if we took the mean F as the current one.

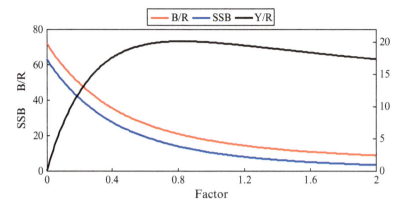

Fig. 20.6 Yield per recruit analysis of *Boops boops* from Egyptian Mediterranean waters

Table 20.2 VIT results

Current Y/R	19.65
Maximum Y/R	20.13
Y/R $_{0.1}$	19.12
F_{max}	0.82
$F_{0.1}$	0.53
Current B/R	14.97
Maximum B/R	72
B/R $_{0.1}$	29.38
SSB$_{0.1}$	21.85
Current SSB	8.59
Maximum SSB	63.1

3.6 Stock Status and Recommendation

The results of the present analysis revealed that the stock of bogue appears overexploited since the current fishing mortality is higher than $F_{0.1}$ and F_{max}. Thus a fishing mortality reduction is necessary in order to avoid future loss in stock productivity and landings.

It is therefore recommended that fishing effort must be controlled and decreased. The available assessment suggests a target reference point of about 30–60 % of the current fishing mortality. Defining nursery areas of important species should be taken into account for recommending closed areas. Also, improving the fisheries data recording system, improving the fishing pattern of the trawl and facilitating data and information exchange are highly recommended.

References

Abdallah M (2002) Length-weight relationship of fishes caught by trawl off Alexandria, Egypt. NagaICLARM Q 25(1):19–20

Abella AJ, Caddy JF, Serena F (1997) Do natural mortality and availability decline with age? An alternative yield paradigm for juvenile fisheries, illustrated by the hake *Merluccius merluccius* fishery in the Mediterranean. Aquat Living Resour 10:257–269

Anato CB, Ktari MH (1986) Age et croissance de *Boops boops* (Linne, 1758) Poisson teleosteen sparidae des cotes Tunisiennes (Age and growth of *Boops boops* (Linne, 1758) sparidae teleostean fish of the Tunisian coast). Bull Inst Natl Sci Tech Oceanogr Peche Salammbo 13:33–54

Djabali F, Boudraa S, Bouhdid A, Bousbia H, Bouchelaghem EH, Brahmi B, Dob M, Derdiche O, Djekrir F, Kadri L, Mammasse M, Stambouli A, Tehami B (1990) Travaux réalisés sur les stocks pélagiques et démersaux de la région de Béni-saf. FAO Fish Rep 447:160–165

El-Agamy A, Zaki MI, Awad GS, Negm RK (2004) Reproductive biology of *Boops boops* (Family Sparidae) in the Mediterranean environment. Egypt J Aquat Res 30(B):241–254

El-Haweet A, Hegazy M, Abu Hatab H, Sabry E (2005) Validation of length frequency analysis for *Boops boops* (Bogue) growth estimation. Egypt J Aquat Res 31(1):399–408

GAFRD (2009) Annual statistical book. General authority for fish resources development, Egypt

Girardin M, Quignard JP (1986) Growth of the *Boops boops* Linne, 1758 (Sparidae) in the Gulf of Lion. J Appl Ichthyol 2:22–32

Gordo LS (1996) On the age and growth of bogue, *Boops boops* (L.), from the Portuguese coast. Fish Manag Ecol 3:157–164

Gulland JA, Holt SL (1959) Estimation of growth parameters for data at unequal time intervals. J Cons Perm Int Explor Mer 25(1):47–49

Hassan MWA (1990) Comparative biological studies between two species of family Sparidae, *Boops boops* and *Boops salpa* in Egyptian Mediterranean waters. M.Sc. thesis, Faculty of Science, Alexandria University, 198 p

Hernandez AV (1989) Study on the age and growth of bogue *Boops boops* (L.) from the central Adriatic Sea. Cybium 13:281–288

Kara A, Bayhan B (2008) Length-weight and length-length relationships of the bogue *Boops boops* (Linnaeus, 1758) in Izmir Bay (Aegean Sea of Turkey). Belg J Zool 138(2):154–157

Karakulak FS, Erk H, Bilgin B (2006) Length-weight relationships for 47 coastal fish species from the northern Aegean Sea, Turkey. J Appl Ichthyol 22:274–278

Lagler KE (1956) Fresh water fishery biology. W. G. Brown co., Dubuque. 421 p

Livadas R (1989) The growth and maturity of bogue (*Boops boops*). Family Sparidae, in the water of Cyprus. In: Savini M, Caddy JF (ed) Report of the second technical consultation of general fisheries council for the Mediterranean on Stock Assessment in the Eastern Mediterranean. 1988, Greece. FAO Fisheries Report no. 412:52–57

Lleonart J, Salat J (1997) VIT: software for fishery analysis. User's manual. FAO computerized information series. Fisheries, 11, 107 p

Mills KH, Beamish RJ (1980) Comparison of fin ray and scale age determination for lake whitefish (*Coregonus clupeaformis*) and their implications for estimate of growth and annual survival. Can J Fish Aquat Sci 37:534–544

Monteiro P, Bentes L, Coelho R, Correia C, Gonçalves JM, Lino PG, Ribeiro J, Erzini K (2006) Age and growth, mortality, reproduction and relative yield per recruit of the bogue, *Boops boops* Linné, 1758 (Sparidae), from the Algarve (south of Portugal) longline fishery. J Appl Ichthyol 22(5):345–352

Panfili J, de Pontual H, Troadec JP, Wright PJ (eds) (2002) Manual of fish sclerochronology. IFREMER-IRD Co-edition, Brest. 464 p

Pauly D (1984) Length-converted catch curves. A powerful tool for fisheries research in the tropics. (Part II). ICLARM Fishbyte 2(1):17–19

Pauly D, Munro JL (1984) Once more on the comparison of growth in fish and invertebrates. ICLARM Fishbyte 2(1):21

Sana K, Gaamour A, Zylberberg L, Meunier F, Romdhane S (2005) Age and growth of bogue, *Boops boops,* in Tunisian waters. Acta Adriat 46(2):159–175

Tesch W (1971) Age and growth. In: Ricker WE (ed) Methods for assessments of fish production in freshwaters. International Biological Programme, Oxford, pp 97–130

Tsangridis A, Filippousis N (1991) Use of length-frequency data in the estimation of growth parameters of three Mediterranean fish species: bogue (Boops boops L.), picarel (Spicara smaris L.) and horse mackerel (Trachurus trachurus L.). Fish Res 12:283–297

von Bertalanffy L (1938) A quantitative theory of organic growth (inquiries on growth laws. 2). Hum Biol 10:181–213

Zoubi A (2001) Etude de la biologie de croissance des principaux stocks demersaux de la mediterranee marocaine. Rapp Comm Int Mer Médit 36:341(only)

Postface

Since 2009, the two international biannual conferences on *"The Integration of Sustainable Agriculture, Rural Development, and Ecosystems in the Context of Food Insecurity, Climate Change, and Energy Crisis"* and on *"Climate Change, Agri-Food, Fisheries and Ecosystems: Reinventing Research, Innovation and Policy Agendas for Environmentally- and Socially-Balanced Growth"* – jointly organized in Morocco by the NRCS and the GIZ – are analyzing the current global situation and actual international research related to sustainable agriculture in the context of local and global environmental change and food security. Both conferences stressed that sustainable agriculture, food security and environmental change should be studied on the basis of linked sciences and economic and resource efficiency. Climate and environmental change are major interferences on the natural balance of efficient resource use worldwide. And additionally human behavior in exploiting land and water resources has resulted in land and vegetation degradation, overexploitation of fisheries, depletion of aquifers, and unsustainable resource use in general.

Currently, seven billion people living on earth, but almost three billion lack access to modern energy for cooking, and some 1.5 billion don't have electricity at all. The fact is directly linked to increased levels of poverty. Prognoses say that the world population is expected to reach about nine billion till 2050. To cover the global natural resource consumption of this population worldwide, a productivity of about 1.5 earths is necessary. Prognoses say that we need two earths to cover our steadily raising resource consumption in 2030. The agricultural production must increase by 70 % globally and 100 % in developing countries. Currently, about one billion people go to bed hungry each night or die of hunger (i.e. in central Africa).

To secure the current world food production, it would be no problem to feed everyone of the current world population of seven billion. Roughly half a billion people are starving worldwide, while only Europe wastes 20 Mio. t food every year. This unequal dispensation is ruining the world's balance in natural resource and also the world's peace.

To support seven billion people on our planet, we have to assure enough resources and water. As this rapid growth mostly happens in developing countries while the industrialized countries almost stay constant, we have to lay the focus of our research on the poorer nations. Growing cities – Megacities and Metacities – are causing serious problems as heavy industries are growing and job possibilities with these industries as well – a serious vicious circle. If current processing technologies and infrastructure systems are not suitable to deal with end-of-life materials of growing cities, other sustainable solutions are still requested.

Scientists, politicians and decision-makers are imposed to the obligation of securing resources – including food and water as basic life essentials – in respect of local and global environmental change. The linkage of the research topics climate, water and natural resources enables us to originate possible future holistic solutions for global food security/food safety, sustainable agriculture, water availability for a secured and environmentally-friendly world.

The biannual *NRCS-GIZ Conferences* have identified and reconfirmed key issues for immediate consideration regarding climate and environmental change adaptation, sustainable agriculture and food security: a need to create a paradigm shift in agricultural policies; reforming food aid policies; investment in agriculture research; development and extension; rewarding the agriculture profession by promoting beneficial price regimes; ensuring research leading to appropriate biotechnology advances and innovations; strengthening of agricultural systems; bringing abandoned lands into production whilst preserving forests; provision of affordable credits to farmers; favorable free trade agreements between developed and developing countries; reforming policies on bio-energy production; improving livestock welfare; combating desertification; and producing scientific soil information.

As a follow up to the NRCS-GIZ Conferences (2009, 2011) and its research outcomes, the international conference on *"Global Environmental Change and Human Security: The Need for a New vision for Science, Policy and Leadership"* (GECS-2012) has been organized on November 22–24, 2012 in Marrakesh, Morocco. This edition has engaged a broad range of audiences and provided an update of the latest understanding of environmental change caused by current development models and schemes, human security implications of this change, and options available for different societies to respond to present and future challenges. Participants have considered how conceptions of security are being transformed in the face of environmental change, and how urgent a shift – in science, policy and leadership – is required to manage efficiently and prudently the current dynamics. The event served as a space to conceive this critically needed roadmap while conceiving future policy and research agendas within the context of post-Durban (2011) and Doha (2012) era.

The Editors
Mohamed Behnassi
Margaret Syomiti
Gopichandran Ramachandran
Kirit N. Shelat

Notes on Contributors

Abinaya, *P*.
Abinaya is a scientific officer in Research program on Markets, Institutions and Policies at International Crop Research Institute for the Semi-Arid Tropics, Hyderabad 502324, India.

Ashok Kumar, *A*.
Ashok Kumar, Ph.D. is a sorghum breeder and senior scientist in Research program on Dryland Cereals at International Crop Research Institute for the Semi-Arid Tropics, Hyderabad 502324, India.

Baig, *Mirza B*.
Dr. Baig is Professor of Agricultural Extension and Rural Society at the King Saud University of Riyadh, Saudi Arabia. He received his education in both social and natural sciences from USA. He earned his MS degree in International Agricultural Extension in 1992 from the Utah State University, Logan, Utah, USA and was placed on the "Roll of Honor". He completed his Ph.D. in Extension for Natural Resource Management from the University of Idaho, Moscow, Idaho, USA and was honored with "1995 Outstanding Graduate Student Award". Previously, he served the School of Rural Extension Studies as the Special Graduate Faculty at the University of Guelph, Ontario, Canada. Dr. Baig has published extensively on the issues faced by agriculture, natural resources, environment and rural communities in national and international journals. He has also presented extension education and natural resource management extensively at various international fora. His areas of interest include areas like degradation of natural resources, deteriorating environment and their relationship with society/community. He has attempted to develop strategies for conserving natural resources, promoting environment and developing sustainable communities through rural development programs. Dr. Baig serves on the editorial boards of many well reputed international research journals published across the globe.

Bantilan, *Cynthia*

Dr. Cynthia Bantilan is the Research Program Director on Markets, Institutions and Policy at the International Crops Institute for the Semi-Arid Tropics (ICRISAT) with headquarters located in India. She has over 250 publications to her credit, including 21 books, 52 book chapters and 40 articles in various international journals. She specializes in agricultural research evaluation and impact assessment, monitoring and evaluation, poverty and income distribution, econometrics, agricultural economics and agricultural statistics, information systems management, and applications for decision support and policy analysis. At ICRISAT, she spearheaded the Research Evaluation and Impact Assessment Program starting early 1990s, covering sub-Saharan Africa and Asia. She currently heads the development and implementation of key research programs on markets, institutions and policy in agriculture, using macro- and micro- (village level) studies in eastern and southern Africa, west and central Africa and South Asia.

Barbieri, *María Ángela*

Fishing engineer, Dr. in Biology Oceanography. I worked in the Instituto de Fomento Pesquero (Fisheries Research Institute) and Pontificia Universidad Católica de Valparaíso as researcher fisheries acoustic, in remote sensing, GIS and modelling applications in fisheries, aquaculture and environmental sciences. I worked as researcher projects with Institute of research for development (IRD), National Centre for Space Studies (CNES), Canadian Centre of Remote Sensing (CCRS), International Research Development Centre-Canadá (IRDC), European Space Agency (ESA), Fund Fisheries Research (FIP), Fishing Subsecretary of Chilean Government (Subpesca), CORFO and Fund for Promotion of Scientific and Technological Development. In 31 years I worked as research in 53 local, regional, national and international projects, 8 consulting services, participating in 52 scientific publications, 126 technical report, and 48 international conferences. In marine science undergraduate, master's and doctoral courses I teach courses Fisheries acoustics, Remote Sensing in Doctorate in Aquaculture, in Chile, Brasil and Argentine. Now I teach Ecosystems Management Approach. I worked too Undersecretaries of Fisheries, advise in the formulation and implementation of the National Fisheries Policy, direct the Fisheries Development Division, on which they depend department of Fisheries Coordination, International Affairs Unit, Innovation and Development Unit, Indigenous Affairs Unit. Now I am National Project Coordinator Chile-Peru, Towards Ecosystem-Based Management of the Humboldt Current Large Marine Ecosystem (HCLME) Project.

Benhamiche, *Nadir*

Benhamiche is State Engineer in Agronomy and 3ed Cycle Doctor in Rural Engineering (Hydrology) Management and Development. He served as Engineer of studies from 1982 to 1988 at the Department of Management and Regional Planning. He then joined the higher education and scientific research where he held teaching/research and administrative functions since 1988 to date. He was a member of several national and international projects including the latest research in progress: PNR entitled 1: Evolution of the water resources of the Soummam watershed : Diagnosis and proposal management tools; Tassili 09MDU785 entitled 2: Contribution to the knowledge of water resources of the watershed Soummam (Algeria) : Climate effect, socio-economic interest in the southern Mediterranean region; AUF-1 BEOM entitled 3: Assessment of water resources in the southern Mediterranean area : Diagnosis and proposal management tools – scientific cooperation projects international (HPIC – BEOM); PHC Maghreb entitled 4: natural and artificial recharge of aquifers; and MISTRALS ENVI-MED entitled 5: Network observations of "badlands" in the Mediterranean area (RESOBAM). These projects address the issue of global warming and human impact on water resources in North African countries.

Bilel, *Weslati*
Bilel Weslati is currently a graduated student at the Mediterranean Agronomic Institute of Zaragoza, Spain. He holds Bachelor Degree in Agricultural Economics from the High Agricultural School of Mograne.

Byjesh, *Kattarkandi*
Mr. Kattarkandi Byjesh is currently working as Consultant with Research Program on Markets, Institutions and Policies at the International Crops Institute for the Semi-Arid Tropics (ICRISAT), Hyderabad, India. He holds a MS from Wageningen University, The Netherlands. His research interests in climate impacts, adaptation strategies, mitigation, long term sustainability of agriculture and environment in the context of rural farming population in the semi-arid tropics of India.

Capouchová, *Ivana*
Capouchová is Engineer and Professor of Plant Production, Department of Plant Production, Faculty of Agrobiology, Food and Natural Resources, Czech University of Life Sciences Prague. Her research focuses on cereal's cropping technologies (especially in organic farming), problems of Fusarium mycotoxins in cereals, technological and nutritional quality of cereals in relation to cropping system, and soil-climatic conditions.

Chebil, *Ali*
Dr. Chebil has a Ph.D. in agricultural economics from the Polytechnic University of Madrid, Spain (2000), and a Master of Science in food and agribusiness marketing from the International Centre for Advanced Mediterranean Agronomic Studies, Zaragoza (1994). He has been at the National Agronomic Research Institute of Tunisia (INRAT) as Associate Researcher from 2001 to July 2003 and at National Research Institute for Rural Engineering, Water and Forest (INRGREF) as Researcher since August 2003. His research primarily focuses on natural resources economics and productivity analysis. He has published numerous journal articles in these fields.

Dhehibi, *Boubaker*
Dr. Dhehibi is an Agricultural Resource Economist and Specialist in the Social, Economics and Policy Research Program (SEPRP) – International Center for Agricultural Research in the Dry Areas (ICARDA). He earned MS degree in Marketing

from the Mediterranean Agronomic Institute of Zaragoza (Spain) – CIHEAM and PhD degree in Quantitative Economics from the University of Zaragoza, Spain with BA degree in Agricultural Economics from the Institute of Agricultural Economics in Moghrane (Tunisia). Dr. Dhehibi is distinguished for his research and teaching on production economics, climate change, economics of water resources, economics of natural resources management, applied micro-econometrics, food demand analysis, international trade, economic modeling, competitiveness and productivity analysis of the agriculture sector in MENA region, growth analysis and economics of development. In addition to many scientific and project reports and working papers, his works are published in ISI scientific journals including New Médit, Cahiers d'Economie et Sociologie Rurales, Applied Economics, Food Policy, African Journal of Agricultural and resources Economics, Journal of Agricultural Science and Technology, Agricultural Economics Review, Food Economics and African Development Review. He has publisher over 25 papers in international refereed scientific journals and has presented more than 25 papers in international conferences.

Fahmy Mehanna, *Sahar*
Dr. Fahmy Mehanna is Professor and Head of Fish Population Dynamics and Stock Assessment Lab, National Institute of Oceanography and Fisheries, Egypt. She has 25 experience in fish population dynamics and aquatic resources assessment and management with 120 published papers in national and international scientific journals. She is an active member in many national and international organizations and societies and joining at least 25 national and international conferences. She supervised 15 MSc and PhD thesis, managed 5 research projects and served as member in another 10 projects.

Hakizimana, *Patrice*
Hakizimana is currently holding a position of Project Manager-Agricultural Specialist in USAID/Rwanda since December 2012. He holds a Research Master in Agriculture and Biology Engineering (crop sciences) from Catholic University of Louvain (UCL/Belgium). He has 13 years of experience in agricultural development (research, extension, education and rural development). Prior to USAID, Patrice worked as an FAO Consultant on Plant Genetic Resources for Food and Agriculture (PGRFA). Previously, he worked for the Government of Rwanda as Researcher on Rice and Sorghum Programs in Rwanda Agriculture Board (RAB)-Directorate of Agricultural Research, and served as the Director General of Rwanda Agricultural Development Authority (RADA) and as the Deputy Vice Chancellor of Higher Institute of Agriculture and Animal Husbandry (ISAE). Hakizimana is passionate about culture and contemporary history, international politics and economics.

Hurd, *Brian H.*
Dr. Hurd is Professor of Agricultural Economics and Agricultural Business at New Mexico State University in Las Cruces, New Mexico. He earned MS and PhD degrees in Agricultural Economics from the University of California, Davis, and graduated magna cum laude from the University of Colorado, Boulder with a BA degree in Economics and Environmental Conservation. Dr. Hurd is distinguished for his research and teaching on the economic assessment of climate change

impacts and adaptation, non-market valuation of natural resources, and the economics of water resources and agro-environmental systems. In addition to many reports and book chapters, his works are published in leading scientific journals including the Journal of the American Water Resources Association, Journal of Agricultural and Resource Economics, Climate Research, Climatic Change, and the American Journal of Agricultural Economics. He has served as Associate Editor for the Journal of Natural Resources Policy Research and is currently President-Elect of the Universities Council on Water Resources (UCOWR) Board of Directors.

Konvalina, *Petr*
Dr. Konvalina is Ph.D. and Engineer. He is also Assistant professor at the University of South Bohemia in České Budějovice, Faculty of Agriculture, Department of Organic Farming. His research focuses on organic plant growing, the use of genetics resources of wheat in organic farming, quality and processing of bioproduction, and problematics of droughtness of wheat. He is Lecturer in courses of "Organic plant growing, Sustainable plant growing, Quality and processing of bioproduction". He is Coordinator of the project ETC SUFA (Sustainable farming) (2009–2012) (sufa.zf.jcu.cz). He is involved in the national research projects, eq: Organic seed multiplication or Use of wheat genetics resources of spring wheat in organic farming. In recent times, he participated in other international projects (AKTION, COST, Interreg, Leonardo da Vinci programme). He published (as author and co-author) more than one hundred scientific publications (reviewed papers, proceeding papers, etc.) and books.

Kumar Pande, *Suresh*
Kumar Pande is born on 15th August 1958 and finished his Bachelor degree (B. Tech.) in Agricultural Engineering from Jawaharlal Nehru Agriculture University, Jabalpur (MP), India during 1980. He did his Master degree in Agricultural Engineering with specialization in Soil and Water Conservation Engineering from Indian Institute of Technology, Kharagpur (WB), India during 1982. He joined the College of Agriculture, Gwalior and then Indore as Assistant Professor during 1982–1983. During December 1983, he joined the Ministry of Agriculture, Government of Madhya Pradesh, India as Assistant Director and worked in the World Bank Aided Watershed Development Project and continued till 1995. From 1995 to 1998 he worked as Associate Professor (Agril. Engg.) in Water and Land Management Institute, Bhopal (MP) on deputation. During 1999, he joined the Higher Institute of Agriculture and Animal Husbandry as faculty member in the department of Soil and Water Management and heading the department since 2007 till date. During his tenure in Rwanda, he supervised 36 students research projects and participated in 7 international conferences for presenting research papers. He also worked as a main consultant to design the hill development strategy for the World Bank aided Rural Sector support Project-2 under the Ministry of Agriculture, Rwanda during 2007. Currently, he is the team lead researcher to help the Rwanda Natural Resource Authority to control floods in lava region of Rwanda.

La Vergne, *Lehmann*
La Vergne Lehmann is PhD Candidate at the University of Ballarat, Victoria, Australia. She has spent the last 10 years working with rural and regional

communities in the Wimmera region in South Eastern Australia in a range of areas focusing specifically on sustainability, biodiversity and water management. As educator, media communicator and more recently as academic researcher, La Vergne has observed the way that rural communities in particular have changed in the face of prolonged drought periods and climate change. To that end, she has worked with rural businesses and communities to develop more environmentally and socially sustainable practices.

Laignel, *Benoit*
Dr. Laignel is Professor at the University of Rouen and Member of the CNRS Research Team UMR 6143 M2C (Continental and coastal Morphodynamics) since 2004 (Prof class 1 since 2009, Assistant Professor 1998–2004) and Associate Professor at the Senghor University of Alexandria, Egypt since 2010. He earned MS and PhD degrees in Geology from the University of Rouen. Dr. Laignel was Assistant Director of the research team UMR M2C and Director of the Laboratory of Geosciences (2004–2010) and member of the Scientific and Administration Board of the University of Rouen (2008–2011). He is member of international scientific networks: FRIEND International Hydrological Program of UNESCO; Expert of IPCC (GIEC); Science Definition Team of the SWOT satellite mission (Surface Water and Ocean Topography; CNES, NASA & ECSA program). Dr. Laignel participated in 32 scientific programs (14 as coordinator), published 58 papers in international journals, and participated in 106 national and international conferences as speaker and 26 as invited speaker. His research program focuses on climate change/water resources, hydrological variability/climate fluctuations, sediment flux in the river, mechanical and chemical erosion, use of the superficial deposits as record of erosion, use of superficial deposits as aggregates, geomorphology of the watersheds.

Madani, *Khodir*
Dr. Madani is Professor of Universities at the research laboratory Biomathematics Biophysics and Biochemistry Scientometry (BBBS), University of Bejaia, Algeria. He earned M.S. and Ph.D. degrees in chemistry and environmental engineering from University of Savoie, France. He is Director of the National Research Agri-Food Centre, Member of the Scientific Council of Natural Sciences, Department of University of Bejaia, Chief of several research projects CNEPRU-MHESR, Head of research PNR-DGRSDT and Co-coordinator of international research projects (CMEP-Tassili, AUF-BEOM, ENVIMED-MISTRAL, PHC-Maghreb, and AUCC). He mainly works on fields of water resources, environment and food. His research and teaching focus on relations between hydrological variability and climate fluctuations, spatio-temporal variability of water and sediment flow in watersheds (surface and underground), geomorphological characterization of watersheds, water requirements of industrial plants of interest, technical extraction, instrumental analysis. In addition to around 30 scientific publications in international journals, he gave a hundred international and national communications, and was invited speaker in many conferences.

Mangora, *Mwita M.*

Mangora is currently working as Assistant Lecturer and registered Ph.D. student at the Institute of Marine Sciences of the University of Dar es Salaam, Tanzania and is specialized in mangrove ecophysiology and ecosystem conservation and management. Shalli is specialized in coastal marine socio-economics, traditional knowledge and fisheries management. Msangamenois is specialized in rocky shore ecology and coastal zone management.

Mason-D'Croz, *Daniel*

Daniel Mason-D'Croz is research analyst in the Environment and Production Technology Division of the International Food Policy Research Institute, Washington, D.C., USA.

Moudrý, *Jan*

Moudrý, CSc. is Engineer and Professor in Plant Production, Faculty of Agriculture, University of South Bohemia, Czech Republic. He is Head of Department of Applied plant Biotechnologies. His main research interests covered rural development, sustainable crop production, and sustainability of landscape management in less favoured areas.

Msangameno, *Daudi J.*

Msangameno is currently working as Assistant Lecturer and registered PhD student at the Institute of Marine Sciences of the University of Dar es Salaam, Tanzania and is specialized in is specialized in rocky shore ecology and coastal zone management.

Mtimet, *Nadhem*

Nadhem Mtimet is Senior Agricultural Economist at the International Livestock Research Institute (ILRI – Kenya). He belongs to the Policy, Trade and Value Chains (PTVC) team. He earned MS degrees in Agrifood Marketing and in Applied Economics from respectively the Mediterranean Agronomic Institute of Zaragoza (Spain) – CIHEAM, and the University of Zaragoza, Spain. He also obtained his PhD degree on Applied Economics from the same University. Before joining ILRI, Nadhem used to work as Assistant Professor at the University of Carthage, Tunisia. Nadhem's research interests include value chain analysis, food demand and food marketing, consumer behavior, and environmental economics. He participated to various reports and book chapters, and has publications in international scientific journals including Agribusiness, Journal of International Food and Agribusiness Marketing, Agricultural Water Management, Spanish Journal of Agricultural Research, African Journal of Agricultural and Resource Economics, and New Médit. He has also various participation to international conferences.

Nedumaran, *Swamikannu*

Dr. Nedumaran is well-trained in agricultural economics and currently working as Scientist in Research Program on Markets, Institutions and Policy at ICRISAT. He has about 7 years of work experience in the field of development policy analysis, natural resource management (NRM) and impact assessment, and has conducted research in collaboration with multiple stakeholders in both Asia and sub-Saharan Africa. Prior to joining ICRISAT in 2010, he worked as a Senior Researcher at the University of Hohenheim, Germany, where he conducted research in Northern Semi-Arid Ghana with multi-stakeholders including the Ghana national program and IFPRI. He also taught postgraduate courses on production economics and farm-level modeling at the University of Hohenheim. His current research work has significantly contributed to strategic foresight analysis, ex-ante impact assessment of promising technologies, research spillover estimation and scenario analysis of ICRISAT mandate crops and targeting and priority setting for Dryland Cereals and Grain Legumes.

Olaniyi, *Wasiu Adekunle*
Olaniyi Wasiu Adekunle is PhD Student at the Department of Animal Sciences. He did his MSc research on Biotechnology and Fisheries/Aquaculture. His research area is Biotechnology with applications to Animal Nutrition, Health and Breeding and Genetics. He is an awardee of World Bank/Federal Government Science and Technology Education Post Basic Project (Cr. 4304-UNI), Innovators of Tomorrow [STEPB (IOT)] Research and Technology Development Grants. He has good understanding of Agroclimatology and has attended many local and international conferences.

Omitogun, *Ofelia Galman*
Omitogun is Professor of Agricultural Biotechnology, Molecular Cytogenetics, and Aquaculture Genetics. She is the pioneer of Biotechnology and Wet Laboratories of her Department where many students are being trained and graduated. She has featured in numerous local and international conferences in France, China, Israel, Japan, Philippine, Tanzania, Kenya, etc. She is a Professor of international repute.

Oyesola, *O.B.*
Dr. Oyesola is an Associate Professor in the Department of Agricultural Extension and Rural Sociology of the University of Ibadan, Ibadan, Oyo State Nigeria.

Plaza, *Francisco*
Fisheries Engineer from the Pontificia Universidad Católica de Valparaíso (PUCV), worked as a researcher since 2006 to 2010, mostly of his work related with environment-resource modeling, using artificial neural networks. During those years, he participated in two international projects (with The University of Huelva, Spain), also national projects (Internal PUCV's projects), and international workshops and conferences presentations, and in 2010 organizing an International Conference related with environment and marine resources. From 2011, he worked in the Instituto de Fomento Pesquero (IFOP, Institute for Fisheries Development) organizing the 7th International Fisheries Observers and Monitoring Conference (7IFOMC), to be held in Chile (April 2013). Currently, Mr. Plaza works in IFOP's Strategical Planning and Management Department. While he works at IFOP, Mr. Plaza is finishing a Statistics M.Sc. Program in Valparaíso, Chile.

Rayar, *Antoni Joseph*
Dr. Rayar was born in Arockiyapuram, Tamil Nadu India in 1943. Graduated with B.Sc. and M.Sc. (Soil Science) in 1967 and 1969, respectively from Agricultural College and Research Institute, Coimbatore, India. He completed his doctoral degree (D.Sc.) from the Catholic University of Leuven, Belgium in 1976. He worked as a Post-Doctoral Scholar in the University of New Brunswick, Canada in 1976–1977, and conducted research on forest soils. He assumed duty as Lecturer in the University of Maiduguri, Nigeria in 1978. He rose to the rank of Associate Professor and Professor of Soil Science in 1984 and 1987, respectively. He served as the Dean, Faculty of Agriculture in 1995–1997. He moved to Rwanda and served as Professor and Head of the Department of Agricultural Engineering in the Higher Institute of Agriculture and Animal Husbandry from 1999. He served as the Dean of the Faculty of Agriculture from 2004 to 2008. He was appointed as the first Director of Quality Assurance from 2009. He has conducted extensive research on soil fertility management in arid and semi-arid regions of Africa. He authored a book titled "Sustainable Agriculture in Sub-Saharan Africa". He moved to Canada in 2011 and serving as a consultant on agricultural projects.

Shahid, *Shabbir A.*
Dr. Shahid is currently Senior Scientist at the Dubai-based International Center for Biosaline Agriculture (ICBA), and Sir William Roberts fellow. He earned Ph.D. degree from the University of Wales, Bangor UK in 1989. He joined ICBA in 2004 and has over 33 years experience (Pakistan, UK, Australia, Kuwait and United Arab Emirates) in agriculture related Research, Development and Extension activities. He has held many positions : Associate Professor (University of Agriculture Faisalabad, Pakistan), Associate Research Scientist in Kuwait Institute for Scientific Research, Kuwait and Manager of Soil Resources Department, Environment Agency – Abu Dhabi, UAE. During his professional career, he has been working in applied agriculture projects in many countries (Pakistan, Kuwait, UAE, Qatar, Oman, Niger, Morocco, Jordan, Spain, Syria etc). In the GCC countries, he has been working for the last 18 years and very much familiar with agriculture activities, climate change challenges, and food security issues. As Technical Coordinator, he completed soil survey of Kuwait and Abu Dhabi emirate. In the latter survey, by the first time in the history of soil research, he has discovered Anhydrite soil in the coastal land of Abu Dhabi emirate. His innovative research on the saline soils of Pakistan highlighted the dominance of thenardite (Na_2SO_4) and mirabillite ($Na_2SO_4.10H_2O$) contrast to older hypothesis of halite (NaCl) dominance. He is a prolific author of over 150 publications in peer-reviewed refereed journals, proceedings, books and manuals. He is currently - life member and Councilor of the World Association of Soil and Water Conservation, and member advisory board "World Forum on Climate Change, Agriculture and Food Security – WFCCAFS". Dr. Shahid enjoyed working as member of scientific committees and advisory boards of international conferences, editor of scientific books, and reviewer of scientific journals.

Shalli, *Mwanahija S.*
Shalli is currently working as Assistant Lecturer and registered PhD student at the Institute of Marine Sciences of the University of Dar es Salaam, Tanzania and is specialized in coastal marine socio-economics, traditional knowledge and fisheries management. Msangamenois is specialized in rocky shore ecology and coastal zone management.

Silva, *Claudio*
Fishing engineer, M.Sc. in Water and Coastal Management, Ph.D. in Marine and Coastal Management. I worked in Pontificia Universidad Católica de Valparaíso as researcher in Remote sensing, GIS and modelling applications in fisheries, aquaculture and environmental sciences. I worked as researcher projects with Canadian Centre of Remote Sensing, International Research Development Centre-Canadá, European Space Agency, Spain Agency of International Cooperation, Fishing Subsecretary of Chilean Government, CORFO and Fund for Promotion of Scientific and Technological Development. In 16 years I worked as research in 35 local, regional, national and international projects, 8 consulting services, participating in 52 scientific publications and 28 international conferences. In faculty and in cooperation with Dra. María Angela Barbieri we gave the Course PES 361-1 in Doctorate in Aquaculture. I taught the course "Remote sensing and GIS in marine pollution analysis" in Master of Integrated Management of Coastal Areas, Erasmus Mundus Master in Water and Coastal Management and Ph.D. in Marine and Coastal Management (MACOMA) in Facultad de Ciencias del Mar y Ambiental, Universidad de Cádiz, Spain. I also taught courses "Tools for assessment of environmental quality in coastal ecosystems" and "Modelling in research: Statistical routines for analysis of marine ecosystems" in the Ph.D. MACOMA.

Singh, *Naveen P.*
Dr. Singh is currently working as Principal Scientist/Agricultural Economist with ICAR at NIASM, Baramati, Pune. Prior to that he served ICRISAT wherein he coordinated the Asian Climate Change program. He has nearly 15 years' of experience working in various countries of Asia, including China. His expertise includes Impact Assessment, Priority Setting, Conservation Agriculture, Participatory research and Adaptation Strategies to Climate Change. He has practical experience and knowledge of a wide range of Agricultural systems, Diversification and other agricultural development issues in Arid and semi-arid tropics of Asia. He is also Associate Fellow of National Academy of Agricultural Sciences (NAAS), India. The experience of working with Asian NARS and knowledge of the intricate issues of socio-economic research critical for uplifting the livelihood of rural poor is a distinct edge. He has published nearly 50 refereed research papers and equivalent numbers of other publications including working papers, proceedings, research reports, policy briefs, popular articles, book chapters.

Sintori, *Alexandra*
Alexandra Sintori graduated from the Agricultural University of Athens (AUA) in 2004 and attended a M.Sc. course in Integrated Rural Development in the same University. Since 2005, she has collaborated with the National Agricultural Research Foundation of Greece, in a number of research programs. She is currently a Ph.D. student in the AUA. Her research focuses mainly on the management and technicoeconomic analysis of livestock production systems.

Stehno, *Zdeněk*
Stehno is Engineer and Researcher in the field of genetics resources of cereals. He is the former Head of Gene Bank, Crop Research institute, Prague. He published more than 40 scientific papers. Some of these publications covered the potential of using genetic resources of cereals for sustainable development of agriculture.

Straquadine, *Gary S.*
Dr. Straquadine earned his BS and MA degree in Agricultural Extension Education from New Mexico State University. He completed his Ph.D. at The Ohio State University, Columbus in Agricultural Education with honors and distinction. He currently serves as the Chair of Agricultural Communication, Education, and Leadership, The Ohio State University, Columbus, USA. Previously he served as the Dean and Executive Director of the Utah State University Regional Campuses. His responsibilities included academic and fiscal oversight for a university outreach system enrolling 1,700 students across the vast geography of Southern and Western Utah. His previous academic experiences involve leadership as the Associate Dean for the College of Agriculture, as Vice Provost for Faculty and Academic Services, and the Department Head for Agricultural Systems Technology and Education (ASTE), where he holds tenure as a Professor. Dr. Straquadine is a recognized scholar in extension education, teaching undergraduate and graduate courses throughout the ASTE curriculum. He has been awarded Teacher of the Year for the College of Agriculture three times, Distinguished Professor in the College and Advisor of the Year for the entire university. While his administrative responsibilities have reduced his teaching and research activities, he still mentors graduate students at both the MS and Ph.D. levels. Dr. Straquadine also serves on the Editorial Boards of numerous highly reputed international scientific research journals.

Suchý, *Karel*
Dr. Suchý is Ph.D. and Assistant Professor at the Department of Live Sciences, Faculty of Agriculture, University of South Bohemia, Czech Republic. He lectures in Plant physiology, biology and production of special plants. His scientific fields cover production potential of grasslands, analysis of physiology and stress risks of plants and accumulation of nitrogen in vegetable. Presently, he is Vice-Dean for Education.

Yamao, *Masahiro*
Dr. Yamao is Professor of Graduate School of Biosphere Science, Hiroshima University, Japan. His fields of specialization include Fisheries Economics, Coastal Resource Management, Community Development, Resource Economics,

Agricultural Economics. His recent research topics consist of Coastal Resource Management in Japan and Southeast Asia, Sustainable Community Development in Coastal Area, Seafood Trade and Fisheries.

Yáñez, *Eleuterio*
Dr. Yáñez is Fisheries Engineer (1974) and PhD in Biological Oceanography and Professor of Fisheries Assessment and Fisheries Oceanography. He began research in demersal fisheries assessment in Chile. Since his first doctoral thesis (1980) on the dynamics of yellow fin tuna in the Atlantic Ocean, he is principally engaged in the assessment and oceanography of pelagic resources, integrating the environment into stock assessment models, including his second doctoral thesis (1998). Recently, being concerned with the ecosystem-based approach to fisheries, he is analyzing fisheries and environmental data using artificial neural networks, and climate change and its likely impacts on fisheries in Chile. He supervised 31 theses on Fisheries Engineering and fisheries assessment and participated in eight doctoral committees. Since 2000, he has edited 6 books, published 39 scientific articles with editorial board and participated in 97 conferences and scientific meetings.

Zamroni, *Achmad*
The author was born in West Java, Indonesia. He received his Graduation in Fisheries in 2002 from Brawijaya University, Malang, Indonesia. After that, he received MSc in Coastal Management from Department of Bioresource Science of Hiroshima University, Japan in 2010. At present, he is continuing his doctoral program in the same department and university. Originally and until now, he served as researcher in Research Center for Marine and Fisheries Socio-Economics, Ministry for Marine Affairs and Fisheries, Republic of Indonesia.